FINITE GRAPHS AND NETWORKS:
AN INTRODUCTION WITH APPLICATIONS

INTERNATIONAL SERIES IN PURE AND APPLIED MATHEMATICS

William Ted Martin and E. H. Spanier
CONSULTING EDITORS

AHLFORS · Complex Analysis
BELLMAN · Stability Theory of Differential Equations
BUCK · Advanced Calculus
BUSACKER AND SAATY · Finite Graphs and Networks
CODDINGTON AND LEVINSON · Theory of Ordinary Differential Equations
DETTMAN · Mathematical Methods in Physics and Engineering
EPSTEIN · Partial Differential Equations
GOLOMB AND SHANKS · Elements of Ordinary Differential Equations
GRAVES · The Theory of Functions of Real Variables
GREENSPAN · Introduction to Partial Differential Equations
GRIFFIN · Elementary Theory of Numbers
HAMMING · Numerical Methods for Scientists and Engineers
HILDEBRAND · Introduction to Numerical Analysis
HOUSEHOLDER · Principles of Numerical Analysis
LASS · Elements of Pure and Applied Mathematics
LASS · Vector and Tensor Analysis
LEPAGE · Complex Variables and the Laplace Transform for Engineers
NEHARI · Conformal Mapping
NEWELL · Vector Analysis
RALSTON · A First Course in Numerical Analysis
ROSSER · Logic for Mathematicians
RUDIN · Principles of Mathematical Analysis
SAATY AND BRAM · Nonlinear Mathematics
SIMMONS · Introduction to Topology and Modern Analysis
SNEDDON · Elements of Partial Differential Equations
SNEDDON · Fourier Transforms
STOLL · Linear Algebra and Matrix Theory
STRUBLE · Nonlinear Differential Equations
WEINSTOCK · Calculus of Variations
WEISS · Algebraic Number Theory
ZEMANIAN · Distribution Theory and Transform Analysis

FINITE GRAPHS AND NETWORKS:
AN INTRODUCTION WITH APPLICATIONS

ROBERT G. BUSACKER
Research Analysis
Corporation

THOMAS L. SAATY
U.S. Arms Control and
Disarmament Agency

McGRAW-HILL BOOK COMPANY
New York St. Louis San Francisco Toronto London Sydney

FINITE GRAPHS AND NETWORKS:
An Introduction with Applications

Library of Congress Catalog Card Number 64-66288

09305

1234567890 MP 7210698765

For HERBERT SCOVILLE, JR.

PREFACE

Graph theory provides simple, accessible, and powerful tools for constructing models and solving problems having to do with discrete arrangements of objects. Technology today poses a great number of problems that require the construction of complex systems through specific arrangements of their components. These problems range from the scheduling of industrial processes to PERT and critical-path analyses, to tactics and logistics, to communication systems, to the study of information transmittal, to the choice of optimum routes and network flows, to electrical networks, to organic-chemical identification, to switching systems, to economics, to social-group structures, to games and puzzles, etc. The applications are numerous. Combinatorial construction techniques for finding appropriate arrangements are considerably different from the arithmetic computational approach to systems through the equations of classical analysis. In addition to the use of graphs, the subject of arrangements can be approached through matrices whose entries are zeros and ones, or simply through the analysis of specific finite families of sets.

It can be truly said that graph theory is one of the simplest, most elegant subjects of modern mathematics which possesses a wide variety of applications. Based on the simple idea of points interconnected by lines, graph theory combines these basic ingredients into a rich assortment of forms and endows these forms with flexible properties, thus making the subject a useful tool for studying many kinds of systems. In addition, graph theory has contributed to the analysis of a wide variety of combinatorial problems.

The term *network* is frequently used instead of *graph*, especially when quantitative characteristics are imparted to the points and lines, in addition to the purely structural relationships that are the defining characteristics of a graph. One speaks, for example, of electrical networks, project networks, and flow networks, in which quantitative measures of energy, effort, and flow, respectively, are associated with the edges.

This book is specially written for the teacher desiring to fulfill the need of the undergraduate and first-year graduate student in mathematics, physical and social sciences, engineering, economics, and operations research to get a relatively broad idea of graph theory and to receive such varied exposure to its application that his creative talent may thereby be inspired. Indeed this has been our experience with students who learned their graph theory from this book and went on to apply it in their work and to conduct research on the many interesting problems in the field. The manuscript is an outgrowth of a semester course taught at the George Washington University, at the Graduate School of the Department of Agriculture, and at various other institutions overseas over a period of several years.

In view of the basic nature and broad applicability of graph theory, it is clearly desirable for anyone engaged in or preparing for work in the general area of science or engineering to become familiar with at least the fundamentals of this topic. It is the objective of this book to provide such familiarity, in a form suitable as the basis for an introductory course in graph theory or for self-study. Despite the abundance of journal articles dealing with specialized topics, the available expository literature does not meet all needs.

In attempting to meet this objective, we have chosen for the most part to place the material in the setting of a relatively informal discussion of central ideas, amplified and illustrated by a variety of applications. By so doing, it is hoped that the interested reader will be left (*a*) in a position to relate and adapt these basic concepts to his own field of application, and (*b*) in a frame of mind to pursue the matter further, convinced of its importance.

The book is organized into two parts. Part I, consisting of Chapters 1 through 5, presents basic theoretical material. Chapter 1 contains the fundamental definitions, terms, and symbols required to describe and classify undirected graphs. It is at the end of this chapter that we have included a large number of references of well-known books on and related to the subject. The bibliography of Moon and Moser which appears at the end of Ref. 8*a* is thorough and indicative of the fantastic range of the subject. Chapter 2 serves a similar function for oriented or directed graphs. In Chapter 3 the basic development of concepts is continued, centering around various ways of partitioning the elements of a graph and the measurement of distances in graphs. Chapter 4 discusses properties and characterizations of the important class of *planar* graphs, as well as the class of problems that have come to be known as *coloring* problems. In Chapter 5, which concludes Part I, the emphasis shifts from geometric to algebraic considerations, with a discussion of the role of matrices in the characterization and structural analysis of graphs.

Part II, consisting of Chapters 6 and 7, illustrates the application of the foregoing theory in a variety of contexts. Chapter 6 in particular (long and rich in variety) contains brief discussions of the utility of graph theory in numerous connections. Chapter 7 considers linear flows through constrained networks from a graph-theoretic point of view. General procedures for maximizing flow and minimizing cost are developed. A brief introduction to stochastic network flows is given, relating the ideas to queueing theory.

The only prerequisites are a familiarity with basic terminology of set theory and a knowledge of elementary concepts related to matrices and vector spaces.

An interesting philosophical observation that one might make is based on Theorem 1–2 given in Chapter 1 of the possibility of representing any graph with a continuum number of vertices and edges in three-dimensional space without intersections. While one might gain in simplicity in representing flows by going to higher dimensions, the analytical concepts of flow would not necessarily be enriched by doing so. The significance of this observation becomes apparent when one considers that man can largely be characterized in terms of various flows in his nervous system (with its infinite possibilities and multidimensional abstractions) which he uses to organize and relate himself to his environment. It would seem that three dimensions is also adequate and economical for the realization of the "flow-system man."

We are grateful to Mr. Jack Edmonds for many helpful suggestions and useful additions, particularly to Chapter 6, and to Dr. Horace Trent for very helpful ideas and profitable discussions on varied topics. For additional suggestions, particularly in Chapter 6, and some help in partial collecting and writing, our thanks also go to Dr. David Rosenblatt (Section 6–2), Mr. Charles MacLean (6–10), Dr. Kurt Fey (6–4), Mr. John Boushka, Miss Susan Meador (6–22), Mr. William D. Murray (6–28), Mr. P. C. Ryan, and Mr. Edward Stern.

ROBERT G. BUSACKER
THOMAS L. SAATY

CONTENTS

PART I

BASIC THEORY

This part of the book contains five chapters. Chapters 1 and 2 contain fundamental definitions and theorems concerning undirected and directed graphs, respectively. In Chap. 3 this basic development is continued, revolving around various methods of partitioning and measuring distances in graphs. Chapter 4 is concerned with planar graphs and coloring problems of which the classical four-color problem is the outstanding example. In Chap. 5 the emphasis shifts from geometric to algebraic considerations, and certain properties of graphs are investigated through their representative matrices.

ONE

BASIC CONCEPTS: UNDIRECTED GRAPHS

1-1 INTRODUCTION

In this chapter the concept of a graph is introduced, first as a geometric structure consisting of points (vertices) in space interconnected by a system of curves (edges) and then in abstract terms. The basic terminology and symbols needed to discuss fundamental local and global structural properties of graphs are then introduced and discussed. Thus this chapter—together with the next, dealing with graphs whose connecting edges are oriented—establishes the basic vocabulary for describing graphs. In order to occasionally relieve the monotony of these necessary preliminaries, a number of basic results which follow rather directly from the definitions are included.

Because the terminology and symbolism currently in use in graph theory is not standardized, the reader is urged to study the conventions of these two chapters carefully before proceeding.

1-2 GEOMETRIC GRAPHS

Before introducing the concept of a graph in its most general form, we consider a class of graphs known as geometric graphs. By doing so, we shall have from the outset a convenient visualization of the various concepts and structures to be considered. It will be shown presently that every graph of interest is abstractly identical (with respect to those properties of interest in graph theory) with an appropriate geometric graph. Thus geometric graphs may be regarded as convenient representations of all graphs, and not merely as specialized examples.

Let $\mathcal{E}^n$ denote n-dimensional euclidean space. (We shall be primarily interested in 2- and 3-dimensional spaces as a result of Theorems 1-1 and 1-2 discussed later.) Euclidean n space is the space of all sequences of n real numbers $x = (x_1, \ldots, x_n)$, with distance between two points $x = (x_1, \ldots, x_n)$, $y = (y_1, \ldots, y_n)$ given by $d(x,y) = \left[\sum_{i=1}^{n} (x_i - y_i)^2\right]^{\frac{1}{2}}$. A *simple open curve* in $\mathcal{E}^n$ is a continuous, non-self-intersecting curve joining two distinct points in $\mathcal{E}^n$ (i.e., a curve obtainable by continuous deformation of a straight-line segment). Similarly, a *simple closed curve* is essentially a continuous, non-self-intersecting curve whose end points coincide.

A *geometric graph* in $\mathcal{E}^n$ is a set $V = \{v_i\}$ of points in $\mathcal{E}^n$ and a set $E = \{e_j\}$ of simple curves satisfying the following conditions:

1. Every closed curve in E contains precisely one point of V.
2. Every open curve in E contains precisely two points of V, and these agree with its end points.
3. The curves in E have no common points, except for points of V.

Thus a geometric graph is simply a geometric configuration or structure in $\mathcal{E}^n$ consisting of a set of points interconnected by a set of non-intersecting continuous curves.

Many familiar structures may (when idealized) be regarded as geometric graphs and studied by applying the methods developed in subsequent pages. A system of highways, for example, can be idealized as a graph by ignoring the width of roads and considering junctions to be points. Many additional examples of graphlike structures in the real world will be introduced in subsequent pages.

Figure 1-1 depicts a geometric graph and illustrates the manner in which graphs will usually be represented. In the context of graph theory, the elements of V and E are respectively called *geometric vertices*

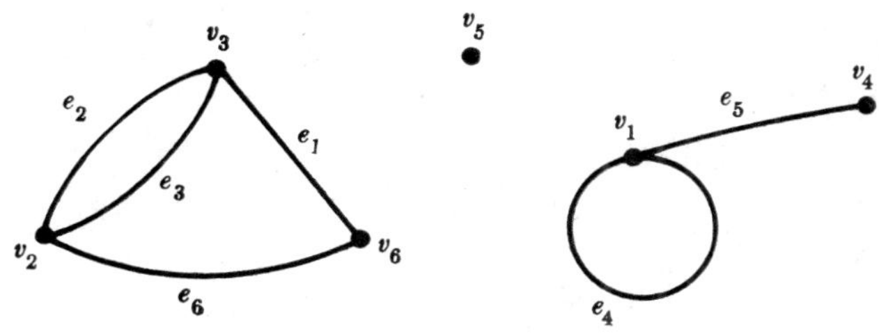

Figure 1-1

and *geometric edges.* We could, at this time, introduce the various descriptive terms which make up the basic vocabulary of graph theory. (For instance, in Fig. 1-1, edges e_2 and e_3 are "parallel," vertex v_5 is "isolated," v_3 and v_6 are "adjacent," etc.) Before doing so, we redefine graphs in general terms, so that these terms need be introduced only once and in a general setting.

1-3 ABSTRACT GRAPHS

While many of the graphs arising in applications are (when idealized) geometric graphs, the only essential structural feature in the context of graph theory is the fact that each geometric edge has associated with it two (possibly coincident) geometric vertices. Graph theory focuses on these associations and disregards the nature of vertices and edges. Thus the enumeration of edges and vertices given in the following table contains all of the essential information concerning the geometric graph of Fig. 1-1.

Edges	*Corresponding vertices*
e_1	v_3, v_6
e_2	v_2, v_3
e_3	v_2, v_3
e_4	v_1
e_5	v_1, v_4
e_6	v_2, v_6
	v_5

To facilitate the general definition of a graph, we first introduce the concept of the unordered product of a set with itself. Recall that the ordered or *cartesian product* of a set S with itself, denoted by $S \times S$, is defined to be the set of all ordered pairs (s,t), where $s \in S$ and $t \in S$. Here (s,t) and (t,s) are considered to be distinct entities except when $s = t$. In a similar vein, the symbol $(s \,\&\, t)$ will denote an unordered pair of elements of S, and the set of all distinct unordered pairs will be denoted by $S \& S$ and called the *unordered product* of S with itself. Here $(s \& t)$ and $(t \& s)$ are equivalent symbols for the same entity and, as in the case of (s,t), it is permissible that $s = t$. Note that if S has k elements, $S \times S$ consists of k^2 ordered pairs while $S \& S$ consists of $k(k+1)/2$ distinct unordered pairs.

An *abstract graph*, or simply a *graph*, may now be defined as follows: A graph consists of a nonempty set V, a (possibly empty) set E disjoint from V, and a mapping Φ of E into $V \& V$. The elements of V and E are called the *vertices* and *edges* of the graph, respectively, and Φ is called the *incidence mapping* associated with the graph.

If $e \in E$ and v and w are the vertices such that $\Phi(e) = (v \& w)$, then edge e is said to be *incident with* each of vertices v and w, and vice versa. All other vertices are regarded as being *nonincident with* e. The vertices incident with an edge are called its *end points*, and are also said to be *joined* by the edge.

While the incidence relationship is fundamental to the concept of a graph, in many contexts there is no need to refer to Φ explicitly. In such cases, the fact that v and w are the end points of e is denoted by $e \sim (v \& w)$, read as "e joins v and w."

A graph will usually be denoted by G or (V,E,Φ), or by (V,E) when the incidence mapping remains implicit. Note that E, but not V, may be empty. While graphs with no edges are uninteresting in themselves, their inclusion is occasionally useful, as, for example, in connection with constructive (or, more accurately, destructive) procedures involving the successive deletion of edges from a given graph. A graph is said to be *degenerate* if and only if it has no edges.

If V and E are both finite sets (the empty set being included as a finite set), G is called a *finite graph*. Otherwise it is said to be *infinite*.

The introduction of the concept of abstract graphs serves a purpose aside from stripping the incidental geometric features away from the essential combinatorial characteristics of a graph. It enlarges the prospects for applications, since many real-world entities have the combinatorial characteristics necessary to be viewed profitably as graphs. For example, the relationships between the component individual activities which make up a complex project can be depicted as a graph. Here the edges (when assigned orientations or directions as in Chap. 2) represent the individual activities, and the incidences within the graph reflect the way in which the execution of certain tasks is contingent on the prior completion of other tasks. We shall return to this important class of project graphs in Sec. 6-4. Note that in this case a geometric visualization of the situation is simply a (convenient) artifice and not an idealization of actual physical connections between objects in space.

1-4 ISOMORPHISMS AND REALIZATIONS

A geometric graph G is clearly a particular instance of a graph, in which the vertices and edges are respectively points and simple curves in $\mathcal{E}^n$ and in which $\Phi(e) = (v \,\&\, w)$ is understood to mean that v and w are the end points of curve e or that v is the sole vertex contained in closed curve e if $v = w$.

It was remarked earlier that every graph is abstractly identical with—or, to use a more customary term, isomorphic to—an appropriate geometric graph. Isomorphism between graphs is formally defined as follows: Graphs $G = (V,E)$ and $G' = (V',E')$ are said to be *isomorphic* to each other if there exist 1-1 correspondences between V and V' and between E and E' which preserve incidences. In other words, edge e is incident with vertex v in G if and only if the corresponding elements e' and v' are incident in G'. If G is isomorphic to a geometric graph G', then G' is called a *geometric realization of* G. (In particular, a geometric graph may be considered as a geometric realization of itself.)

A graph is said to be *planar* if and only if it has a geometric realization in $\mathcal{E}^2$. For example, the graph G shown in Fig. 1-2*a* is planar, since it is isomorphic to the graph G' shown in (*b*). This figure illustrates an evident but important fact: A geometric graph may be planar even though it cannot be deformed by a continuous transformation into a planar graph. While G and G' have important distinctive features from the standpoint of topology, they are equivalent from the standpoint of graph theory.

Figure 1-3 depicts a *nonplanar* graph: one having no geometric realization in $\mathcal{E}^2$. This graph is, in fact, one of two fundamental configurations which characterize all nonplanar finite graphs. This fact, embodied in an important theorem due to Kuratowski, is established in Chap. 4, where the distinctive features of planar graphs are investigated. In con-

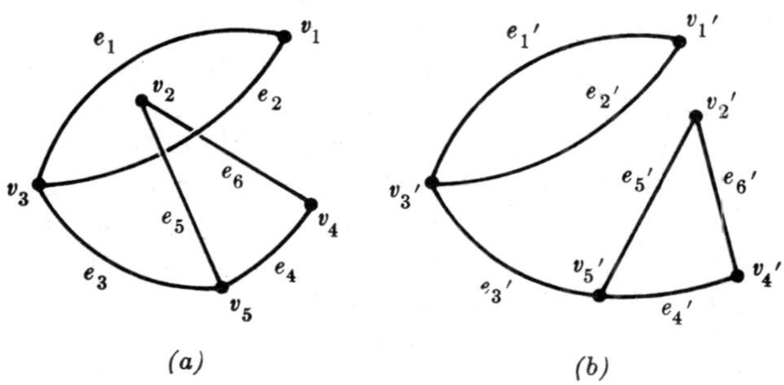

Figure 1-2

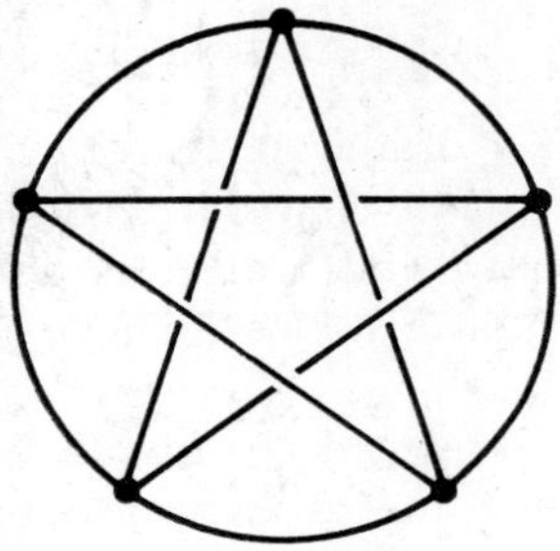

Figure 1-3

trast to the fact that only a restricted class of finite graphs have realizations in $\mathcal{E}^2$, we have the following:

Theorem 1-1 Every finite graph $G = (V,E)$ has a geometric realization in $\mathcal{E}^3$.

Proof: Let L denote an arbitrarily chosen line in $\mathcal{E}^3$. Corresponding to each vertex $v \in V$ select a distinct point v' on L, and corresponding to each unordered pair $(v \,\&\, w) \in V \,\&\, V$ select a distinct half plane $H_{v\&w}$ having L as its boundary. For each edge e in G such that $e \sim (v \,\&\, w)$ construct in $H_{v\&w}$ a simple curve e' joining v' and w' and having no other points in common with L or with other curves in $H_{v\&w}$. (This is clearly possible.) The resulting structure is a geometric realization of G.

While the validity of Theorem 1-1 is self-evident, we have indicated this construction primarily because it can be extended in a straightforward manner to establish the following less obvious theorem, which completely characterizes realizable graphs:

Theorem 1-2 A graph $G = (V,E)$ has a geometric realization in $\mathcal{E}^3$ if and only if the elements of V and also those of E can be placed in 1-1 correspondence with a subset of the set of all real numbers.

(For the reader familiar with the concept of cardinal number, G is realizable in $\mathcal{E}^3$ if and only if neither V nor E has cardinality exceeding that of the continuum.)

In this book we shall be almost exclusively concerned with finite graphs, and never with graphs which do not meet the conditions of Theorem 1-2. Moreover, the structural properties to be studied are all preserved by isomorphisms, since they all rest, ultimately, on the concept of incidence. It follows that we can, without any real loss of generality, think entirely in terms of geometric graphs in $\mathcal{E}^3$. In particular, all illustrations of graphs will be geometric graphs.

Sketch of proof of Theorem 1-2: The points in $\mathcal{E}^n$ can be placed in 1-1 correspondence with the real numbers.

Clearly, a graph $G = (V,E)$ is not realizable in $\mathcal{E}^3$ (or, in fact, in $\mathcal{E}^n$ for any positive integer n) if either the vertices or the edges cannot be placed in 1-1 correspondence with a subset of the points in $\mathcal{E}^3$. On the other hand, if such correspondences exist, it is possible (referring to the terminology in the proof of Theorem 1-1) to select a distinct point v' on L for each vertex and a distinct half plane $H_{v\&w}$ for each unordered pair of vertices. This having been done, the curves in $H_{v\&w}$ can be constructed according to the scheme shown in Fig. 1-4. Each point on the segment

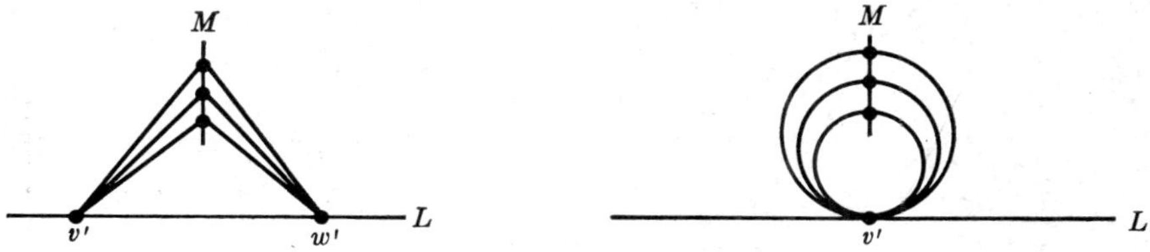

Construction when $v \neq w$

(a)

Construction when $v = w$

(b)

Figure 1-4

M determines a simple curve (a broken line segment or circle) disjoint from other such curves except for v' and w'.

1-5 TERMS DESCRIBING LOCAL STRUCTURE

In order to describe various structural features of graphs efficiently, the following terms are useful. They are all particularly transparent in the context of geometric graphs.

If $e \sim (v \,\&\, w)$, v and w are called the *end points* of e, whether the graph under consideration is geometric or not. If $v = w$, then v is the sole end point of e and e is called a *loop*. If $e_1 \sim (v \,\&\, w)$ and $e_2 \sim (v \,\&\, w)$, then e_1 and e_2 are called *parallel edges*. In particular, two loops incident with the same vertex are parallel. Vertices v and w are said to be *adjacent vertices* if $e \sim (v \,\&\, w)$ for at least one edge e. In particular, v is adjacent with itself if there is a loop incident with v, but not otherwise. Similarly, edges e_1 and e_2 are *adjacent edges* if they have at least one common end point. Note that adjacency is a relationship between two like elements (either vertices or edges), while incidence is a relationship between unlike elements.

The number of edges incident with v, with loops counted twice, is called the *degree* of v and denoted by $\delta(v)$. A vertex v is said to be *isolated* if $\delta(v) = 0$. Thus, in particular, a degenerate graph is one for which every vertex is isolated.

If S is any finite set, $|S|$ denotes the number of elements in S. In particular, $|V|$ and $|E|$ denote the number of vertices and edges, respectively, of the finite graph $G = (V,E)$. Since every edge contributes 1 to the degree of two vertices (or, in the case of a loop, 2 to the degree of a single vertex), we have

$$\sum_{v \in V} \delta(v) = 2|E|$$

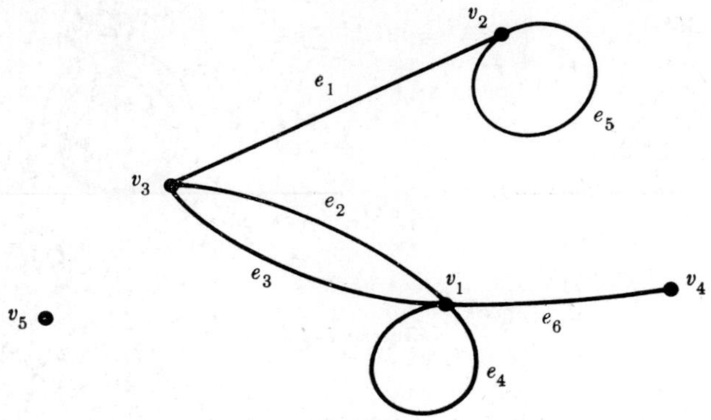

Figure 1-5

If V_0 and V_1 are the sets of vertices having even and odd degree, respectively, then clearly $\sum_{v \epsilon V_0} \delta(v)$ is even, since it is a finite sum of even numbers. It follows that

$$\sum_{v \epsilon V} \delta(v) - \sum_{v \epsilon V_0} \delta(v) = \sum_{v \epsilon V_1} \delta(v)$$

is also necessarily even, which establishes the following:

Theorem 1-3 In a finite graph, the number of vertices of odd degree is even.

In order to fix the terminology of this section firmly in the reader's mind, these terms are now illustrated in the context of the graph of Fig. 1-5. The end points of e_1 are v_3 and v_2, and the loop e_4 has the single end point v_1. Edge e_2 is adjacent with e_1 and parallel with e_3. (Note that parallel edges are also adjacent.) Vertex v_1 is adjacent with v_4 and with itself, but v_4 is not adjacent with itself. Vertex v_5 is an isolated vertex. Four vertices (namely, v_1, v_2, v_3, and v_4) have odd degree.

Given a graph $G = (V,E,\Phi)$, the system $G_1 = (V_1,E_1,\Phi_1)$ is called a *subgraph* of G if and only if the following conditions are met:

1. $V_1 \subset V$ and $E_1 \subset E$.
2. $\Phi_1(e) = \Phi(e)$ for every $e \epsilon E_1$.
3. If $e \epsilon E_1$ and $\Phi(e) = (v \,\&\, w)$, then $v \epsilon V_1$ and $w \epsilon V_1$.

Expressed differently, a subgraph of G consists of selected edges and vertices of G, with the same incidences as those of G, and with the reasonable requirement that the selected vertices must include all of the end points of the selected edges. The graph G is also called a *supergraph* of G_1.

Exercise 1-1 Interpret the various terms connected with the local structure of a graph in the context of a graph whose vertices and edges represent the junction points and individual segments of a system of city streets. (The reader may also wish to anticipate certain global characteristics to be introduced shortly. For example, what would be a reasonable definition for the property of connectedness in graphs?)

1-6 EDGE PROGRESSIONS, CHAINS, AND CIRCUITS

In a geometric graph one can visualize starting at a given vertex, traversing a sequence of geometric edges in a continuous manner, and ultimately arriving at another specified vertex or returning to the original vertex. Sequences of edges which form continuous routes play a fundamental role in graph theory. In particular, the familiar structures which we shall call *chains* and *circuits,* in which no edge is traced more than once, enter in some way into nearly all aspects of graph theory. The above intuitive procedure suggests the following formal definition:

A finite sequence $e_1, e_2, \ldots, e_n$ of (not necessarily distinct) edges of a graph is said to constitute an *edge progression* of length n if there exists an appropriate sequence of $n + 1$ (not necessarily distinct) vertices $v_0, v_1, \ldots, v_n$ such that

$$e_i \sim (v_{i-1} \,\&\, v_i) \qquad \text{for } i = 1, 2, \ldots, n$$

The edge progression is said to be *closed* if $v_0 = v_n$ and *open* if $v_0 \neq v_n$. In the latter case the progression is said to extend *from* v_0 *to* v_n. Note that an individual edge may be considered as an edge progression of length 1.

(Edge progressions are frequently referred to as edge sequences in the literature. However, since most sequences of edges fail to be edge sequences in this special sense, we have avoided this terminology.)

Referring to Fig. 1-6, the sequence $e_7, e_1, e_8, e_3, e_4, e_5$ constitutes an open edge progression from v_2 to v_4 of length 6, the corresponding sequence of vertices being $v_2, v_5, v_5, v_6, v_1, v_5, v_4$. If we replace e_5 by e_7 as the last

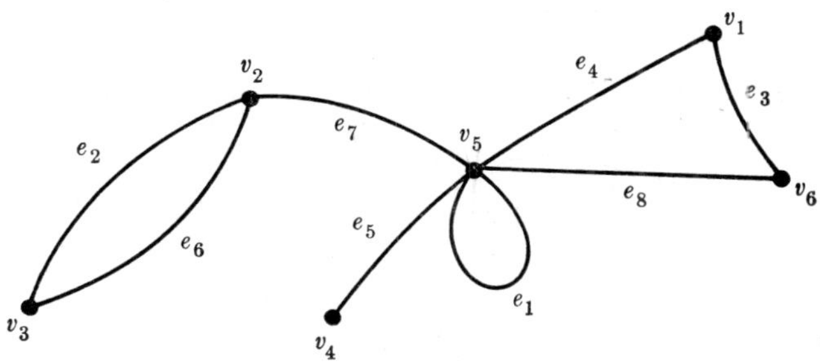

Figure 1-6

term of the progression, we have an example of a closed edge progression of length 6.

If all elements of the progression represent distinct edges, the progression is called a *chain progression* if it is open and a *circuit progression* if it is closed. The set of edges itself, without regard to sequencing, is said to constitute a *chain* in the former case and a *circuit* in the latter. For example, the set of edges $\{e_3,e_4,e_7,e_8\}$ in Fig. 1-6 constitutes a chain, since the sequence e_4, e_3, e_8, e_7 is a chain progression having v_5, v_1, v_6, v_5, v_2 as the corresponding sequence of vertices. This same chain is also associated with the progression e_8, e_3, e_4, e_7, which has v_5, v_6, v_1, v_5, v_2 as the associated sequence of vertices.

In some contexts it is important to distinguish between alternative ways of ordering the edges of a chain or circuit to form a progression; in others the ordering is immaterial so long as at least one such ordering exists. Both situations occur often enough to warrant the double terminology (e.g., chain progression vs. chain) introduced here.

If all the $n + 1$ vertices $v_0, v_1, \ldots, v_n$ are distinct (in which case the edges are necessarily also distinct), the edge progression is called a *simple chain progression* and the set of edges—without regard to order—is called a *simple chain*. If $v_0 = v_n$ but the vertices are otherwise distinct, the progression is called a *simple circuit progression* and the unordered set of edges is said to constitute a *simple circuit*. Note that in a geometric graph simple chains determine simple open curves, as defined earlier, while simple circuits determine simple closed curves.

It is instructive to interpret the concepts of progression, chain, and circuit in various specific application contexts. Consider, for example, a graph whose vertices represent the individuals in an organization and whose edges correspond to pairs of individuals between whom direct communication is possible. Describe the movement of information within such a structure in terms of progressions, chains, and circuits.

Exercise 1-2 Prove that every open edge progression from v to w includes, as a subsequence, a simple chain progression from v to w. In particular, every nonsimple chain progression contains such a simple chain progression.

Exercise 1-3 Show, with an example, that the following is *not* true in general: If P is an edge progression from v_1 to v_2 which passes through v_3, where $v_3 \neq v_1$ or v_2, then P includes, as a subsequence, a simple chain progression from v_1 to v_2 which also passes through v_3.

Exercise 1-4 Why is the following characterization inadequate? An edge progression is a sequence of edges such that adjacent terms of the sequence correspond to adjacent edges of the graph under consideration.

Exercise 1-5 Prove that if v_1, v_2, and v_3 are distinct vertices such that v_1 and v_2 are joined by a chain and v_2 and v_3 are joined by a chain, then v_1 and v_3 are joined by a chain.

Exercise 1-6 Under the assumptions of Exercise 1-5, give an example to show that there does not necessarily exist a simple chain joining v_1 and v_3 which passes through (i.e., is incident with) v_2.

Exercise 1-7 Prove that if the edges of a chain can be ordered in more than two ways to form a chain progression, then the chain is not a simple chain.

Exercise 1-8 The converse of the statement in Exercise 1-7 is not true in general. That is, there exist chains which are not simple but which cannot be ordered in more than two ways to form a chain progression. Provide an example, and try to characterize all such chains.

Exercise 1-9 Prove that every nonsimple circuit can be partitioned into two or more simple circuits.

Exercise 1-10 Prove that every nonsimple chain joining v and w can be partitioned into a simple chain joining v and w together with one or more simple circuits. Thus only simple chains are minimal in the sense that they contain no proper subset of edges connecting their end points.

Exercise 1-11 If "nonsimple chain" is replaced by "open edge progression" in Exercise 1-10, or if "nonsimple circuit" is replaced by "closed edge progression" in Exercise 1-9, the resulting statement is not true in general. Provide a counter-example for each.

Exercise 1-12 Show that any finite set of nonnegative integers can be realized as the degrees of the vertices of an appropriate graph, provided that an even number of them are odd.

If loops are not permitted, a realization such as that of Exercise 1-12 is not generally possible. One must place more stringent requirements on the given set of integers. Tutte [The Factors of Graphs, *Can. J. Math.*, **4:** 314–328 (1952)] has developed necessary and sufficient conditions under which a given graph contains a subgraph having vertices of prescribed degrees.

1-7 CONNECTIVITY

A graph is said to be *connected* if every pair of distinct vertices are joined by at least one chain. Other graphs are said to be *disconnected.* For finite geometric graphs, these definitions conform with conventional usage; i.e., a finite geometric graph is connected in this graph-theoretic sense if and only if it is connected when considered as a point set. However, this is not always the case for infinite graphs. Consider the geometric graph $G = (V,E)$ in $\mathcal{E}^2$, where V consists of all points with coordinates (x,y), $x = 0$ or 1 and $0 \leq y \leq 1$, and where for each y the vertices $(0,y)$ and $(1,y)$ are joined by a straight edge. Viewed as a point set, G consists of a unit square in $\mathcal{E}^2$ and is connected. As a graph, however, it is highly disconnected, since vertex $(0,y)$ is joined to $(1,y)$ but to no other vertex by a chain.

As an alternative characterization of connectivity we have the following theorem:

Theorem 1-4 A graph $G = (V,E)$ is connected if and only if its vertices cannot be partitioned into two nonempty subsets V_1 and V_2 in such a way that both end points of every edge are in the same subset.

Proof: If G is not connected, select a vertex v_1 and let V_1 consist of v_1 together with all vertices joined with v_1 by chains. Since G is not connected, $V_1 \neq V$ (why?), so the complementary set $V_2 = V - V_1$ is nonempty. Considering the way in which V_1 was defined, no edge joins a vertex in V_1 with one in V_2, so a partitioning of the desired type has been obtained.

Conversely, if such a partitioning exists, arbitrarily select $v_1 \in V_1$ and $v_2 \in V_2$. A chain joining v_1 and v_2 would necessarily contain at least one edge having end points in both V_1 and V_2. Since no such edge exists, G is not connected, and the proof is complete.

If $G = (V,E)$ is any graph, consider the binary relation ρ defined between certain ordered pairs of vertices as follows: $v \rho w$ if and only if either $v = w$ or else v and w are joined by a chain. The relation ρ is clearly reflexive ($v \rho v$ for every v), symmetric ($v \rho w$ implies $w \rho v$), and transitive ($u \rho w$ whenever $u \rho v$ and $v \rho w$). Thus ρ is an equivalence relation, and it partitions V into uniquely determined equivalence classes of mutually related vertices. In Fig. 1-1, for example, these are $\{v_2,v_3,v_6\}$, $\{v_1,v_4\}$, and $\{v_5\}$. Each equivalence class of vertices, together with those edges of E incident with these vertices, constitutes a connected subgraph called a *component* of G. A component G_1 of G is readily seen to be a maximal connected subgraph, in the sense that no proper supergraph of G_1 is connected.

Exercise 1-13 Prove that a connected graph remains connected after removal of an edge if and only if the edge is contained in some circuit.

Exercise 1-14 Give an example of a connected graph which becomes disconnected if any one edge is removed. (In view of Exercise 1-13, the graph can have no circuits.) Such graphs, called *trees*, are considered in the next section.

Exercise 1-15 Prove that a connected graph having k vertices must have at least $k - 1$ edges.

Exercise 1-16 Prove that the set of all edges of a connected finite graph constitutes a simple circuit if and only if every vertex has degree 2. Formulate and prove a similar characterization of connected graphs whose edges constitute a simple chain.

Exercise 1-17 Show that all the edges of a finite connected graph can be included in an appropriately chosen edge progression. (In Chap. 3 we shall determine the circumstances under which this can be done with a chain or circuit progression.)

1-8 TREES AND FORESTS

A graph is said to be a *tree* if it is connected and has no circuits. More generally, a graph which is free of circuits and consists of k connected components is sometimes referred to as a *forest* of k trees. The concept of a tree plays a central role in many aspects of graph theory. A graph is a tree if and only if every pair of distinct vertices are joined by precisely one chain. (Connectivity implies the existence of at least one chain, while the absence of circuits implies the existence of at most one chain.)

The removal of any one edge of a tree yields a disconnected graph, since the edge removed constituted the unique chain joining its end points. On the other hand, given any connected graph which is not a tree, it is possible to remove certain edges (namely, any edge included in a circuit) without destroying connectivity. Consequently, a tree may also be characterized as a minimal connected graph in the sense that it contains no proper subgraph which includes all of the vertices and is connected.

If a tree T is a subgraph of a graph G, the edges of G which appear in T are called *branches* relative to T and the edges not included in T are called *chords* relative to T. If all vertices of G are included in T, then T is said to *span* G. It is clear that only connected graphs contain spanning trees, and of these only trees contain unique spanning trees.

In Fig. 1-7 two alternative spanning trees of the same graph are indicated by bold edges. The fact that each of these trees contains four edges is an instance of the following general property:

Theorem 1-5 Every tree with n vertices has precisely $n - 1$ edges.

Proof: Removal of any one edge divides a tree into two components, i.e., into a forest of two trees. (Why never more than two?) Similarly, removal of a second edge yields a forest of three trees. In general, after removal of any $k - 1$ edges, we are left with a forest of k trees. On the other hand, after removal of all edges, the forest clearly consists of n trees (each being an isolated vertex). It follows that the number of edges removed was necessarily $n - 1$. This completes the proof.

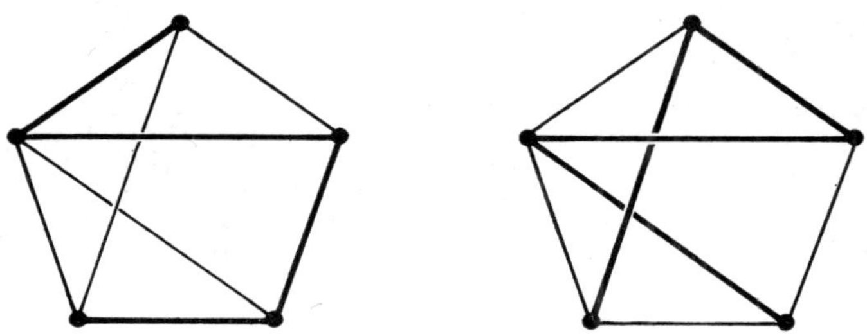

(a) (b)

Figure 1-7

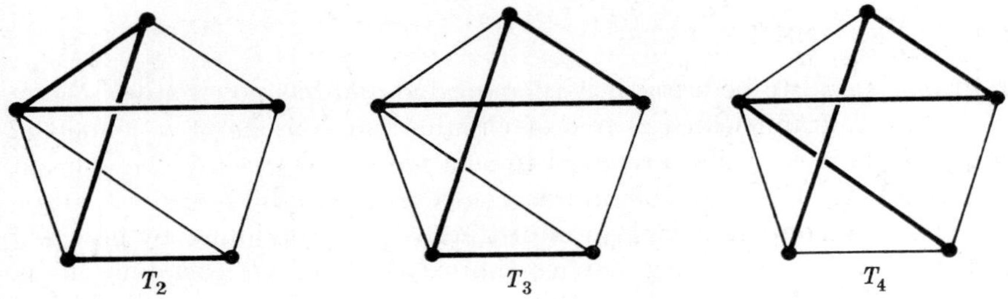

Figure 1-8

By applying the preceding result to each tree of a forest, we obtain the following "generalized" statement:

Theorem 1-6 A forest of k trees which has n vertices has precisely $n - k$ edges.

Given two trees which span the same graph, it is always possible to transform one into the other by constructing a sequence of spanning trees each of which has all but one edge in common with the preceding one. For example, consider the sequence T_1, T_2, T_3, T_4, where T_1 is the tree shown in Fig. 1-7*a* and the remaining elements are shown in Fig. 1-8. (Note that T_4, Fig. 1-8, is the tree shown in Fig. 1-7*b*.)

The four trees constitute a "monotonic" transition from T_1 to T_4, in the sense that each succeeding tree has one more edge in common with the final tree. In general, such a transition from one spanning tree T to another T' can always be achieved as follows: Let e_1 be any edge contained in T' but not in T. Then the (unique) chain in T which joins the end points of e_1 must contain at least one edge $\bar{e}_1$ not appearing in T', since T' has no circuits. Construct the graph T_1, which is identical with T except that it includes e_1 and excludes $\bar{e}_1$. Then T_1 is necessarily also a spanning tree. (Why?) If T_1 and T' are not identical, repeat the process. If the underlying graph has k vertices, then each spanning tree has $k - 1$ edges and it will be necessary to generate at most $k - 1$ intermediate trees before T' is produced.

1-9 DISCONNECTING SETS AND CUT-SETS

In situations where the emphasis is on ways of separating sets of vertices from one another, the concept of a cut-set becomes important. For example, in the study of flows through networks, cross sections of the network separating flow origins from destinations are considered in order to find a restrictive cross section which constitutes a bottleneck. Such bottlenecks determine the flow capacity of the network as a whole.

Before formally defining cut-sets, we introduce a more general con-

cept. If $G = (V,E)$ is a connected graph and if $F \subset E$, then F is called a *disconnecting set* if and only if the subgraph $G' = (V, E - F)$ is not connected. Here $E - F$ denotes the set of edges which are in E but not in F. Disconnecting sets always exist (if G has at least two vertices), since we may take $F = E$. (We assume that G has no loops, since they could play no role in connecting or disconnecting distinct vertices.)

The dashed edges in Fig. 1-9 indicate two disconnecting sets in a graph G, the second of which is a subset of the first. The disconnecting set depicted in Fig. 1-9*a* divides the remaining subgraph into three components, one of which contains the vertices of the set W encircled in the figure. It is clear that in order to disconnect the graph, it would be sufficient to remove only those edges which join a vertex in W to one in $W' = V - W$. These are the edges shown in Fig. 1-9*b*.

In general, if $G = (V,E)$ is connected and if V is partitioned into nonempty sets W and $W' = V - W$, the set of edges joining W with W' is called a *cut-set*. (For every W this set must be nonempty, since G is connected.) For any given graph, the totality of cut-sets determined by various W's forms a subclass of the class of all disconnecting sets and, moreover, every disconnecting set contains at least one cut-set as a subset.

Of particular interest are disconnecting sets which are minimal in the sense that they contain no proper subsets which also disconnect the graph. Minimal disconnecting sets are called *proper cut-sets*. In view of the above observations, proper cut-sets are necessarily cut-sets. However, not every cut-set is proper. The one shown in Fig. 1-9*b*, for instance, is not. In general, if removal of a cut-set F divides the graph into three or more components, the cut-set cannot be proper. For if we return any one edge of F, it can at most reunite two components, and the resulting graph still has at least two components.

If $G = (V,E)$ is a connected graph having at least two vertices and if $v \in V$, then the set of all edges (excepting loops) incident with V is a cut-set corresponding to the partition $\{v\}$, $\{V - v\}$. It is important to note the complementary nature of the concept of a spanning tree and that of a

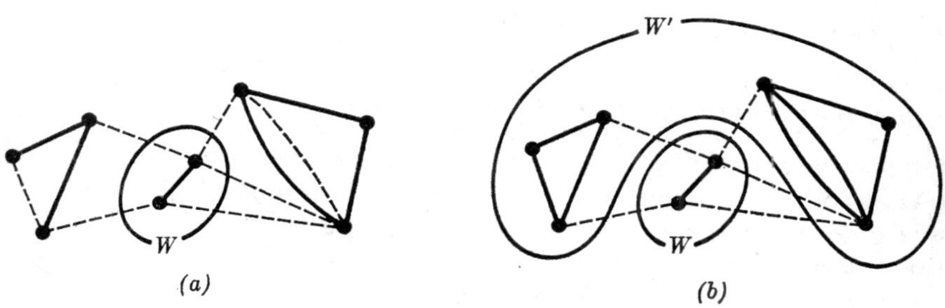

Figure 1-9

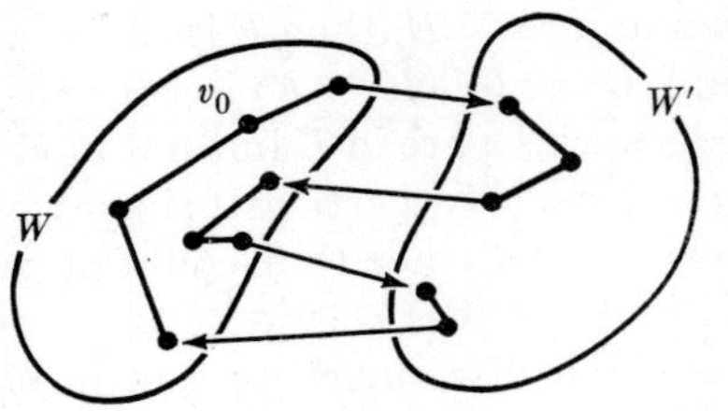

Figure 1-10

cut-set. The former represents a minimal set of edges which connects all vertices of the graph, while the latter is a minimal set of edges which disconnects some vertices from others. The following result combines these notions and is an immediate consequence of our definitions:

Theorem 1-7 Every spanning tree has at least one edge in common with every cut-set of a graph.

Suppose the vertices of a connected graph G are partitioned into nonempty sets W and W' and that $P = e_1, e_2, \ldots, e_n$ is a closed edge progression which starts at and returns to v_0. Assume, without loss of generality, that $v_0 \in W$ (see Fig. 1-10). In traversing P, we either remain in W at all times or else cross between W and W' an even number of times. Formalizing this observation yields the following theorem.

Theorem 1-8 Every closed edge progression in a connected graph has an even number (possibly zero) of elements in common with every cut-set. Consequently, every circuit has an even number of edges in common with every cut-set.

Exercise 1-18 The disconnecting set shown in Fig. 1-9*a* contains two cut-sets which are different from (but not disjoint from) the one shown in Fig. 1-9*b*. Identify these.

Exercise 1-19 The cut-set shown in Fig. 1-9*b* is actually the union of disjoint proper cut-sets. Determine whether this is coincidental or a general property possessed by cut-sets which are not proper cut-sets.

Exercise 1-20 Given a connected graph and two vertex sets W_1 and W_2, where $W_2 \neq W_1$ and $W_2 \neq W_1'$, is it possible for W_1 and W_2 to generate the same cut-set? (Here W_1' denotes the set of all vertices of the graph which are not in W_1.)

Exercise 1-21 Connected graphs which contain cut-sets consisting of single edges can be characterized in terms of circuits. Give such a characterization.

Exercise 1-22 Prove that a cut-set F in a connected graph $G = (V,E)$ is proper if and only if $G' = (V, E - F)$ has precisely two components.

1-10 SOME SPECIAL CLASSES OF GRAPHS

Graphs can be classified in a variety of ways depending on the structural features which are used as the basis for classification. We have already seen graphs classified on the basis of connectivity (or lack of it) and on the basis of whether they are planar or nonplanar. We now introduce a number of other useful classifications.

A graph is said to be *simple* if it has no loops and no parallel edges.

Note that we could also characterize simple graphs as graphs having no circuits composed of fewer than three edges. In many situations it is sufficient to restrict consideration to simple graphs. For example, connectivity (or lack of it) is not affected if loops and parallel edges are removed. Similarly, if the edges have nonnegative lengths associated with them, a search for shortest chains is not affected if loops, as well as all but the shortest edge in each set of parallel edges, are removed.

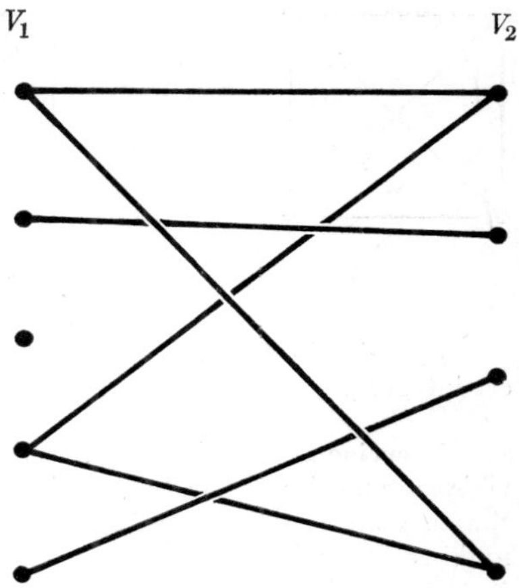

Figure 1-11

A graph is said to be *complete* if every two distinct vertices are adjacent, i.e., joined by an edge. Usually this term is applied to simple graphs. For such graphs there is essentially only one complete graph having a specified number of vertices, since all such graphs are isomorphic. Consequently, we speak of *the* complete graph having k vertices. Figure 1-3, for example, represents the complete graph having five vertices.

A graph is said to be *bipartite* if its vertices can be partitioned into two disjoint sets V_1 and V_2 in such a way that every edge has one end point in V_1 and the other in V_2. More generally, a graph is *k-partite* if its vertices can be partitioned into k disjoint sets $\{V_1, V_2, \ldots, V_k\}$ in such a way that no edge joins two vertices of the same set. It is often convenient to emphasize the structural character of bipartite graphs by displaying the vertices of V_1 and V_2 in separate columns (or rows) as indicated in Fig. 1-11.

A graph is said to be *k-connected* if every pair of distinct vertices v and w are joined by at least k chains which have no common vertices except, of course, v and w. For example, any simple circuit (except a loop) constitutes a 2-connected graph. When $k = 1$, this concept reduces to ordinary connectivity.

If $\delta(v) = k$ for every vertex of a graph, the graph is said to be *regular of degree k* or simply *k-regular*. For example, Fig. 1-12 illustrates several connected 3-regular graphs. Note that the (simple) complete graph having k vertices is $(k - 1)$-regular. Also, according to Theorem 1-3, a k-regular graph has an even number of vertices, if k is odd.

Exercise 1-23 What is the largest integer p such that the (simple) complete graph having q vertices is p-connected?

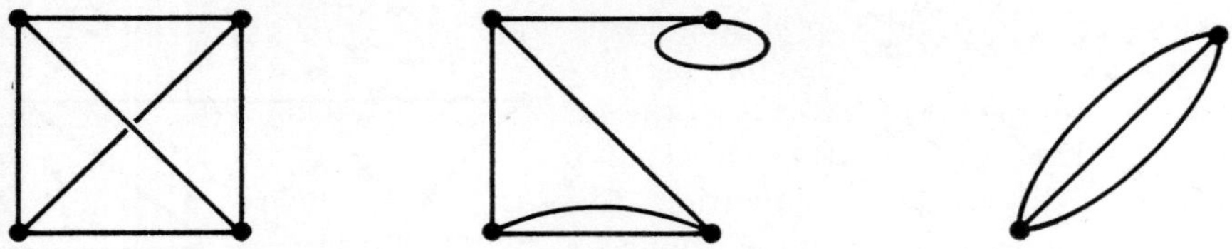

Figure 1-12

Exercise 1-24 Are any complete graphs bipartite? In general, what is the smallest integer p such that the complete graph having q vertices is p-partite?

Exercise 1-25 Figure 1-12 illustrates two connected graphs having four vertices which are not isomorphic although both are 3-regular. However, one of these is not a simple graph. Construct two simple graphs which are connected, have the same number of vertices, and are k-regular for the same k, but which are not isomorphic. Note that this cannot be done for $k = 2$. (Why not?)

Exercise 1-26 Determine whether the following characterization of bipartite graphs is equivalent to the defining one: A graph is bipartite if and only if the totality of its edges forms a cut-set.

Exercise 1-27 Formulate several specific application contexts in which the appropriate graph to consider is bipartite owing to the nature of the problem. (In general, such graphs arise when the vertices represent two distinct types of objects and when the edges serve to describe a relationship which by its nature holds only between unlike objects such as men and jobs or theatergoers and tickets.)

Remark: In the preceding pages, the reader may have noticed the number of structural concepts which were related to some type of optimality. To recall a few of them: a simple chain is a minimal set of edges connecting two given vertices. A spanning tree is a maximal subgraph of a connected graph which contains no circuits and is a minimal subgraph which joins all vertices. A proper cut-set is a minimal set of edges whose removal disconnects a connected graph. A circuit (other than a loop) is a minimal 2-connected graph. Here minimal (maximal) means that no proper subset (superset) or subgraph (supergraph) has the property also.

This theme is strongly present in the remaining pages also, both in the definitions of additional concepts and in the nature of many of the exercises which are considered. We shall be concerned, for example, with shortest and longest routes, maximal network flows, flows having minimal cost, partitioning the vertices of a graph into the fewest "independent" sets, etc. It may be of interest to note the various manifestations of optimality in the subsequent chapters.

Exercise 1-28 List 10 activities with which you are familiar to which the concept of a graph is applicable. Point out some aspect of each which can be studied by means of the graph.

REFERENCES

1 Aleksandrov, P. S.: "Combinatorial Topology," Graylock Press, Rochester, N.Y., 1956.

2 Arnold, B. H.: "Intuitive Concepts in Elementary Topology," Prentice-Hall, Inc., Englewood Cliffs, N.J., 1962.

3 Avondo-Bodino, G.: "Economic Applications of the Theory of Graphs, Gordon and Breach, Science Publishers, Inc., New York, 1962.

4 Ball, W. W. R.: "Mathematical Recreations and Essays," The Macmillan Company, New York, 1960.

5 Berge, C.: "Theory of Graphs and Its Application," John Wiley & Sons, Inc., New York, 1962.

6 Courant, R., and H. Robbins: "What Is Mathematics?" Oxford University Press, London, 1941.

7 Dirac, G. A., and M. D. Stojakovic: The Four-Colour Problem. *Matematicka Biblioteka*, **16:** (1960). MR 22-1946

8 Dynkin, E. B., and W. A. Uspenski: "Multicolor Problems," D. C. Heath and Company, Boston, 1952. (Original in German.)

8a Fiedler, M.: "Theory of Graphs and Its Applications," *Proc. Symp. Smolenice, Czech.*, June 17–20, 1963; Academic Press Inc., New York, 1964.

9 Flament, C.: "Applications of Graph Theory to Group Structure," Prentice-Hall, Inc., Englewood Cliffs, N.J., 1963. Translated by Maurice Pinard, Raymond Breton and Fernand Fontaine.

10 Ford, L. R. Jr., and D. R. Fulkerson: "Flows in Networks," Princeton University Press, Princeton, N. J., 1962.

11 Franklin, P.: The Four Color Problem, *Scripta Mathematica*, 1961.

12 Gale, D.: "The Theory of Linear Economic Models," McGraw-Hill Book Company, New York, 1960.

13 Hilbert, D., and S. Cohn-Vossen: "Geometry and the Imagination," Chelsea Publishing Company, New York, 1952.

14 Kim, W. H., and R. T. Chien: "Topological Analysis and Synthesis of Communication Networks," Columbia University Press, New York, 1962.

15 Koenig, H. E., and W. A. Blackwell: "Electromechanical System Theory," McGraw-Hill Book Company, New York, 1961.

16 König, D.: "Theorie der endlichen und unendlichen Graphen," Acad. Verl. M.B.H., Leipzig, 1936. Reprint, Chelsea Publishing Company, New York, 1950, Zbl.

17 Le Corbeiller, P.: "Matrix Analysis of Electric Networks," John Wiley & Sons, Inc., New York, 1950.

17a Lorens, C. S.: "Flowgraphs," McGraw-Hill Book Company, New York, 1964.

18 Mason, W. P.: "Electromechanical Transducers and Wave Filters," D. Van Nostrand Company, Inc., Princeton, N.J., 1942.

19 Olson, Harry F.: "Dynamical Analogies," D. Van Nostrand Company, Inc., Princeton, N.J., 1943.

20 Ore, O.: "Graphs and Their Uses," Random House, Inc., New York, 1963.

21 Ore, O.: "Theory of Graphs," Colloquium Publications, vol. 38, American Mathematical Society, Providence, R.I., 1962.

22 Reed, M. B.: "Foundation for Electrical Network Theory," Prentice-Hall, Inc., Englewood Cliffs, N.J., 1961.

23 Ringel, G.: "Färbungsprobleme auf Flachen und Graphen," VEB Deutscher Verlag der Wissenschafteng, Berlin, 1959.

24 Riordan, J.: "An Introduction to Combinatorial Analysis," John Wiley & Sons, Inc., New York, 1958.

25 Ryser, H. J.: "Combinatorial Mathematics," Carus Mathematical Monograph No. 14, John Wiley & Sons, Inc., New York, 1963.

26 Sainte-Lague, A.: "Les Reseaux (au Graphes)," Memorial des Sciences Mathematiques, vol. 18, Paris, 1926.

27 Seshu, S., and M. B. Reed: "Linear Graphs and Electrical Networks," Addison-Wesley Publishing Company, Inc., Reading, Mass., 1961.

28 Veblen, O.: "Analysis Situs," American Mathematical Society, Providence, R. I., 1931.

TWO

BASIC CONCEPTS: DIRECTED GRAPHS

2-1 INTRODUCTION

In this chapter we introduce the basic concepts and terms associated with directed graphs, which are graphs having the additional characteristic that every edge has been oriented or given a direction. The discussion is somewhat abbreviated, compared to that of Chap. 1, since many of the concepts are directly analogous to those presented for undirected graphs. On the other hand, some new concepts that by their nature have no undirected counterparts are introduced.

2-2 DIRECTED GRAPHS

In many contexts it is necessary to associate with each edge of a graph an orientation or direction. In the case of a geometric graph this can be interpreted as a direction of traversal of the edge, whereas in the case of an abstract graph it means merely that the end points of each edge are distinguished from each other by ordering them. Thus the only structural difference between an undirected graph and a directed graph (also called *digraph*) is that the end points of an edge constitute an ordered rather than an unordered pair of vertices.

In applications of graph theory the need for introducing directions on the edges arises in two ways. In some situations, the edges represent a relationship between pairs of vertices which is not symmetric. In the context of the structure of a system of city streets, for example, it becomes necessary to represent one-way streets. Or in the context of communications being transmitted between men or machines, devices that are essentially unidirectional may be employed. In other situations it is necessary to introduce directions in order to establish a frame of reference and thus avoid ambiguities. For example, in a connection between electrical devices, it is necessary to designate one direction as "positive" in order to unambiguously describe current flow, even though the actual direction of flow may not be essentially restricted.

Defined formally, a *directed graph* consists of a nonempty set V, a set A disjoint from V, and a mapping Δ of A into $V \times V$. The elements of V and A are respectively called *vertices* and *arcs* (or directed edges), and Δ is called the *directed incidence mapping* associated with the directed graph. If $a \in A$ and $\Delta(a) = (v,w)$, then arc a is said to have v as its *initial vertex* and w as its *terminal vertex*. The symbolism $a \simeq (v,w)$ will be employed to convey the same information in contexts where Δ is not explicitly represented. (As in the undirected case, it is seldom necessary to symbolize the incidence mapping itself. Its existence is, however, fundamental to the concept of a directed graph.) For our purposes the number of vertices and arcs is again assumed to be finite.

Directed graphs will usually be denoted by D or (V,A,Δ), or by (V,A) when Δ is not used explicitly. Given a directed graph $D = (V,A,\Delta)$, the *associated undirected graph* is the graph $G = (V,A,\Phi)$ whose incidence mapping is defined as follows: $\Phi(a) = (v \,\&\, w)$ whenever $\Delta(a) = (v,w)$. Thus G is obtained by disregarding the ordering of the end points of each arc. The structural terms introduced in Chap. 1 are also applied to a directed graph, with the understanding that the associated undirected graph is being described. For example, two arcs of D are said to be parallel (adjacent) if the corresponding edges of the associated undirected graph are parallel (adjacent).

Two directed graphs are said to be *isomorphic* if their associated undirected graphs are isomorphic in the undirected sense and if, also, the end points of each pair of corresponding arcs are ordered the same way. Expressed formally, directed graphs $D = (V,A,\Delta)$ and $D' = (V',A',\Delta')$ are said to be isomorphic if the elements of V and A can be placed in 1-1

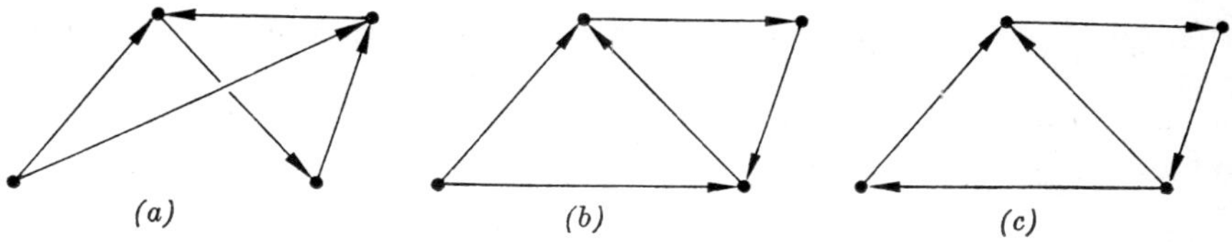

Figure 2-1

correspondence with those of V' and A', respectively, in such a way that

$$\Delta'(a') = (v',w')$$

if and only if

$$\Delta(a) = (v,w)$$

where a', v', and w' denote the images of a, v, and w, respectively. Thus the two directed graphs shown in Fig. 2-1a and b are isomorphic. On the other hand, the directed graph shown in (c) is not isomorphic with them, despite the fact that the associated undirected graphs of all three are isomorphic in the undirected sense.

If D and D' are isomorphic directed graphs and if D' is a directed geometric graph, then D' is called a *geometric realization* of D. Theorems 1-1 and 1-2, which characterize the realizability of undirected graphs, are equally applicable to directed graphs, since the possibility of making the necessary constructions in $\mathcal{E}^3$ does not depend on whether or not the curves are to be considered as oriented. Similarly, a directed graph is planar (realizable in $\mathcal{E}^2$) if and only if its associated undirected graph is planar.

2-3 TERMINOLOGY DESCRIBING LOCAL STRUCTURE

The following terms are useful for describing certain structural features of a directed graph which have no undirected counterpart. If $a_1 \simeq (v,w)$ and $a_2 \simeq (v,w)$, arcs a_1 and a_2 are said to be *strictly parallel*. [If $a_3 \simeq (w,v)$, then a_1 and a_3 are parallel but not strictly parallel.] If $a \simeq (v,w)$, arc a is said to be directed *away from* vertex v and *toward* vertex w. Arc a is also said to be *positively incident* with its initial vertex v and *negatively incident* with its terminal vertex w. The number of arcs which are positively incident with v is called the *positive degree* of v and denoted by $\delta^+(v)$. The *negative degree* of v is defined analogously and denoted by $\delta^-(v)$. (A directed loop incident with v is considered to be both positively and negatively incident with v.) The signed degrees $\delta^+(v)$ and $\delta^-(v)$ are related to the unsigned degree $\delta(v)$ introduced earlier by the following equation:

$$\delta(v) = \delta^+(v) + \delta^-(v)$$

(For infinite graphs this relationship remains true if interpreted as an equation relating cardinal numbers.) Since every arc is positively incident with one vertex and negatively incident with one vertex, it is evident that

$$\sum_{v \in V} \delta^+(v) = \sum_{v \in V} \delta^-(v) = |A|$$

where $|A|$ denotes the number of arcs of the graph, which is the directed counterpart of the relationship

$$\sum_{v \in V} \delta(v) = 2|A|$$

A directed graph is said to be *simple* if it has no strictly parallel arcs and no loops. Note that if a simple directed graph has two parallel but oppositely oriented arcs, the associated undirected graph is not simple in the undirected sense, since it has parallel edges. The arcs of a simple directed graph can be represented, without ambiguity, by ordered pairs of vertices, since at most one arc joins a given pair of vertices in a specified direction. (Note that this is a specialization of, and not a deviation from, the concept of having a set V of vertices and a second set A of arcs. No restriction was placed on the nature of the elements of A, and so in particular they may be elements of $V \times V$.)

2-4 ARC PROGRESSIONS, PATHS, AND CYCLES

An *arc progression of length n* is a sequence of (not necessarily distinct) arcs $a_1, a_2, \ldots, a_n$ such that for an appropriate sequence of $n + 1$ vertices $v_0, v_1, \ldots, v_n$ we have $a_i \simeq (v_{i-1}, v_i)$ for $i = 1, 2, \ldots, n$. For example, in Fig. 2-2 the sequence a_1, a_5, a_5, a_3 forms an arc progression of length 4 with v_3, v_2, v_2, v_2, v_1 as the corresponding sequence of vertices. An arc progression is said to be *closed* if $v_0 = v_n$ and *open* if $v_0 \neq v_n$. In the latter case it is called an arc progression *joining* v_0 and v_n and, more specifically, leading *from* v_0 *to* v_n.

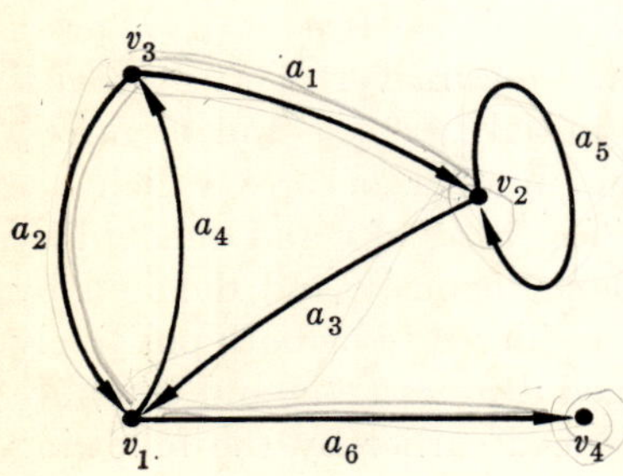

Figure 2-2

It is clear that an open (closed) arc progression in a directed graph determines a corresponding open (closed) edge progression in the associated undirected graph. The converse is not generally true, however. For example, in Fig. 2-2, the sequence a_1, a_2, a_6 determines an open edge progression joining v_2 and v_4, but it fails to constitute an arc progression for lack of consistent directions.

An arc progression in which no arc is repeated is called a *path progression* or *cycle progression*, depending on whether it is open or closed. The corresponding set of arcs, without regard to sequencing, is called a *path* or *cycle*, respectively. If the vertices $v_0, v_1, \ldots, v_n$ are all distinct (in which case the arcs are also distinct), the path progression or cycle progression, as well as the corresponding path or cycle, is said to be *simple*. A directed graph is said to be *cyclic* if it contains at least one cycle, and *acyclic* otherwise. (Note that a loop is a special instance of a cycle.)

Exercise 2-1 Show that the arcs of a path can be arranged in only one way to form a path progression if the path is simple. Show also that the converse is not true in general.

Exercise 2-2 Prove that if there exist paths from v_1 to v_2 and from v_2 to v_3, where $v_1 \neq v_3$, then there exists a path from v_1 to v_3.

Exercise 2-3 Prove that a nonsimple cycle can be partitioned into two or more simple cycles.

Exercise 2-4 Prove that a nonsimple path from v to w can be partitioned into a simple path from v to w and one or more simple cycles.

Exercise 2-5 Is it true in general that if $\delta^+(v) \geq 1$ and $\delta^-(v) \geq 1$ for every vertex v of a finite directed graph, then every vertex is incident with at least one cycle?

Exercise 2-6 Prove that if D contains a path from w to v and one from v to w, then D contains a cycle. However, give an example to show that there may be no cycle incident with both v and w.

Exercise 2-7 Prove that if $\delta^+(v) > 0$ for every vertex of a directed graph D, then D is necessarily a cyclic graph. [The same is true if $\delta^-(v) > 0$ for every vertex v.]

Exercise 2-8 Show that there is no path in Fig. 2-3 which passes through each junction vertex once and only once, no matter how the edges are oriented.

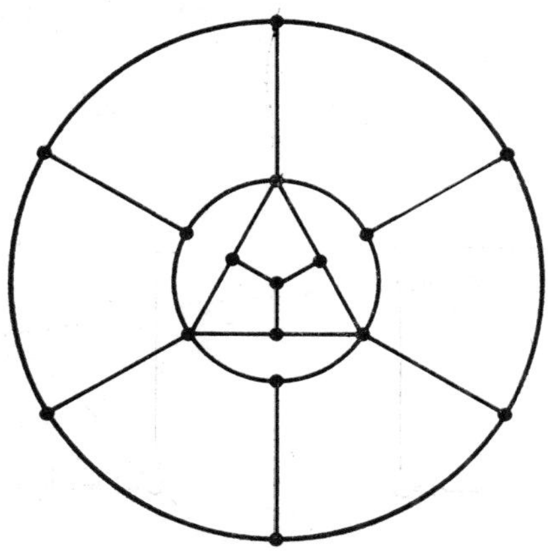

Figure 2-3

Exercise 2-9 Use an argument similar to that of Exercise 2-8 to prove that it is not possible for a rook which starts at the southwest square of a chessboard to end at the northeast square and pass through every square once and only once. Note that in each move a rook can travel any number of squares along the row or the column of the square on which it is located.

2-5 STRONG CONNECTIVITY

A directed graph is said to be *strongly connected* if, for every pair of distinct vertices v and w, there exists a path from v to w as well as one from w to v. It is evident that strong connectivity of a directed graph implies connectivity of the associated undirected graph. The converse, of course, is not generally true. In Fig. 2-4, D_1 is strongly connected but D_2 is not. In Chap. 3 we shall develop necessary and sufficient conditions for an undirected graph to have the property that an appropriate orientation of its edges will produce a strongly connected graph.

A directed graph is said to be strongly k-connected if, for every pair of distinct vertices v and w, there exist at least k paths from v to w which have no vertices (and hence also no arcs) in common, except, of course, for v and w. For a directed graph to be strongly k-connected, it is clearly necessary, but not sufficient, that the associated undirected graph be k-connected in the undirected sense.

Consider a directed graph whose arcs represent the (directed) communication channels between a group of individuals. If the graph is strongly connected, then it is possible for every individual to communicate with every other individual in at least one way (i.e., by means of at least one path). If the graph is strongly k-connected, then there are at least k disjoint communication paths from any one individual to any other individual. Thus one would have to disrupt communications at at least k points in order to deny communication between any specific pair of individuals completely.

Exercise 2-10 Prove that a finite directed graph is strongly connected if and only if there exists a closed arc progression which includes every arc at least once.

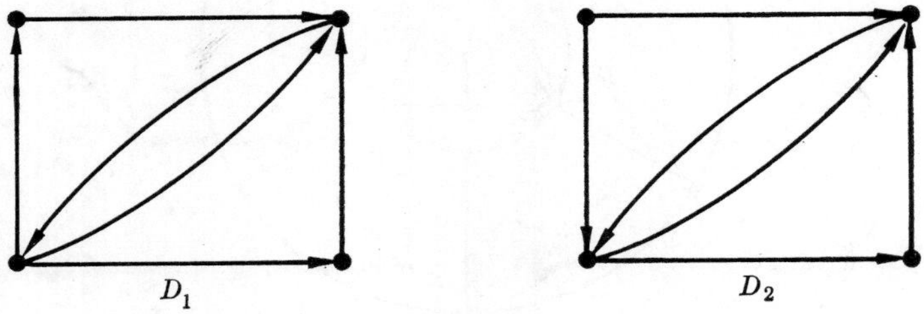

Figure 2-4

Exercise 2-11 Prove that a directed graph $D = (V,A)$ is strongly connected if and only if, for every vertex partitioning $\{W, V - W\}$, the corresponding cut-set in the associated undirected graph includes at least one arc directed from W to $V - W$ and at least one arc directed from $V - W$ to W.

Exercise 2-12 Prove that if $C_1, C_2, \ldots, C_k$ is a sequence of cycles in a directed graph D, where every two consecutive cycles have at least one common vertex, then the subgraph determined by the union of these cycles is strongly connected.

Exercise 2-13 Let D be a directed graph having a vertex v such that every other vertex of D is joined to v by an arc in each direction and such that there is at least one arc which is not incident with v. Prove that for any integer $k \geq 4$ and for any two (not necessarily distinct) vertices v and w there exists an arc progression from v to w having precisely k terms.

2-6 TREES AND CUT - SETS

When the terms "tree," "forest," "disconnecting set," "cut-set" and "proper cut-set" are used without qualification, it is to be understood that arc directions are ignored and the associated undirected graph is being described. However, if we do take cognizance of arc directions, several additional notions emerge.

A directed graph is said to be a *directed tree* rooted at v_0 if (1) it forms a tree in the undirected sense and (2) the unique chain determined by v_0 and any other vertex w is in fact a path from v_0 to w. It is clear that a tree can be rooted at at most one vertex and that, in general, trees in directed graphs may not be directed trees. In Fig. 2-5a, for instance, there can be no directed tree which spans the graph. (Why not?) In Fig. 2-5b, on the other hand, the bold arcs illustrate a directed tree rooted at v_0 which spans the directed graph shown. In Chap. 3 we shall develop a systematic procedure for finding a directed tree, rooted at a specified vertex v_0, whose (unique) paths to all other vertices represent shortest paths relative to a specified distance function defined on the arcs.

Recall that cut-sets, and proper cut-sets in particular, are sets of edges which join W and W', where $\{W,W'\}$ is a partitioning of the vertices

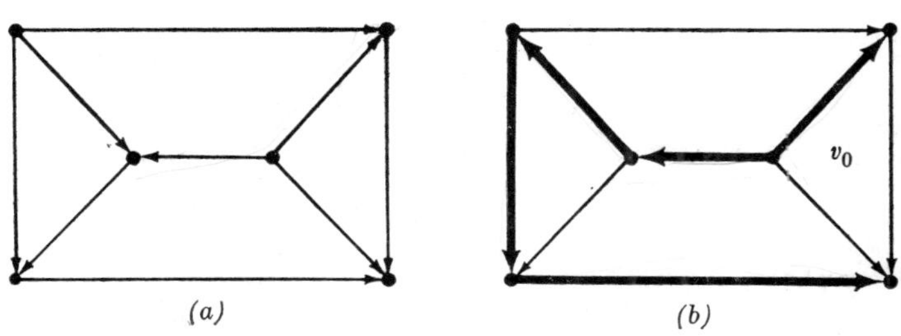

Figure 2-5

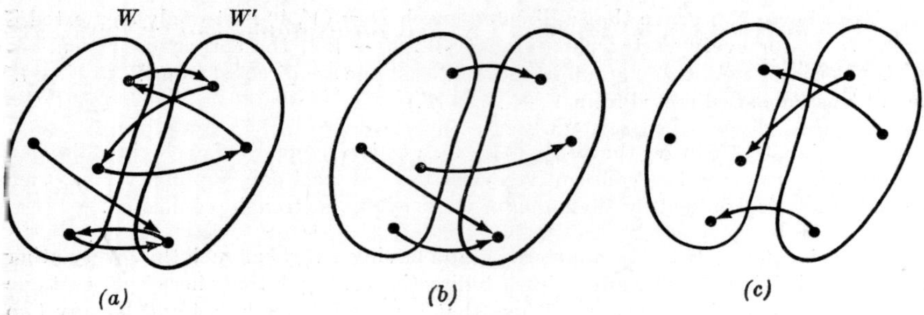

Figure 2-6

of a connected graph into two disjoint, nonempty sets. The deletion of a cut-set breaks the graph into two or more components. The deletion of a proper cut-set results in precisely two components.

In a directed graph the arcs of a cut-set can be classified into two groups: those directed from W to W' and those directed from W' to W. Removal of the former set disconnects all paths from W to W', while removal of the latter disconnects all paths from W' to W. We shall have occasion later (particularly in connection with network flows) to consider these two sets individually. The former will be called a $W \to W'$ (read "W to W'") cut-set, and the later a $W' \to W$ cut-set. Figure 2-6*b* and *c* exhibits these subsets of the cut-set shown in (*a*).

Note that if D is strongly connected, then for every partition $\{W,W'\}$ both directed cut-sets are necessarily nonempty. (See Exercise 2-11 concerning the converse of this statement.)

Exercise 2-14 If the edges of a tree having n edges are randomly assigned directions, what is the probability that the result will be a directed tree rooted at some vertex? (The answer depends on n, but not on the specific structure of the tree.)

Exercise 2-15 Prove that if the edges of a complete simple graph are randomly assigned directions, the resulting directed graph necessarily contains a rooted directed tree which spans the graph.

Exercise 2-16 Prove that if a simple graph is not complete, there is at least one way to direct its edges so that it does not contain a spanning tree rooted at any vertex.

2-7 DIRECTED GRAPHS AND BINARY RELATIONS

If ρ is a binary relation in a set S, the *graph of* ρ, denoted by D_ρ, is the directed graph whose vertices are the elements of S and is such that there is an arc $a \simeq (s,t)$ if and only if $s \; \rho \; t$. Conversely, if $D = (V,A)$ is a directed graph having no strictly parallel arcs, the *binary relation associated with D*, denoted by ρ_D, is the binary relation in V such that $v \; \rho_D \; w$ if and only if $a \simeq (v,w)$ for some arc a.

The directed graph shown in Fig. 2-7, for example, can be considered as representing a certain binary relation defined in a set of eight elements. In fact, any binary relation R defined in a set S can be represented by an appropriate directed graph whose vertices correspond to the elements of S. The graph exhaustively catalogs the relation by enumerating all pairs of related elements. Thus in a sense a directed graph is the "extensive form" of a relation, i.e., it is a complete tabulation of those pairs for which the relation holds. Given a particular directed graph, there may or may not be a characterization of the same binary relation given by a simple rule that avoids complete tabulation. The graph of Fig. 2-7, for instance, is actually an alternative representation of the inclusion relation $\subset$ defined on the subsets of a set having three elements. This is apparent from Fig. 2-8, where the vertices are identified with appropriate subsets of the set $\{x,y,z\}$.

We wish to point out that it is generally not profitable to take a binary relation, originally given in some "closed form," and generate the corresponding graph in order to study the properties of the relation. Nor is it often possible, when given a directed graph, to find a simple closed form for the binary relation which it represents. But at times this may be feasible, and the tools of one theory can be applied to the problems of the other. In any case, the conceptual relationship between the two notions should be kept in mind. Note that in one sense the notion of a directed graph is more general, in that varying numbers of strictly parallel arcs can be used to reflect a quantitative degree of relatedness.

We shall borrow some of the terminology of binary relations to describe special classes of directed graphs having no strictly parallel edges. A *reflexive graph* is one having a loop at each vertex. A *symmetric graph* is one for which every arc $a \simeq (v,w)$ is matched by an arc $a' \simeq (w,v)$. A

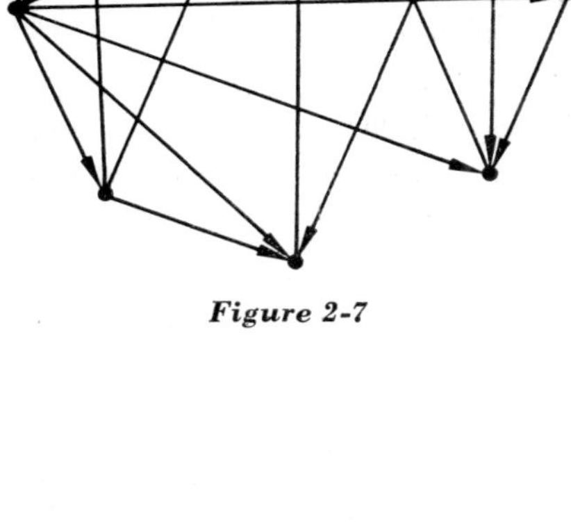

Figure 2-7

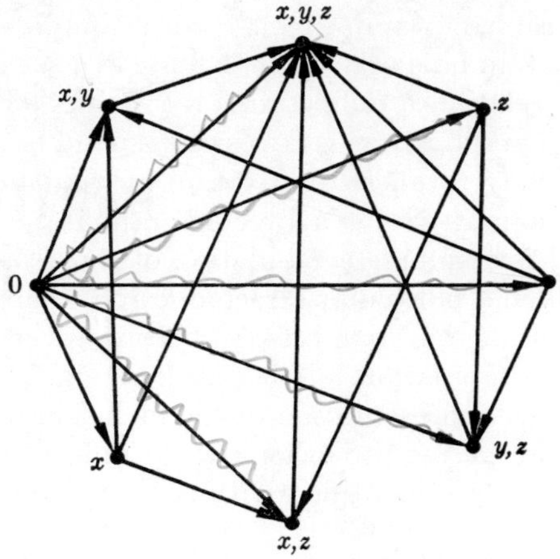

Figure 2-8

transitive graph is one for which the existence of arcs $a \simeq (v,w)$ and $b \simeq (w,x)$ implies the existence of an arc $c \simeq (v,x)$. It follows from this that the existence of any path from v to another vertex y implies the existence of an arc from v to y.

The graph shown in Fig. 2-8 is transitive, but it fails to be either reflexive or symmetric. In fact, it is *asymmetric*, meaning that the existence of an arc $a \simeq (v,w)$ precludes the existence of an arc $a' \simeq (w,v)$.

Exercise 2-17 Referring to the definition of a transitive graph in the text, vertices v, w, and x were not required to be distinct. Use this fact to show that if D is transitive but not reflexive, then there exists a vertex which is not incident with any cycle.

Exercise 2-18 A graph can be symmetric and transitive, yet fail to be reflexive. (*a*) Produce an example to show this. (*b*) What property characterizes all such graphs?

Exercise 2-19 Let vertices correspond to the integers 0 to 12 and construct graphs characterizing the binary relation R, where vRw means: (*a*) $v = w^2$; (*b*) v and w are congruent mod 4 (that is, $v - w$ is a multiple of 4); (*c*) v and w are relatively prime.

Exercise 2-20 A "family tree" T is a directed graph displaying the relation "is a parent of" defined for a selected group of individuals. Develop a systematic procedure for deriving from T the graph of the relation "is a great-grandparent of" defined for the same group of individuals.

Remark: In the remainder of this book, except where otherwise noted, all graphs mentioned are assumed to be finite graphs.

THREE

PARTITIONS AND DISTANCES IN GRAPHS

3-1 INTRODUCTION

The topics developed in this chapter are related to two general themes. The first concerns partitioning the edges, arcs, or vertices of a graph into sets of some specified structural type. For example, the famous problem of the seven bridges of Königsberg involves partitioning the edges of a particular graph into the smallest number (hopefully, 1) of circuits or chains. The second theme is the measurement of distances in graphs. The determination of a longest or "critical" path through a PERT network exemplifies this type of problem. Here one determines the shortest time within which some total project can be completed by finding a most time-consuming path of component activities.

3-2 EDGE PARTITIONS

The first general problem to be considered is that of partitioning the edges of a graph into the fewest disjoint subsets each of which

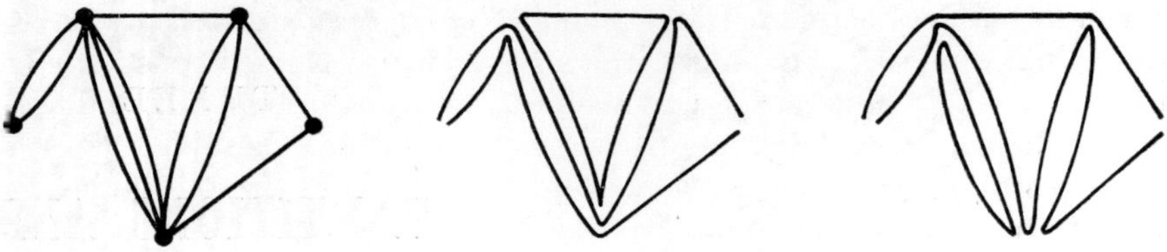

Figure 3-1

is either a chain or a circuit (which need not be simple). Partitions of this type will be called *coverings,* and coverings involving the fewest possible chains and circuits will be called *minimal coverings.* Note that each individual edge is either a chain or (in the case of a loop) a circuit. Consequently, every graph has coverings. Moreover, since we are considering finite graphs, clearly minimal coverings always exist. For example, Fig. 3-1 shows two different coverings of the same graph. Each consists of two chains and one circuit. In either of these, we could incorporate the circuit into one of the chains, thus forming a nonsimple chain and reducing the number of sets in the covering by 1.

Isolated vertices are clearly of no consequence when considering coverings. Also, a minimal covering for a disconnected graph can be obtained by finding a minimal covering for each component having edges. Consequently, we shall here concern ourselves only with connected graphs. Of particular interest are graphs which can be covered by a single chain or circuit, that is, graphs the totality of whose edges form a chain or a circuit. Such graphs are said to be *unicursal,* inasmuch as it is possible to traverse all edges in a continuous manner without repeating any edge.

Returning to the general problem, the characterization of minimal coverings depends essentially on whether or not any vertices have odd degree. (Recall that when such vertices exist, they occur in pairs.) Graphs—whether connected or not—all of whose vertices are of even degree are called *Euler graphs.* For connected Euler graphs, minimal coverings are characterized as follows:

Theorem 3-1 If $G = (V,E)$ is a connected Euler graph, then E is a circuit and hence constitutes the unique minimal covering of G.

Proof: If G has any loops, we remove them initially and consider the reduced graph, all of whose vertices are still of even degree. Starting at any vertex v_1, we proceed along any edge to its other end point, say, v_2. Since v_2 is of even degree, it is incident with a second edge, which leads us to a vertex v_3. We can always proceed in this manner so long as $v_n \neq v_1$, since the evenness of all degrees assures us that we can always "leave" any other vertex after having "entered" it, using an edge not previously traversed. Since there are a finite number of edges, ultimately we have $v_n = v_1$ for some n. The edges which have been traversed constitute a circuit. If every edge has been traversed, we are through. If not, we remove the traversed circuit and consider the resulting subgraph. Every vertex is still of even degree, (why?), and hence we can repeat the preceding construction, finding and removing a second circuit. After a finite number of steps, removal of the next circuit will leave no remaining edges. This set of circuits, together with the loops removed at the outset, constitutes a covering of G. But since G is connected, the union of these circuits, that is, E, is again a circuit. (Verify this in detail.) This completes the proof.

A circuit which covers a graph is called an *Euler circuit*. Hence, according to Theorem 3-1, every connected Euler graph possesses an Euler circuit. Conversely, it is easily seen that a graph which possesses an Euler circuit is a connected Euler graph (provided that it has no isolated vertices). In contrast to the preceding result, minimal coverings of connected graphs having vertices of odd degree consist solely of chains. (Of course, these need not be simple chains.) More precisely, we have the following:

Theorem 3-2 Given a connected graph $G = (V,E)$ having $2n$ vertices of odd degree, where $n \geq 1$, every minimal covering of G consists of n chains, each of which joins two vertices of odd degree.

Proof: We initially augment G by adding n edges, each of which joins two vertices of odd degree, in such a way that every such vertex is incident with one of these edges. Let F denote this set of edges. Then

$$G' = (V, E \cup F)$$

is a connected Euler graph, and hence $E \cup F$ is a circuit. When the edges of F (no two of which are adjacent) are removed, the circuit is broken into n chains, which cover G and which have as end points the $2n$ vertices of odd degree. Hence coverings of the type described in the statement of the theorem exist.

Now consider an arbitrary covering of G. It must include at least n chains in order to generate $2n$ vertices of odd degree. (Why?) If it includes more than n chains, or any circuits, it cannot be minimal, since a covering with n members has been shown to exist. If the covering consists of precisely n chains, they must be of the stated variety. (Why?) Hence only such sets of n chains constitute minimal coverings. This completes the proof.

A characterization of unicursal graphs follows immediately from Theorems 3-1 and 3-2:

Theorem 3-3 A connected graph is unicursal if and only if it has either 0 or exactly 2 vertices of odd degree. In the former case it is

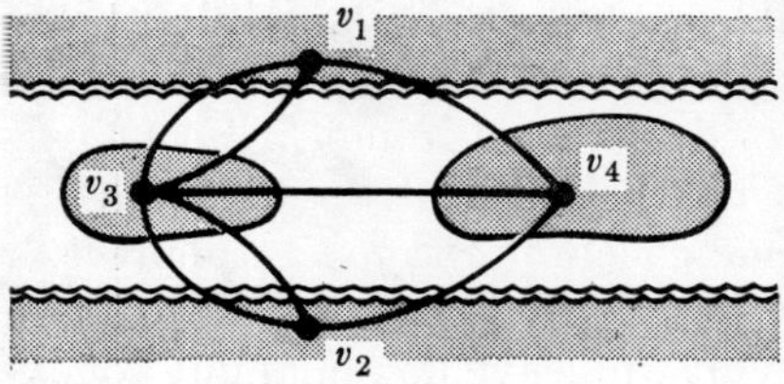

Figure 3-2

covered by a circuit; in the latter case, by a chain joining the two vertices of odd degree.

A classic problem concerning unicursal graphs is the Königsberg bridge problem. It arose with regard to a promenade to be taken across seven bridges which interconnected two islands with the two banks of the Pregel river and with each other, in Königsberg. Letting v_1 and v_2 denote the two banks and v_3 and v_4 the islands, the graph of Fig. 3-2 indicates the manner in which the bridges joined the land masses. The problem, studied in 1736 by L. Euler, was this: Is it possible to describe a continuous route which traverses each bridge precisely once; i.e., is the graph of Fig. 3-2 unicursal? Since all four vertices have odd degree, it is not. According to Theorem 3-2, at least two chains are required to cover this graph. Figure 3-3 exhibits two minimal coverings. (In each case, one chain is shown dashed and the other solid.)

Exercise 3-1 Find minimal coverings for the graphs shown in Figs. 1-2, 1-3, 1-5, and 1-6.

Exercise 3-2 If a connected graph is not unicursal (and hence has at least four vertices of odd degree), prove that it has at least two distinct minimal coverings.

Exercise 3-3 If a minimal covering of a connected graph G has k chains and/or circuits, with $k > 1$, then removal of an edge produces a subgraph whose minimal coverings may have $k - 1$, k, or $k + 1$ elements. Give examples illustrating all three possibilities.

Exercise 3-4 Prove that every cut-set of a connected Euler graph has at least two edges and that it necessarily has an even number of edges.

3-3 ARC PARTITIONS

We next consider an analogous general problem for directed graphs. In this case a *covering* means a partitioning of the arcs into disjoint paths and cycles. The problem is formally the same: to characterize the minimal coverings of an arbitrary connected directed graph.

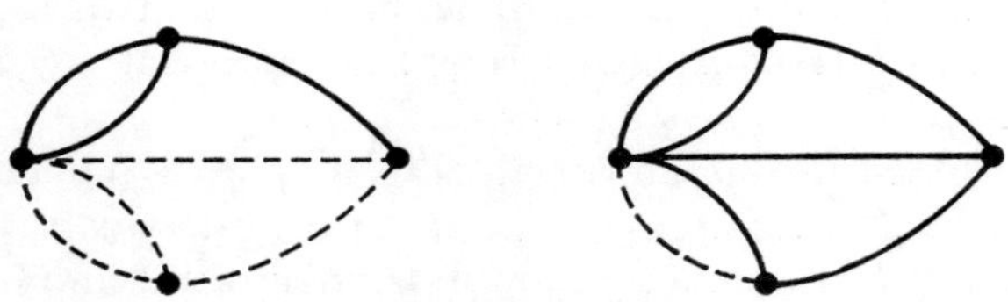

Figure 3-3

It is convenient here to partition the vertices of an arbitrary directed graph $D = (V,A)$ into disjoint sets R, S, and T as follows:

$$R = \{v \in V | \delta^+(v) = \delta^-(v)\}$$
$$S = \{v \in V | \delta^+(v) > \delta^-(v)\}$$
$$T = \{v \in V | \delta^+(v) < \delta^-(v)\}$$

The symbols R, S, and T retain these meanings throughout this section.

A directed graph is said to be *pseudosymmetric* at v if $\delta^+(v) = \delta^-(v)$. A directed graph which is pseudosymmetric at every vertex is called a *directed Euler graph.* (Note that the associated undirected graph is then necessarily an Euler graph in the undirected sense.) Thus a directed Euler graph is one such that $R = V$. It was seen earlier (Chap. 2) that for any directed graph (V,A) we have

$$\sum_{v \in V} \delta^+(v) = \sum_{v \in V} \delta^-(v)$$

It follows that S is nonempty if and only if T is nonempty and moreover that

$$\sum_{v \in S} [\delta^+(v) - \delta^-(v)] = \sum_{v \in T} [\delta^-(v) - \delta^+(v)]$$

if S and T are nonempty. For directed graphs, the characterization of minimal coverings depends essentially on whether or not S and T are empty. If they are empty, we have the following:

Theorem 3-4 Given a directed Euler graph $D = (V,A)$ which is connected, A is a cycle and hence constitutes the unique minimal covering of D.

The proof, which is entirely analogous to that of Theorem 3-1, is omitted. For other directed graphs, we have the following:

Theorem 3-5 If a directed graph $D = (V,A)$ is connected but not pseudosymmetric, then every minimal covering of D consists of k paths each of which joins a vertex in S to one in T, where

$$k = \sum_{v \in S} [\delta^+(v) - \delta^-(v)] = \sum_{v \in T} [\delta^-(v) - \delta^+(v)]$$

Proof: As a first step, we augment D by adding k arcs in such a way that the enlarged graph is a directed Euler graph. Specifically, for each vertex $w \in T$ we add $\delta^-(w) - \delta^+(w)$ arcs having w as initial vertex and some vertex of S as terminal vertex. We can do this in such a way that, for every vertex $v \in S$, precisely $\delta^+(v) - \delta^-(v)$ arcs have v as terminal vertex. Let B denote this set of k arcs, and let $D' = (V, A \cup B)$

Since D' is a directed Euler graph, $A \cup B$ is a cycle. If an arc $b_1 \in B$ is removed, the remaining arcs form a path from some $v \in S$ to some $w \in T$. If a second arc $b_2 \in B$ is now removed, this path is broken into two paths, each joining a vertex in S to one in T. This would fail to be the case only if b_1 and b_2 were adjacent arcs, but that is impossible. (Why?) Proceeding in this manner, after all k arcs are removed, we are left with k paths having the stated property, which form a covering of D.

It remains for us to prove that a covering of this type is minimal and that no other coverings are minimal. To do this, we note that an arbitrary covering includes at least $\delta^+(v) - \delta^-(v)$ paths having $v \in S$ as initial vertex. (There will be more than this number if and only if some paths of the covering have v as terminal vertex.) Hence, at least k paths originate in S. For similar reasons, at least k paths terminate in T. If the covering under consideration has more than k paths or any cycles, it cannot be minimal, since we have previously produced a covering having k members. If the covering has precisely k paths and no cycles, then the k paths originating in S and those terminating in T are the same. In this case the covering has the stated characteristics, and the proof is complete.

The following interesting class of problems, considered by Berge (Ref. 5 of Chap. 1), involves an application of Theorem 3-4. This application illustrates the fact that the selection of an appropriate graph to analyze is sometimes the crucial step in applying graph theory and that this graph is not always evident from the nature of the problem.

Let the finite set $X = \{x_1, x_2, \ldots, x_p\}$ be regarded as an "alphabet," whose elements are "letters." A sequence $y_1, y_2, \ldots, y_q$, where each y_i is a letter, will be regarded as a "q-letter word" in the alphabet X. A cyclic sequence of letters $z_1, z_2, \ldots, z_r$ is a sequence of letters together with the convention that z_1 is regarded as the successor of z_r (so that, for example, z_{r-1}, z_r, z_1, z_2 is considered to be a sequence of four consecutive letters of the cyclic sequence).

The problem to be considered is this: Given an alphabet X having p letters and given a positive integer q, find a shortest cyclic sequence S of letters such that every q-letter word appears at least once as q consecutive letters of S. Note that there are precisely p^q distinct q-letter words which are formable in X, so S necessarily contains at most $q \cdot p^q$ terms (since we could simply itemize all words, one after another) and at least p^q terms. We shall see that for $p \geq 2$ and $q \geq 2$ the latter bound is precise; i.e., we shall produce a cyclic sequence containing every q-letter word and having only p^q terms.

We construct the directed graph D whose vertices are the p^{q-1} different $(q-1)$-letter words formable from X and whose arcs are the p^q different q-letter words formable from X. The initial and terminal vertices of the arc $y_1, y_2, \ldots, y_k$ are taken to be $y_1, y_2, \ldots, y_{k-1}$ and $y_2, y_3, \ldots, y_k$, respectively. It is readily verified that D is a connected Euler graph and consequently has at least one Euler cycle, which necessarily has precisely p^q terms.

Figure 3-4 illustrates the construction when $p = 3$ (the specific alphabet being $X = \{a,b,c\}$) and $q = 3$. The arc from *ba* to *ac*, for example, is the arc *bac*. (For simplicity, we have omitted commas between letters.) The arcs are numbered 1 to 27 to indicate an Euler cycle.

In general, having found an Euler cycle, we form the cyclic sequence (having p^q terms) $z_1, z_2, \ldots, z_k$, where z_i is the first letter of the *i*th arc in the Euler cycle. In the example, this sequence (without commas) is

$$aabcababbbcbcccacaaacbaccbb$$

The cyclic sequence S constructed in this manner has the required characteristics. To see this, note that the $q - 1$ arcs following any arc $y_1, y_2, \ldots, y_q$ in the Euler cycle necessarily have the form

$$\begin{array}{l} y_2y_3 \cdots y_{q-1}y_qx_1 \\ y_3y_4 \cdots y_qx_1x_2 \\ \cdots\cdots\cdots\cdots \\ y_qx_1 \cdots x_{q-2}x_{q-1} \end{array}$$

for appropriate letters $x_1, x_2, \ldots, x_{q-1}$. The corresponding terms of

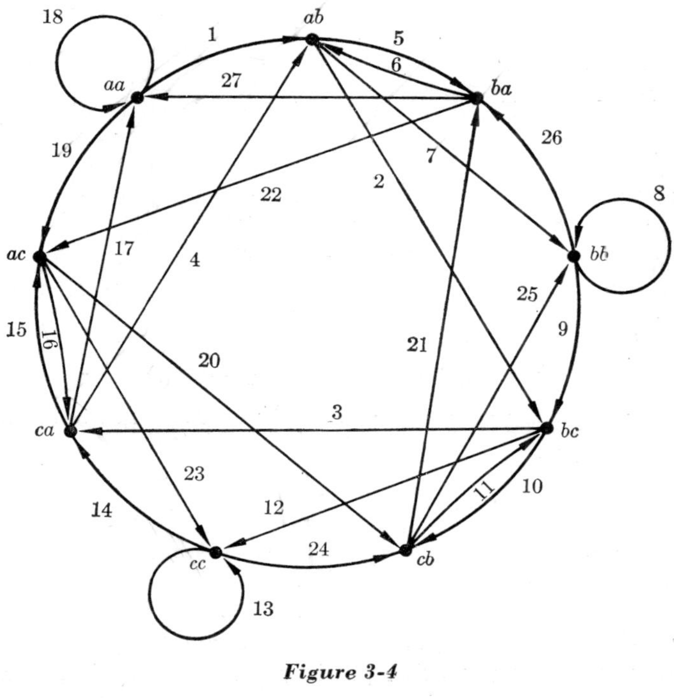

Figure 3-4

S are

$$y_1y_2 \cdot \cdot \cdot y_q$$

which is the arc (that is, q-letter word) with which we arbitrarily started. Thus S clearly contains every q-letter word and is as short as possible. S also is a solution to the problem of finding a longest cyclic sequence in which no q-letter word occurs more than once. (Why?)

Note that a shortest cyclic sequence containing all q-letter words has p^q terms also in the cases when $q = 1$ and/or $p = 1$. (If $q = 1$, any sequence itemizing the p letters will do; while if $p = 1$, the single letter of the alphabet constitutes the appropriate sequence.) In particular, the above construction is trivially applicable when $p = 1$ and $q > 1$.

Exercise 3-5 Construct the directed graph and find an Euler cycle and corresponding cyclic sequence for the case when $X = \{a,b\}$ and $q = 4$.

Exercise 3-6 Repeat Exercise 3-5 for the case when $X = \{a,b,c,d\}$ and $q = 2$.

Exercise 3-7 What is the nature of the graph when $X = \{a\}$ and $q > 1$?

Exercise 3-8 Prove that if every q-letter word occurs precisely once in a cyclic sequence, then every $(q - 1)$-letter word occurs precisely p times and, in general, every $(q - k)$-letter word occurs precisely p^k times, for $k = 1, 2, \ldots, p - 1$.

By use of Theorems 3-4 and 3-5 and by interpretation of unicursal to mean capable of being covered by a single path or cycle, the following theorem characterizes unicursal directed graphs:

Theorem 3-6 Given a directed graph $D = (V,A)$ which is connected, D is unicursal if and only if it is either a directed Euler graph or is such that $|S| = |T| = 1$ and $\delta^+(v) - \delta^-(v) = 1$ for the single vertex $v \in S$.

Theorems 3-5 and 3-6 play a basic role in connection with the network flows to be discussed later. In this connection, the following specialized theorem will also be useful:

Theorem 3-7 Let $D = (V,A)$ be a directed graph which is such that $|S| = |T| = 1$. Let s and t denote the single vertices in S and T, respectively, and let

$$k = \delta^+(s) - \delta^-(s) = \delta^-(t) - \delta^+(t)$$

Then D can be covered by k simple paths from s to t, together (possibly) with some cycles.

Proof: If D is connected, Theorem 3-5 asserts that D can be covered by k paths of the stated variety. We have seen earlier that any nonsimple path can be partitioned into a simple path joining the same vertices, together with one or more simple cycles. If D is not connected, let D_1, $D_2, \ldots, D_n$ denote its components. Vertices s and t are necessarily in the same component. (Why?) This component can be covered by k simple paths from s to t together possibly with some cycles, as we have just seen. Each of the remaining components is a directed Euler graph which is connected. As such, its arcs form a cycle. If this cycle is non-simple, it can be partitioned into simple cycles. This completes the proof.

Exercise 3-9 Find minimal coverings for the directed graphs in Figs. 2-2, 2-4, and 2-7.

Exercise 3-10 Given an Euler graph, show that it is always possible to direct its edges in such a way that a directed Euler graph results.

Exercise 3-11 If a directed Euler graph is connected, is it necessarily strongly connected?

Exercise 3-12 Give a detailed proof of Theorem 3-4.

A somewhat related problem is the following: Under what conditions can the edges of an undirected graph be oriented in such a way that the resulting directed graph is strongly connected? For example, when can all the streets of a city be restricted to one-way traffic and still permit travel from any point to any other point in the city? Such graphs are characterized in the following theorem:[11]†

Theorem 3-8 A connected graph $G = (V,E)$ is orientable (in the above sense) if and only if every edge of G is contained in at least one circuit.

Proof: Note first that if the theorem is true for simple graphs, it is true for all graphs We therefore assume that the graph is simple. It is clear that G cannot be orientable if some edge $e \sim (v \;\&\; w)$ is not contained in any circuit; for such an edge represents the only simple chain in the graph joining v and w, and as soon as e is oriented, one of these vertices will not be accessible from the other.

To prove the converse, we shall describe a construction which produces appropriate orientations. To start, take any simple circuit $\{e_1,e_2, \ldots ,e_n\}$ incident with vertices $\{v_1,v_2, \ldots ,v_n\}$. If we orient the circuit in either direction, we produce a strongly connected subgraph consisting of these n arcs and vertices.

Assume that at least one other vertex w besides $v_1, v_2, \ldots , v_n$ exists. (Otherwise, we are through.) Then at least one such w is adjacent to one of the v_i's. Let e be such that $e \sim (v_i \;\&\; w)$. By assumption, e is contained in some simple circuit C. Orient those edges of C which have not already been oriented; to do so, use the orientation induced by traversing C, in either direction. The resulting, enlarged directed graph is also strongly connected (verify this in detail) and contains at least one new vertex, namely, w. If not all vertices are included, a new w is selected and the process is repeated until all vertices are included.

† Superscript reference numbers pertain to the list of references at the end of the chapter.

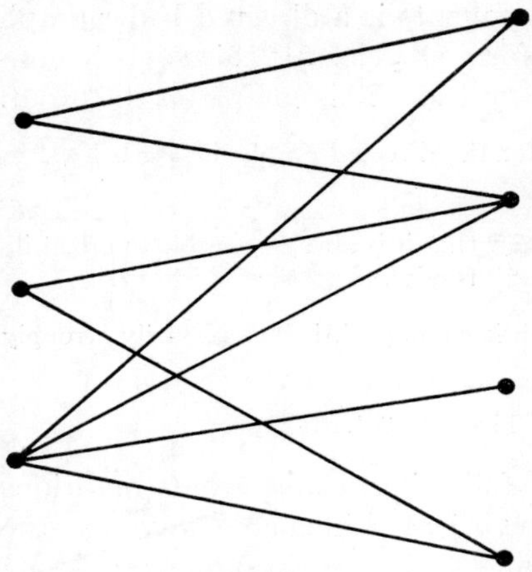

Figure 3-5

3-4 HAMILTONIAN CHAINS AND CIRCUITS

We have already considered the problem of determining whether or not a given finite graph is unicursal. It is natural also to formulate the following superficially related problem: Under what circumstances does a finite connected graph contain a chain or circuit which is incident with every vertex? If, in addition, the chain or circuit is required to be simple, it is correspondingly called a *hamiltonian chain* (often referred to as a *hamiltonian line*) or a *hamiltonian circuit.*

If a graph possesses a hamiltonian circuit S, then it clearly possesses a hamiltonian chain. In contrast, the bipartite graph of Fig. 3-5 possesses several hamiltonian chains but no hamiltonian circuit. And, of course, many graphs possess neither. For some special classes of finite graphs, the existence or nonexistence of hamiltonian chains or circuits is easily settled. For example, a tree cannot possess a hamiltonian chain unless the tree itself constitutes a chain. (The details are left to the reader.) Similarly, Fig. 3-5 exemplifies the general fact that no bipartite graph having an odd number of vertices possesses a hamiltonian circuit. (Every simple circuit in a bipartite graph has an even number of edges and hence is incident with an even number of vertices.) On the other hand, every complete graph clearly has numerous hamiltonian circuits.

While partial results for special classes of graphs exist, the problem has not been adequately settled in general. For instance, there is no efficient constructive procedure for finding a hamiltonian chain in an arbitrary graph or else providing information which proves that none exists.

The celebrated mathematician Hamilton introduced a commercial game, whose object was to find a hamiltonian circuit in the graph determined by the vertices and edges of a certain polyhedron. An account of this may be found in Ref. 20 of Chap. 1.

Remark: The problem of finding a hamiltonian circuit can be viewed as a special case of the following problem: In a graph G, each of whose edges e has a positive length $L(e)$, find a simple circuit whose total length is maximal. By setting $L(e) = 1$ for every edge, the length of a simple circuit is simply the number of edges which it contains and, equivalently, the number of vertices with which it is incident. If G has n vertices, then

G has a hamiltonian circuit if and only if a longest simple circuit has length n.

Exercise 3-13 The planar graph shown in Fig. 3-6 represents the vertices and edges of a dodecahedron. (A dodecahedron is a solid having 20 vertices and 12 pentagonal faces. The outside region also represents a face.) Find a hamiltonian circuit for this graph. This is the graph involved in Hamilton's game mentioned earlier.

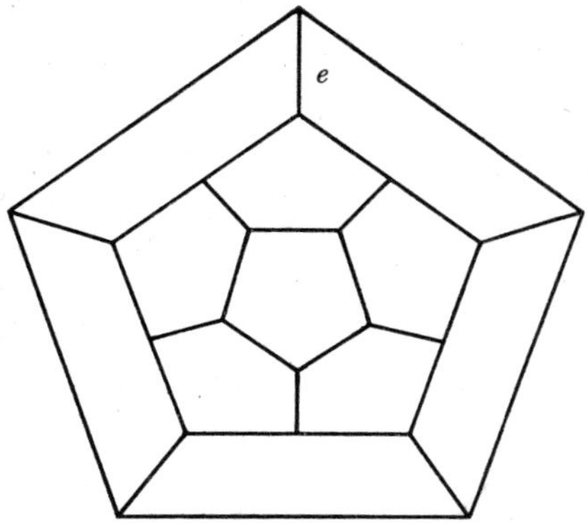

Figure 3-6

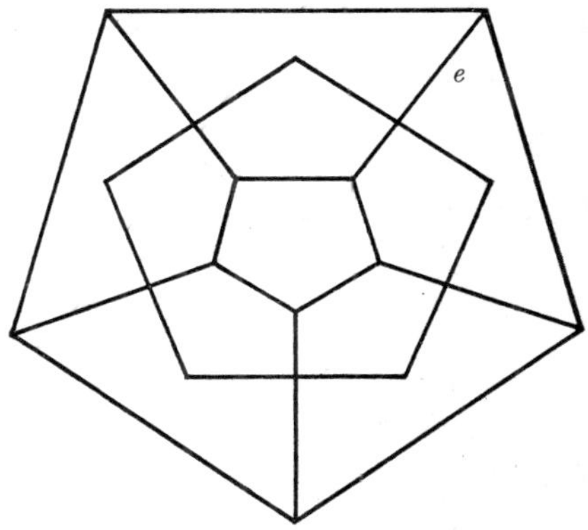

Figure 3-7

Exercise 3-14 The graph shown in Fig. 3-7 is obtained from that of Fig. 3-6 by rotating the outer pentagon 36° and modifying the end points of five edges (such as edge e) accordingly. Note that this graph also has 20 vertices and 12 pentagonal faces (including the outside region). Prove, however, that there exist no hamiltonian circuits in this graph. (*Hint:* Exploit the five vertices of degree 2).

Exercise 3-15 The graph G associated with a rhombic dodecahedron has eight vertices of degree 3 and six of degree 4. Moreover, it is bipartite: no two vertices of the same degree are adjacent. Prove that G has no hamiltonian chain (and hence also no hamiltonian circuit).[13]

Exercise 3-16 Use Exercise 3-15 as a guide to formulate a more general necessary condition for a connected bipartite graph to have a hamiltonian circuit or chain.

We turn now to a particularly interesting topic related to the concept of hamiltonian circuits. An undirected graph G is *hypohamiltonian* (HH for brevity) if it has no hamiltonian circuit but every subgraph of immediately smaller order has such a circuit.[8] The problem is to find an HH graph of minimum order and show that the solution is unique. We assume that the graph has n vertices and m edges, and is a simple graph.

Lemma 3-9 If G is HH, then $n \geq 3$.

Lemma 3-10 If G is HH, $d(v) \geq 3$ for every v.

Proof: Every vertex is in a simple circuit of length $n - 1$, and hence $d(v) \geq 2$. If w is adjacent to v, then in the subgraph obtained by suppressing w we also have $d(v) \geq 2$. Thus $d(v) \geq 3$.

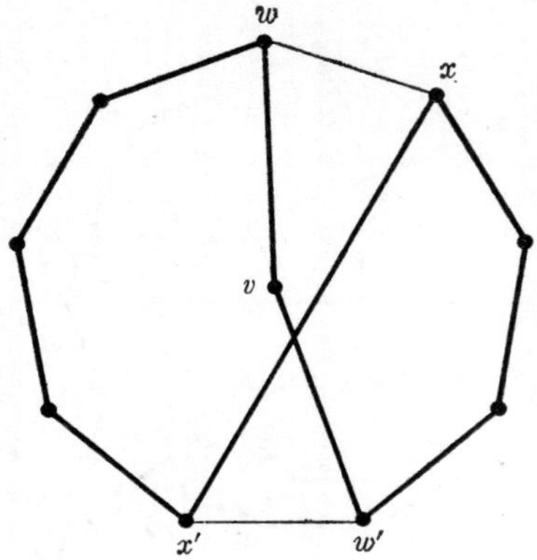

Figure 3-8

Lemma 3-11 If G is HH and if w and x are consecutive vertices of a simple circuit of length $n - 1$ in the graph in which a vertex v is suppressed, then v is not adjacent to both w and x.

Proof: Otherwise the insertion of v yields a hamiltonian circuit for the entire graph.

Lemma 3-12 If G is HH, then $d(v) \leq (n - 1)/2$ for every v.

Proof: Follows from Lemma 3-11.

Lemma 3-13 If G is regular of constant degree $d(v) = d$, then $d \cdot n = 2m$.

Lemma 3-14 If v, w, x, w', x' are distinct vertices of G such that v is adjacent to w and w', and x is adjacent to x', and if (w,x) and (w',x') are two edges of the same hamiltonian circuit of the graph obtained by suppressing v, then G has a hamiltonian circuit.

Proof: The circuit is indicated in Fig. 3-8.

Theorem 3-15 If G is HH, then $n \geq 7$.

Proof: Follows from Lemmas 3-10 and 3-12.

Theorem 3-16 If G is HH, then $n \neq 7$.

Proof: By Lemmas 3-10 and 3-12, $n = 7$ implies that $d(v) = 3$ for every v, contradicting Lemma 3-13.

Theorem 3-17 If G is HH, then $n \neq 8$.

Proof: By Lemmas 3-10 and 3-12, $n = 8$ implies $d(v) = 3$. If G is HH of order 8, by Lemma 3-11 it is one of two graphs, Figs. 3-9 and 3-10, each of which has a hamiltonian circuit by Lemma 3-14.

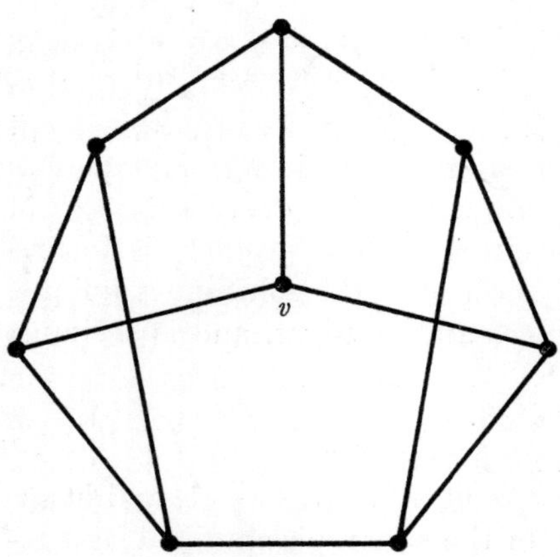

Figure 3-9

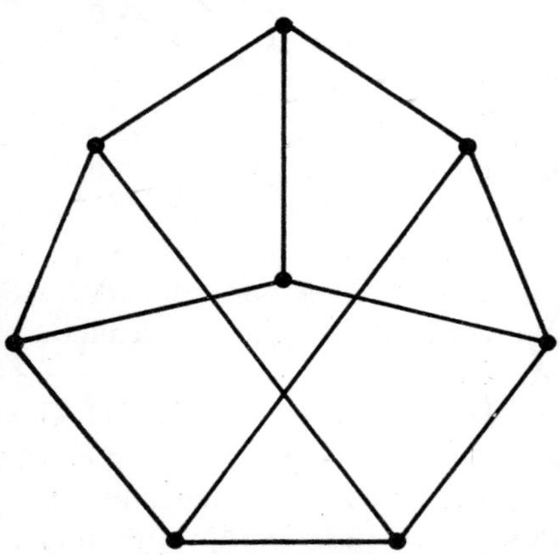

Figure 3-10

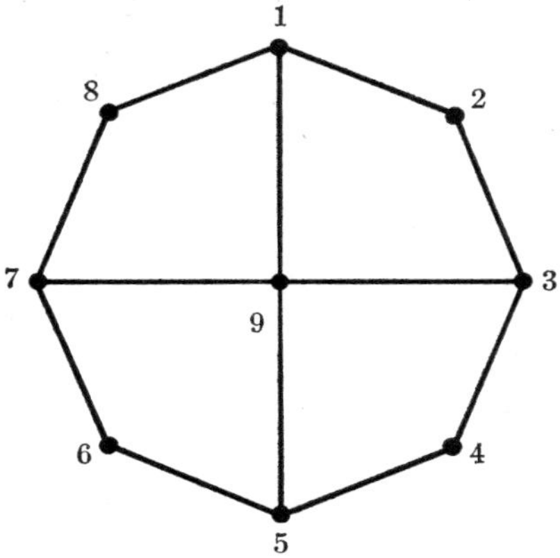

Figure 3-11

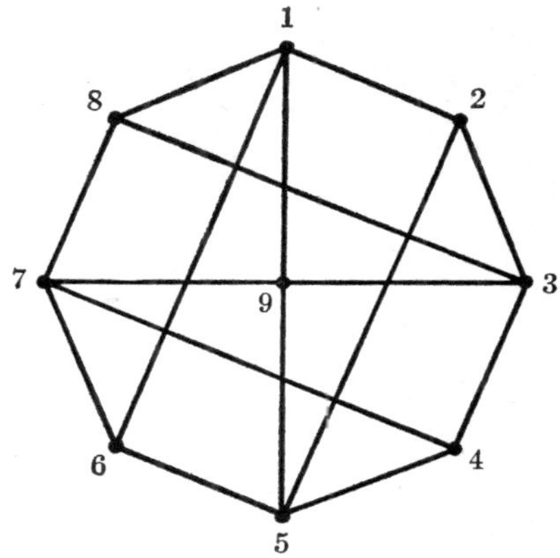

Figure 3-12

Theorem 3-18 If G is HH, then $n \neq 9$.

Proof: By Lemma 3-13, if G is HH of order 9, it cannot be regular of degree 3; and by Lemma 3-12, it has at least one vertex of degree 4. By Lemma 3-11, it has the partial graph of Fig. 3-11. Vertex 2 must be adjacent to at least one of the vertices 4 to 8, since $d(2) \geq 3$. But the edges 2–4, 2–6, and 2–8 each yield a hamiltonian circuit (by Lemma 3-14). If 2 is adjacent to 5, then 8, which by the same reason cannot be adjacent except to 3 or 5, is not adjacent to 5 without having $d(5) > 4$ contrary to Lemma 3-12. Thus 8 is adjacent to 3. Similarly, 6 is adjacent to 1 and 4 to 7, and there can be no other edges since the vertices 9, 1, 3, 5, 7 are saturated. The same figure (see Fig. 3-12) would be obtained if 2 were adjacent to 7. Note that every edge joins two vertices whose numbers have opposite parity. Thus the graph obtained from G by suppressing vertex 1 has no hamiltonian circuit, and therefore G cannot be HH.

Theorem 3-19 If G is HH of order 10, it is homogeneous of degree 3.

Proof: Let some vertex have degree 4. G admits the partial graph of Fig. 3-13. By Lemma 3-14 each of 2, 4, 5, 7, 9 cannot be adjacent except to one or two of 1, 3, 6, 8. Since each of the latter cannot be adjacent except to at most one of the former vertices, we have a contradiction.

Theorem 3-20 The graph of Fig. 3-14 is HH.

Proof: Suppressing vertex 1, we have the circuit (0,4,3,2,6,5,9,8,7,10); and suppressing vertex 2, we have the circuit (0,1,9,8,3,4,5,6,7,10). The other cases are obtained by symmetry. The graph has no hamiltonian circuit. Thus the maximal elementary chains which start with 10, 1, 2 are (6,5,4,3,8,7), (6,5,4,3,8,9), (6,5,9,8,3,4), (6,5,9,8,7), (6,7,8,3,4,5,9), (6,7,8,9,5,4,3), (3,4,5,6,7,8,9), (3,4,5,9,8,7,6), (3,8,7,6,5,4), (3,8,7,6,5,9), (3,8,9,5,4), (3,8,9,5,6,7) and none of them gives a hamiltonian circuit. By symmetry, this is the case for all other chains.

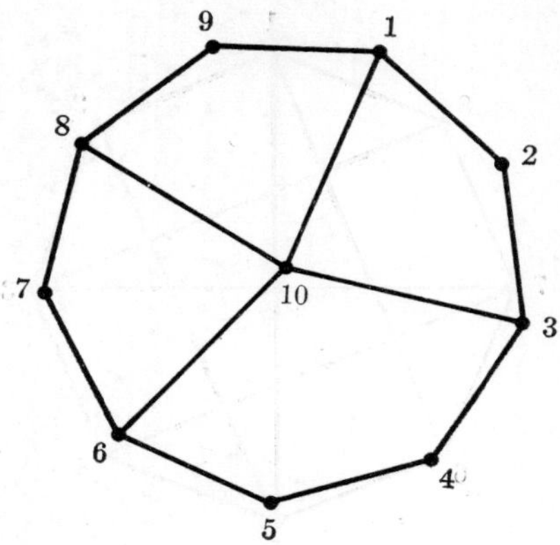

Figure 3-13

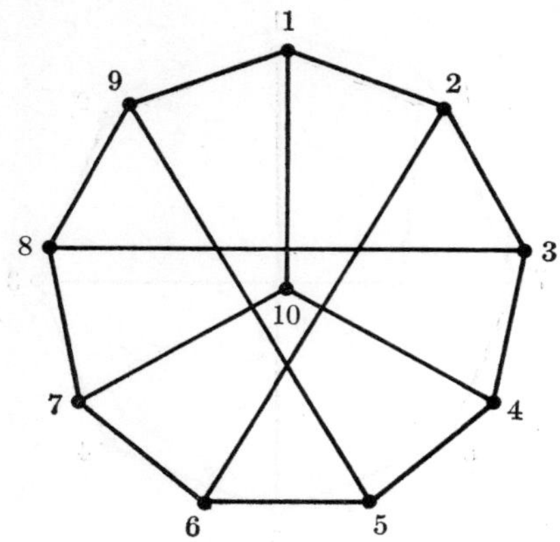

Figure 3-14

Theorem 3-21 If G is HH of order 10, it is isomorphic to the graph of Theorem 3-20.

Proof: By numbering the vertices 1, 2, . . . , 9 of the circuit in the graph without vertex 10 there are three types of graphs possible for G:

1. The type in which 10 is adjacent to 1, 4, 7, which by Lemma 3-14 and Theorem 3-19 gives the graph of Fig. 3-13 or the graph of Fig. 3-15. But the latter has the hamiltonian circuit (10,1,2,9,8,7,6,5,3,4,10).
2. That in which 10 is adjacent to 1, 3, 6, which leads to the graph in Fig. 3-16 in which we have the hamiltonian circuit (10,6,7,5,4,3,2,8,9,1,10).
3. That in which 10 is adjacent to 1, 3, 5, which leads to the graphs shown in Figs. 3-17 and 3-18, which respectively have the hamiltonian circuits (10,1,2,3,4,7,8,9,6,5,10) and (10,1,2,7,8,9,6,5,4,3,10).

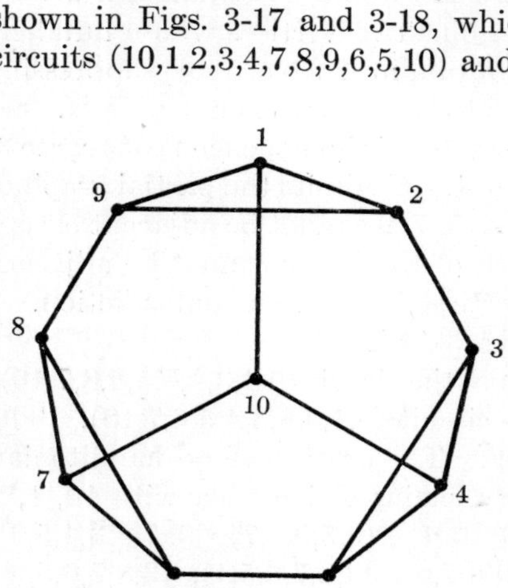

Figure 3-15

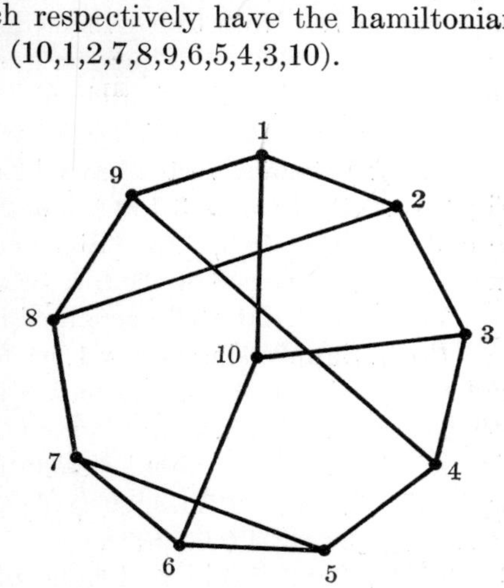

Figure 3-16

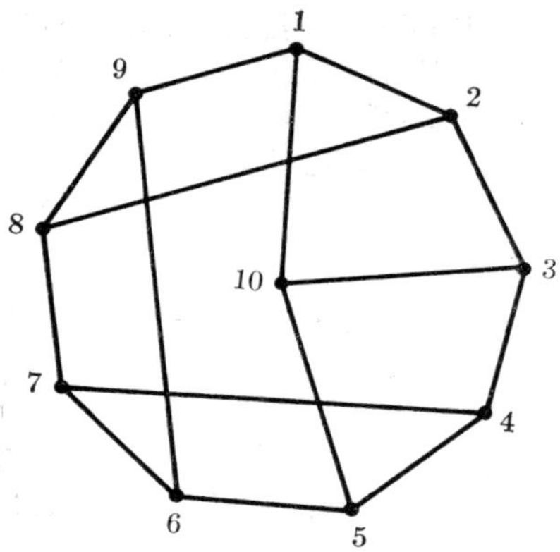

Figure 3-17

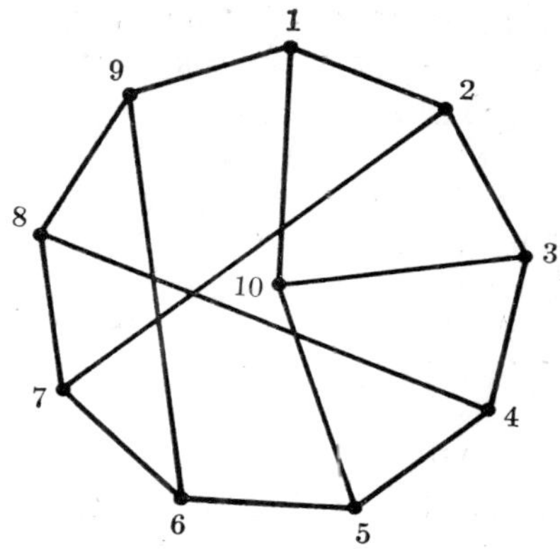

Figure 3-18

The concepts of hamiltonian chains and circuits can be extended in a straightforward manner to directed graphs. Specifically, a simple path or cycle which is incident with every vertex is called a *hamiltonian path* or *hamiltonian cycle*, respectively. (A hamiltonian path is sometimes referred to as a *directed hamiltonian line.*) Hamiltonian paths and cycles in a directed graph determine corresponding hamiltonian chains and circuits, respectively, in the associated undirected graph. They require, in addition, consistency of arc orientations; hence we can expect their existence to be still more rare. We shall indicate several results which are applicable to a special class of directed graphs.

A directed graph will be said to be *d-complete* if every pair of distinct vertices are joined by precisely one arc. Thus one can obtain a d-complete graph by taking a complete undirected graph (which has no parallel edges) and arbitrarily directing its edges. In contrast to the fact that every complete graph has numerous hamiltonian circuits, a d-complete graph may have no hamiltonian cycles. For example, if there exists a vertex v for which either $\delta^+(v) = 0$ or $\delta^-(v) = 0$, then there can exist no hamiltonian cycles. More generally, it is clear that the existence of a hamiltonian cycle implies that the graph is strongly connected. (Why?)

If we do not insist on returning to the starting vertex (i.e., if we will settle for a hamiltonian path), the prospects are a good deal brighter. König has proved that every d-complete graph G contains at least one hamiltonian path. For a proof, see pages 30 and 31 of König's book (Ref. 16 of Chap. 1). Rédei[12] has shown that there are in fact an odd number of hamiltonian paths. Camion[2] investigated uniqueness and obtained the following theorem:

Theorem 3-22 A d-complete graph G contains a unique hamiltonian path if and only if G contains no cycles.

Proof: If G contains two different hamiltonian paths, they determine two permutations of the vertices of G, and these differ by at least one inversion. Hence there are two vertices each of which is joined to the other by a path. This implies the existence of a cycle. On the other hand, if a cycle exists, then the method of construction employed by König and described in his book, cited earlier, has options which lead to different hamiltonian paths.

A directed graph is said to be *total* if every pair of distinct vertices is joined in at least one direction by a path. Note that a d-complete graph is a special type of total graph in which the above paths are single arcs. All strongly connected graphs are also total. The following result, also due to Camion and stated here without proof, applies to all total graphs.

Theorem 3-23 A necessary and sufficient condition that a directed graph without cycles have precisely one hamiltonian path is that it be total.

An interesting related problem seeking the shortest hamiltonian circuit is that of a traveling saleman who must pass through each of n cities (every two of which are connected) precisely once and return to his starting point. Thus the salesman must travel the shortest distance. It is clear that computing the length of all hamiltonian circuits to determine the shortest one involves $(n-1)!/2$ possible circuits, which is an astronomical number for large n. The problem is to find a more convenient algorithm for the solution of the problem. No such general algorithm exists as yet. For a number of computational algorithms dealing with this problem see Refs. 3, 7, and 9.

Exercise 3-17 Let d_{ij} be the distance between the ith and jth cities. Give a matrix interpretation of the problem. One is tempted to select the smallest distance connecting any pair of cities, cross out the corresponding row and column, and repeat the process. But this does not in general yield an optimum solution. Now take five cities, set up the matrix assigning distances, compute an optimal hamiltonian circuit by exhausting the possibilities, and compare with the plausible procedure mentioned above. Show that in some cases the solution does not contain the shortest edge and may contain a longest one.

In this regard it is interesting to note that the shortest closed route connecting n points in a plane not all on a straight line is a simple polygon. In particular, if the convex hull of the set contains none of the n points in its interior, then its boundary is the shortest closed route. (Thus there need not be any crossing of routes in the corresponding traveling salesman problem.) To prove this fact,[15] we connect the n points by any closed path and note that a shorter path is obtained by connecting the points in the same order by straight line segments with the given points as the only vertices. If the segments v_iv_{i+1} and v_jv_{j+1} intersect in a point

v, we suppose that the path followed is $v_i vv_{i+1} \cdots v_j vv_{j+1} \cdots v_i$. If v is not one of the given points, then $v_iv_j \cdots v_{i+1}v_{j+1} \cdots v_i$ is a shorter polygonal path which does not contain the intersection v. If, on the other hand, v is in the set of given points, then $v_ivv_j \cdots v_{i+1}v_{j+1} \cdots v_i$, is a polygonal path which does not contain v as an intersection point. (A sketch will clarify the argument.)

For the interested reader, we present an interesting and simple approach to the traveling salesman problem due to P. C. Ryan of the Naval Research Laboratory. This approach is not any more efficient than other algorithms, but it provides some insight.

Recall that the problem is to find a shortest hamiltonian circuit when each edge of the graph is assigned a length. The graph need not be complete, in which case the existence of one or more hamiltonian circuits is not assured, and consequently the problem may have no solution. If parallel edges are present, we can clearly disregard all except the shortest of each parallel group. The present algorithm will either find a shortest hamiltonian circuit or indicate that none exists—in either case in a finite number of steps.

We will assume that the graph under consideration is planar, and the fact (established in Chap. 4) that a simple planar graph has at least one vertex of degree 5 or smaller. In addition, the edge lengths will be assured positive.

Starting with an arbitrary vertex v_1 of minimum degree $d \leq 5$, it is possible to determine $\binom{d}{2} \leq 10$ distinct subgraphs, each consisting of all vertices of the original graph, all edges that are not incident with v_1, and two of the d edges that are incident with v_1. Clearly, the problem of finding all hamiltonian circuits of the original graph is equivalent to finding all hamiltonian circuits in these subgraphs, each of which has a vertex (v_1) of degree 2.

Now consider one of these subgraphs $G_i = (V,E_i,\Phi_i)$, where $|V| = n$, $\Phi_i = \Phi/E_i$ (the restriction of Φ to E_i), and degree $v_1 = 2$. Starting with the edge e_1 incident with v_1 and the adjacent vertex of least degree (if both adjacent vertices are of the same degree, either will suffice), we construct a hamiltonian circuit C_n as the end result of a sequence of n simple chains $C_k = \{e_1, \ldots ,e_k\}$, where $e_j \sim (v_j \;\&\; v_{j+1})$ for $j = 1, \ldots , k$, with the understanding that $v_{n+1} = v_1$. (This notation is meant to imply that the edges and vertices are labeled sequentially as they occur in the construction of the chain.) An edge e_k may be adjoined to C_{k-1} if the following conditions are satisfied:

1. $v_{k+1} \neq v_j$ ($j = 1, \ldots , k$), unless $k = n$, in which case the single exception noted above applies.

2. $C_k = C_{k-1} \cup \{e_k\}$ does not separate the remaining graph, in the sense that every pair of vertices $v_p, v_q \in \{v_{k+2}, \ldots, v_{n+1} = v_1\}$ can be connected by a chain none of whose edges is incident with some vertex in $\{v_2, \ldots, v_{k+1}\}$.

Conditions 1 and 2 ensure that the final chain will be a simple circuit.

It is likely that at some stage more than one edge will qualify for addition to the chain at the same vertex. When this occurs, an arbitrary incident edge of least length is selected and unselected qualified edges are noted. Later each of the unselected edges must be investigated.

If no edge satisfying conditions 1 and 2 can be found at the terminal vertex v_{k+1} of a chain $C_k(k < n)$, the chain is not part of a hamiltonian circuit and is abandoned. At this point, a return is made to the most recent arbitrary selection and an alternate chain is investigated. When all such selections have been checked, any hamiltonian circuits which exist will have been found. Following this procedure for each subgraph G_i yields all hamiltonian circuits.

Let s_i be the length of the shortest edge in the ith subgraph [that is, $s_i = \min_{e \in E_i} \lambda(e)$, where λ is the length function defined on E]. After the first hamiltonian circuit C_n has been found (in any subgraph), its length $|C_n|$ may be calculated. Let C'_k denote any subsequent chain under construction in the ith subgraph, with length $|C'_k|$. In order for C'_k to result in a hamiltonian circuit of length shorter than C_n, we must have

$$|C'_k| + (n - k)s_i < |C_n|$$

at every step of the process (i.e., for every value of $k = 1, \ldots, n$). This inequality provides another criterion for rejecting edges which are under consideration for addition to the chain.

To illustrate, we apply the method to the graph G of Fig. 3-19. The integer labels represent the lengths associated with the edges. Figure 3-20a, b, and c gives the three subgraphs that result from the (semi-arbitrary) choice of v_1 and the corresponding values of s_i.

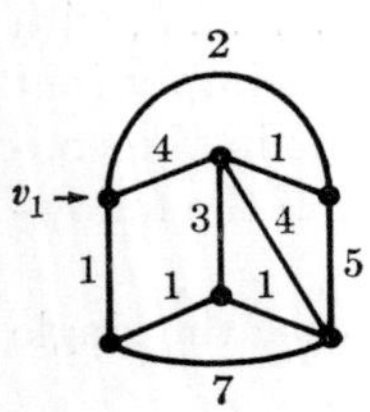

Figure 3-19

Figure 3-21 illustrates the series of steps which are required to construct the first hamiltonian circuit (in G_1). The edges shown by heavy lines are the circuit edges. Condition 2 of the method is equivalent to requiring that the graph obtained by deleting from G_1 the heavy and dotted edges and the circled heavy vertices (at each step) be connected. Numbers in parentheses at a vertex refer to the number of unselected (but qualified) edges remaining to be investigated there.

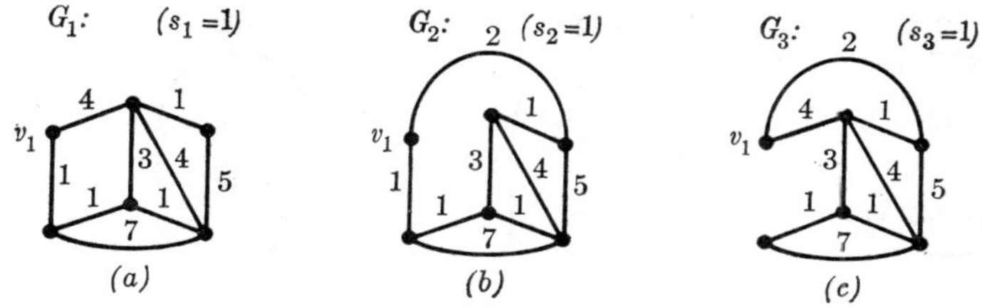

Figure 3-20

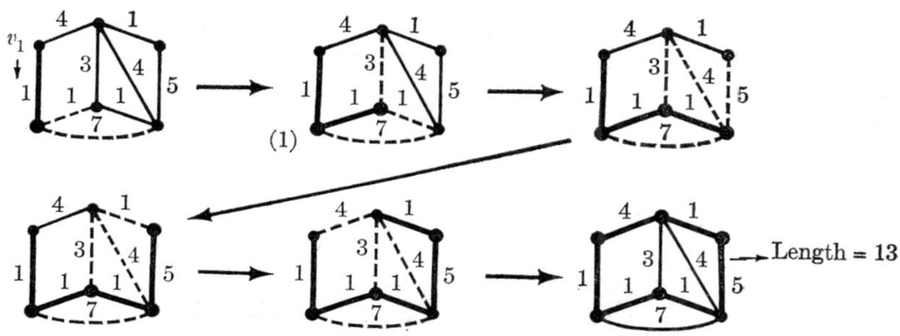

Figure 3-21

Returning to consider the alternative at step 2 (above), the situation is as depicted in Fig. 3-22. Note that the edge of length 5 is not a permissible alternative in step 2 (Fig. 3-22), since we already have a complete hamiltonian circuit of length 13. Since no further edges may be added to it, this chain is abandoned and G_1 may be dropped from consideration.

Figures 3-23 and 3-24 illustrate the steps necessary for full consideration of G_2 and G_3. Note that the unused alternative in the second sketch of Fig. 3-23 becomes an illegitimate alternative (by the chain length condition) when the hamiltonian circuit of length 10 is found (see Fig. 3-23).

Although this method works well for small-scale graphs like the one of the example and quite possibly would be acceptable (if done graphically as above) for somewhat larger-scale planar graphs, it would not be easy to mechanize, especially since condition 2 (which is easy to visualize graphically) would be difficult to formulate efficiently on a computer.

Exercise 3-18 In the directed graph of Fig. 3-25 start at A, visit each intersection only once, and return to A. There is a route. Find it.[4]

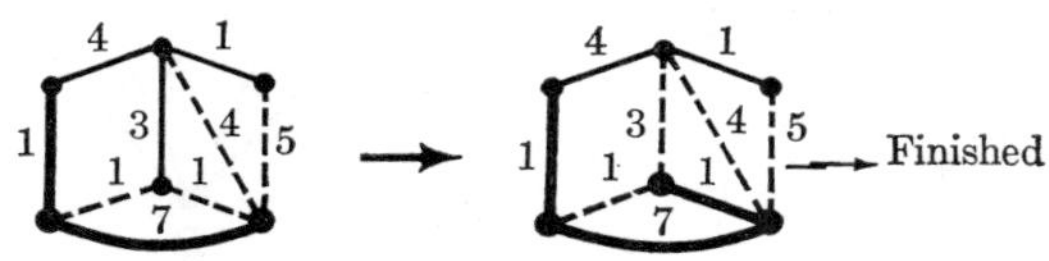

Figure 3-22

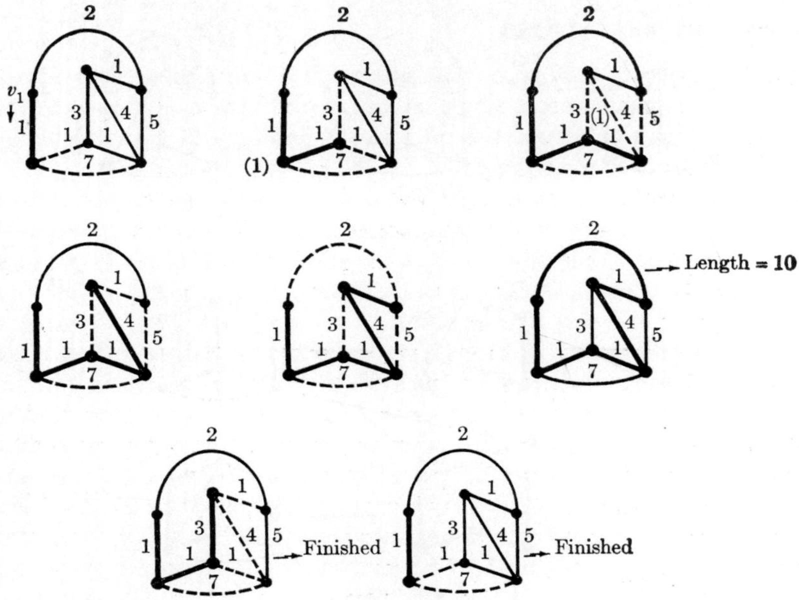

Figure 3-23

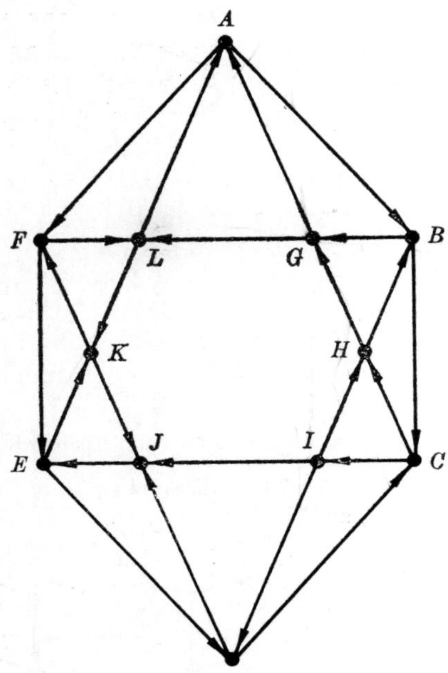

Figure 3-25

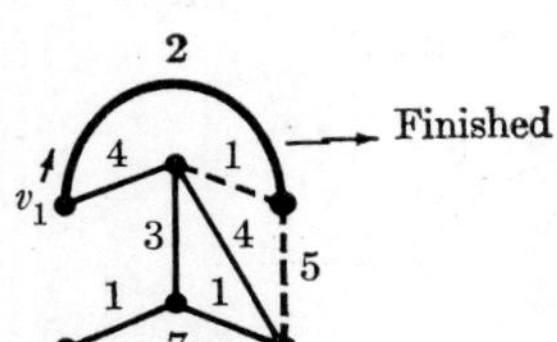

Figure 3-24

3-5 VERTEX PARTITIONS

A set of vertices is said to be *independent* if it contains no two adjacent vertices. In particular, a single vertex constitutes an independent set. One sometimes encounters the problem of finding a largest independent set of vertices in a given graph.

For example, suppose that we wish to form the largest possible committee from a group of individuals certain pairs of whom cannot be jointly assigned. This problem is of the above type, where the vertices of the graph represent the various individuals and the edges indicate conflicting pairs. We may as well assume that the graph determined in this manner is connected. (If it is not, we can select a maximal subcommittee in each component and combine the subcommittees to solve the problem.)

A reasonable approach would be to start with one individual and successively add others, at each stage selecting one who does not conflict with any previously selected individual. Such a procedure will ultimately yield a maximal independent set W, in the sense that W is not a subset of any larger independent set. We cannot be assured, however, that W will have the largest possible number of vertices (the maximum maximorum), since the size of W depends on the particular vertices selected at each stage of the process.

As an extreme case consider the graph shown in Fig. 3-26. If we start with any vertex other than v_1 and at each stage select any vertex other than v_1, a maximal independent set W_1 that includes all but one vertex of the graph is ultimately obtained. On the other hand, if we have the misfortune to initially select v_1, we are through, and we have discovered a maximal (though very small) independent set W_2 consisting of a single vertex. In this illustration, the complement of one maximal independent set is another maximal independent set—a coincidence. In Fig. 3-27 the independent set $W = \{v_3,v_4\}$ is maximal, but its complement is not independent.

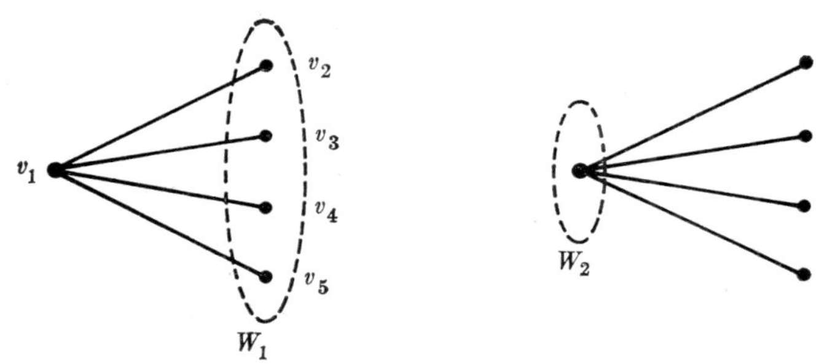

Figure 3-26

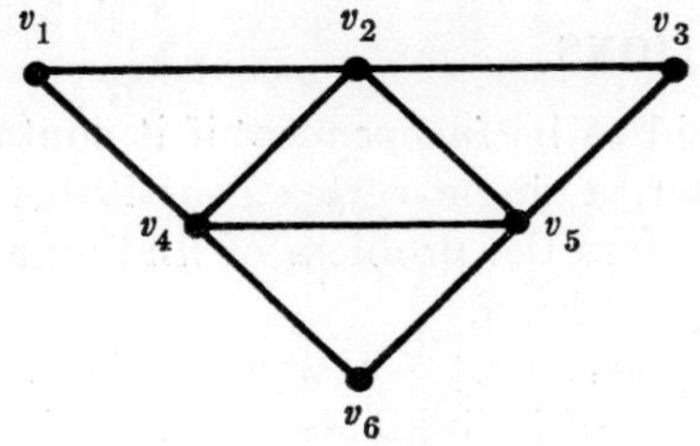

Figure 3-27

A maximal independent set W has the property that it dominates the graph in the sense that every vertex is either a member of W or is adjacent to a member of W. For if an independent set does not dominate the graph, there exists at least one vertex which is neither in the set nor adjacent to it. We can adjoin this vertex to the set without destroying independence. Thus the set is not maximal.

A set of vertices having this dominance property is called a *dominating set,* whether it is independent or not. In view of the preceding remarks, the number of vertices in a largest independent set is at least as great as the number in a smallest dominating set. (In Fig. 3-26, W_2 is a smallest dominating set and W_1 is a largest independent set.)

In a simple graph G a vertex v dominates itself and the $\delta(v)$ other vertices with which it is adjacent. Consequently, a set of k vertices $S = \{v_1, \ldots, v_k\}$ dominates at most $k + \sum_1^k \delta(v_i)$ vertices. If G has n vertices, it follows that a necessary condition for S to be a dominating set is that

$$k + \sum_1^k \delta(v_i) \geq n$$

In particular, a dominating set for a simple graph which is p-regular $[\delta(v_i) = p$ for all $i]$ must contain at least $n/(p+1)$ vertices.

Exercise 3-19 Produce several graphs for which a largest independent set coincides with a smallest dominating set. Can you think of any classes of graphs which have this property?

Exercise 3-20 In the text the procedure for producing a maximal independent set of vertices was seen to fail (sometimes) to produce a largest independent set. Consider the following modification: (1) start with a vertex of smallest degree and (2) at every stage select as the next vertex to enter the set one which is adjacent with the fewest possible vertices not in the set (and, of course, with none already chosen). This procedure will correctly find a largest independent set for the graph shown in Fig. 3-26. However, construct a graph for which it fails to yield a largest independent set.

Exercise 3-21 We have seen that a maximal independent set is necessarily a dominating set. A minimal dominating set may be independent (see, for example, Fig. 3-26). Give an example to show that it need not be.

A related class of problems, and one which has been the object of much research, is the following: Partition the vertices of a given graph into the smallest number of independent sets. Problems of this type, frequently called *coloring problems*, have the alternative formulation: Assign to each vertex a color (or abstract label) in such a way that adjacent vertices always have different colors (labels) and the number of different colors (labels) employed is as small as possible.

This smallest number is called the *chromatic number* of the graph, in keeping with the above coloring formulation. The chromatic number of a simple graph G having n vertices is less than n if G is not complete and is greater than 1 if G has any edges at all. The chromatic number of a given class of graphs is defined to be the largest chromatic number realized by any member of the class. Expressed differently, it is the smallest number of colors which suffices to color every graph in the class. This number can be determined for various classes. As an example, the reader may verify the following theorem:

Theorem 3-24 Every tree (except the degenerate one consisting of a single isolated vertex) has chromatic number 2.

It is important to recognize the close relationship between the chromatic number of a graph and the concept of a k-partite graph. The chromatic number of G is simply the smallest k such that G is k-partite. In particular, Theorem 3-24 asserts that every tree is bipartite. Two representations of the same tree are shown in Fig. 3-28. The second shows the bipartite property clearly, while connectivity and the absence of circuits are more apparent in the first representation.

We shall return later to the topic of coloring problems and, in particular, to the problem of determining the chromatic number of the class of planar graphs. This is the celebrated *four-color problem* (since it is conjectured that the chromatic number is four).

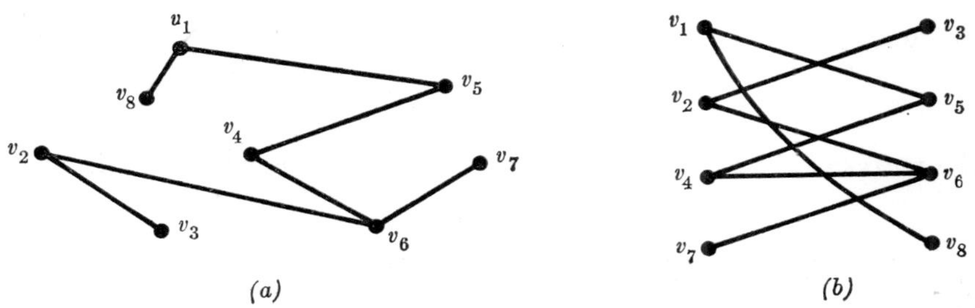

Figure 3-28

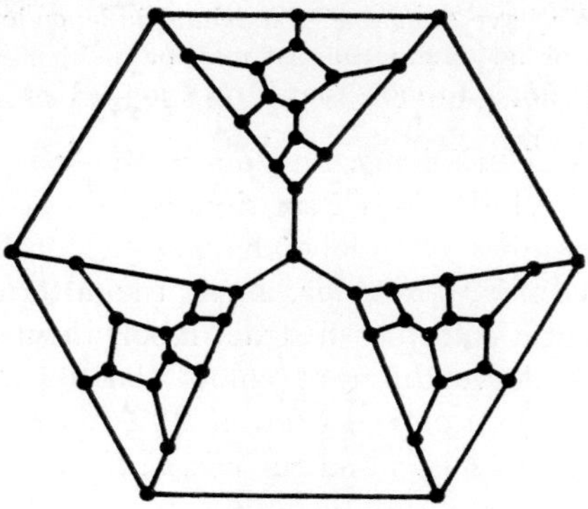

Figure 3-29

Remark: Tait conjectured that the graph of every convex polyhedron contains a hamiltonian circuit, and the four-color conjecture would follow immediately from that. It is known that a graph is a skeleton of a convex polyhedron if and only if it is planar and it is 3-connected. Thus Tait's conjecture is equivalent to the fact that every 3-connected planar graph has a hamiltonian circuit. Tutte[14] has proved that the planar graph of Fig. 3-29 has no hamiltonian circuit.

3-6 RADIUS AND DIAMETER

Let $G = (V,E)$ be a connected graph. We can reasonably define a distance between vertices as follows: If $v \neq w$, $d(v,w)$ denotes the minimum number of edges contained in any chain joining v and w, whereas $d(v,v) = 0$ for all v.

For a fixed vertex v, the integer

$$R(v) = \max_{w \in V} d(v,w)$$

measures the distance from v to the vertex (or vertices) most remote from v. It is intuitively clear that a vertex is relatively central if $R(v)$ is relatively small. Thus it is natural to call

$$R_0 = \min_{v \in V} R(v)$$

the *radius* of the graph and to refer to any vertex v_0 as a *center* if

$$R(v_0) = R_0$$

Figure 3-30*a* and *b* shows graphs having radii 1 and 3, respectively. A graph may have many centers. In Fig. 3-30*b*, for example, every vertex is a center.

An application of another measure of centrality is found in Sec. 6-16, in connection with communication networks.

The *diameter* T of a connected graph is the maximum distance between pairs of vertices. Symbolically,

$$T = \max_{v,w \in V} d(v,w)$$

The graphs shown in Fig. 3-30 have diameters 2 and 3, respectively. Note from the second example that the radius and diameter of a connected graph may be equal.

Exercise 3-22 Let G be the graph determined by a simple chain having $2n$ edges. Determine its radius, diameter, and number of centers.

Exercise 3-23 Repeat Exercise 3-22 for a simple chain having an odd number of edges and for simple circuits having even and odd numbers of edges.

Exercise 3-24 Characterize simple complete graphs in terms of the concept of diameter.

Exercise 3-25 (*a*) Show that the radius and diameter of a connected graph necessarily satisfy the relationship $T/2 \leq R \leq T$. (*b*) Produce examples which realize each extreme.

Exercise 3-26 Show that in a tree having at least two edges, the radius R and diameter T always satisfy $R < T$.

Exercise 3-27 Show that $d(v,w)$ possesses the usual properties of a metric, or distance function:

$$\begin{aligned} d(v,v) &= 0 \\ d(v,w) &> 0 \qquad \text{if } w \neq v \\ d(v,w) &= d(w,v) \\ d(u,v) + d(v,w) &\geq d(u,w) \end{aligned}$$

The concepts of radius, center, and diameter can be extended to directed graphs of certain types. If we redefine $d(v,w)$ to mean the number of arcs in a shortest path from v to w, then we must either assume that there exists a path from v to w, for every v and w, or else consider the radii and diameters of certain directed graphs to be undefined. We can avoid

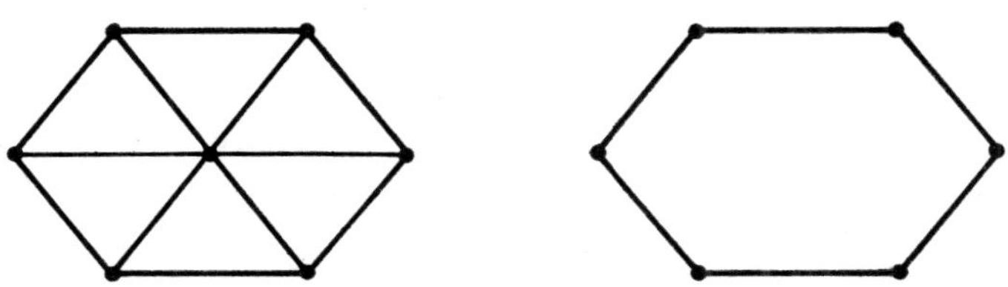

(*a*) (*b*)

Figure 3-30

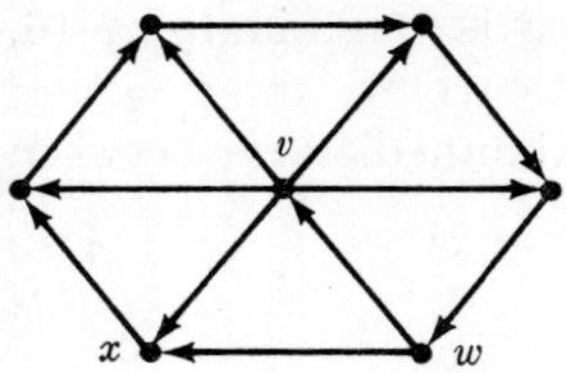

Figure 3-31

the latter possibility by restricting consideration to strongly connected graphs.

If $D = (V,A)$ is a strongly connected graph, we formally define the radius, diameter, and centers precisely as in the undirected case, using the modified definition of $d(v,w)$. In order to avoid confusion, the radius and diameter in the directed case will be denoted by $\vec{R}$ and $\vec{T}$.

In Fig. 3-31 we have a directed graph D obtained from the undirected graph of Fig. 3-30*a* by a particularly unfortunate orientation of edges. It is easily seen that this graph is strongly connected and that

$$\vec{R} = 2$$

the vertices v and w being centers. Moreover,

$$\vec{T} = 6$$

this distance being realized by the (unique) simple path from x to v. So in the directed case, the radius may be less than half of the diameter, which ultimately stems from the fact that the distance function is no longer symmetric. [Note, for example, that in Fig. 3-31 we have $d(x,v) = 6$, while $d(v,x) = 1$.]

These unusual characteristics can be avoided by restricting consideration still further to the class of directed graphs which are connected and symmetric. But for such graphs the directed and undirected meanings of radius and diameter are essentially the same. Every chain joining v and w in the associated undirected graph gives rise to a path from v to w (as well as one from w to v) having an equal number of arcs. Thus the two meanings of $d(v,w)$ produce numerically identical distance functions.

3-7 MINIMAL - LENGTH PROBLEMS

Consider a graph each of whose edges (or arcs, if the graph is directed) has an associated real number, which we shall call its "length." Depending on the application, this number may be a measure of physical distance, of time consumed, of cost, or of some other factor of interest. Although we shall use terminology appropriate for physical distances, the other possible interpretations should be borne in mind. The two essential characteristics of length, for the present discussion, are these:

1. It must be *additive* in the sense that the length of a collection of edges or arcs may be taken to be the sum of the lengths of the individual edges or arcs.

2. It must be a measure which it is inherently desirable to either minimize or maximize, subject to restrictions concerning which sets of edges or arcs are to be considered "admissible."

The traveling salesman problem is an illustration in which the admissible sets of edges are those that form hamiltonian circuits. In Sec. 6-31, spanning trees will be the admissible sets.

The problem to be considered now is that of finding a shortest path between two specified vertices in a directed graph. More precisely, let v_0 and v_1 be distinct vertices in a directed graph $D = (V,A)$ and let $\lambda(a)$ denote the length of arc a. Assuming that there exists at least one path from v_0 to v_1, the problem is to find a path $P = \{a_1,a_2, \ldots ,a_n\}$ from v_0 to v_1 whose length is minimal.

Note first of all that a shortest path from v_0 to v_1 necessarily exists even if some or all arcs have negative length, since there are only a finite number of paths from v_0 to v_1 (and, by assumption, there is at least one). However, in a very real sense there may be no shortest distance from v_0 to v_1 under certain circumstances. This is the case if there is a cycle C such that $\lambda(C) < 0$ and if there is a path P from v_0 to v_1 via a vertex w incident with C. For then we could construct arc progressions whose lengths are (algebraically) as small as we please by traversing P from v_0 to w, then going around C a sufficient number of times, and finally traversing P from w to v_1. For this reason, we shall assume that

$$\lambda(C) \geq 0$$

for every cycle C. Under this assumption, while a nonsimple path P may be a shortest path if $\lambda(C) = 0$ for each cycle in P, it is sufficient to consider simple paths in order to find a shortest path.

It turns out that in the course of finding a shortest path from v_0 to v_1, we shall as a by-product produce a shortest path from v_0 to every vertex $w \neq v_0$ which is "reachable" from v_0 via a path. So we may as well suppose that we wish to solve the more general problem of finding a path from some fixed "reference vertex" v_0 to every vertex w which is reachable from v_0. Toward this end, we present the following definitions:

A *shortest-distance tree* T relative to reference vertex v_0 is a directed tree, rooted at v_0, such that the (unique) tree path from v_0 to each tree vertex $w \neq v_0$ is in fact a shortest path from v_0 to w. A *maximal shortest-distance tree* relative to v_0 is a shortest-distance tree which includes every vertex reachable from v_0. We shall show that, for every directed graph D having a distance function λ which satisfies

$$\lambda(C) \geq 0$$

for every cycle C, and for any fixed reference vertex v_0, D does in fact contain as a subgraph at least one maximal shortest-distance tree relative to v_0. We first establish the following theorem:

Theorem 3-25 Let T be a rooted tree in a directed graph $D = (V,A)$ rooted at v_0 and containing all vertices reachable from v_0. If v is any vertex of T, let $L(v)$ denote the tree distance from v_0 to v, with $L(v_0) = 0$. Then T is a shortest-distance tree, relative to v_0, if and only if every chord (v,w) whose end points are both in T satisfies

$$L(w) \leq L(v) + \lambda(v,w)$$

Proof: If $L(w) > L(v) + \lambda(v,w)$ for some chord, then the tree path from v_0 to v, together with the chord (v,w), has length $L(v) + \lambda(v,w)$. Since this is smaller than $L(w)$, the length of the tree path to w is clearly not minimal.

Conversely, suppose that T is not a shortest-distance tree. Let v be a vertex of T for which the tree path is not a shortest path and let $P = a_1, a_2, \ldots, a_n$ be a shortest path progression from v_0 to v. Let

$$a_{i+1} = (v_i,v_{i+1})$$

be the first (i.e., lowest indexed) arc of P which is not an arc of T. Since P is a shortest path progression from v_0 to v, $P_j = a_1, a_2, \ldots, a_j$ must be a shortest path progression as far as v_j for $j = 1, 2, \ldots, n$. (The verification of this statement is left to the reader.) But $\lambda(P_i) = L(v_i)$, since P_i and the tree path agree as far as vertex v_i. By assumption, $L(v_{i+1}) > \lambda(P_{i+1}) = L(v_i) + \lambda(a_{i+1})$. Thus we have produced an arc, namely, a_{i+1}, such that its end points satisfy

$$L(w) > L(v) + \lambda(v,w)$$

This completes the proof.

Theorem 3-25 provides a theoretical basis for an iterative procedure which will actually yield a maximal shortest-distance tree, relative to a given reference vertex v_0, provided that

$$\lambda(C) \geq 0$$

for every cycle of the graph under consideration. The procedure is as follows:

1. Take as T_0 any tree, rooted at v_0, that includes all vertices which are reachable from v_0.
2. In general, having obtained tree T_i, let $L_i(v)$ denote the distance from v_0 to v through T_i. If every chord (v,w) satisfies

$$L_i(w) \leq L_i(v) + \lambda(v,w)$$

then, by Theorem 3-25, T_i is a maximal shortest-distance tree and we are through. If not, let (v^*,w^*) be a chord such that

$$L_i(w^*) > L_i(v^*) + \lambda(v^*,w^*)$$

Let T_{i+1} be the graph obtained from T_i by adding arc (v^*,w^*) and deleting the arc of T_i whose terminal vertex is w^*. Given that T_i is a tree rooted at v_0 and that

$$\lambda(C) \geq 0$$

for every cycle, it follows that T_{i+1} constructed as above is necessarily also a tree rooted at v_0. (This assertion is left as an illuminating exercise for the reader.) Step 2 is now repeated, with $i + 1$ in place of i.

For any vertex v which is reachable from v_0, the sequence of values $L_0(v)$, $L_1(v)$, $L_2(v)$, . . . is a nonincreasing sequence. Moreover, at each stage i, $L_i(v) < L_{i-1}(v)$ for at least one vertex (namely, the vertex w^* above). Finally, the values of $L_i(v)$ are bounded below, since there do not exist paths from v_0 to v having indefinitely small lengths. The observations of this paragraph, taken together, imply that at some stage i we must have

$$L_i(w) \leq L_i(v) + \lambda(v,w)$$

for every arc (v,w) which is a chord relative to T_i. When this occurs, T_i is the desired maximal shortest-distance tree according to Theorem 3-25.

Example: Figure 3-32 illustrates this method for finding a shortest-distance tree. In this example, the reference vertex is v_0. Part (*a*) shows the arc lengths. In part (*b*) the solid arcs represent an initial tree T_0 rooted at v_0 which spans the graph. The values $L_0(v)$ are also shown. The dotted arc in (*b*) is an "improvement" chord (since $3 > 0 + 1$). When this arc is added and the appropriate arc is removed, the new spanning tree T_1 shown in part (*c*) is obtained. In a similar manner, T_2 and T_3 are obtained; they are shown in parts (*d*) and (*e*), respectively. By checking all chords relative to T_3, we can verify that T_3 is indeed a shortest-distance tree. Note that there may be several "improvement" chords at any stage. In that case any one may be selected. Different choices may lead to different final trees (and possibly in a different number of steps) since the shortest-distance tree is not in general unique.

By defining $\lambda(a) = 1$ for all $a \in A$, we can find paths which are shortest in the sense that they contain the fewest possible arcs. A variety of techniques for finding shortest-distance trees in directed graphs have been developed. See Ref. 10 for a comparative discussion of several algorithms; the one described here is based on the "labeling" or "index-reduction" approach described in Refs. 5 and 6.

Exercise 3-28 Repeat the preceding example, but select a different improvement chord to go from T_1 to a different tree T_2'.

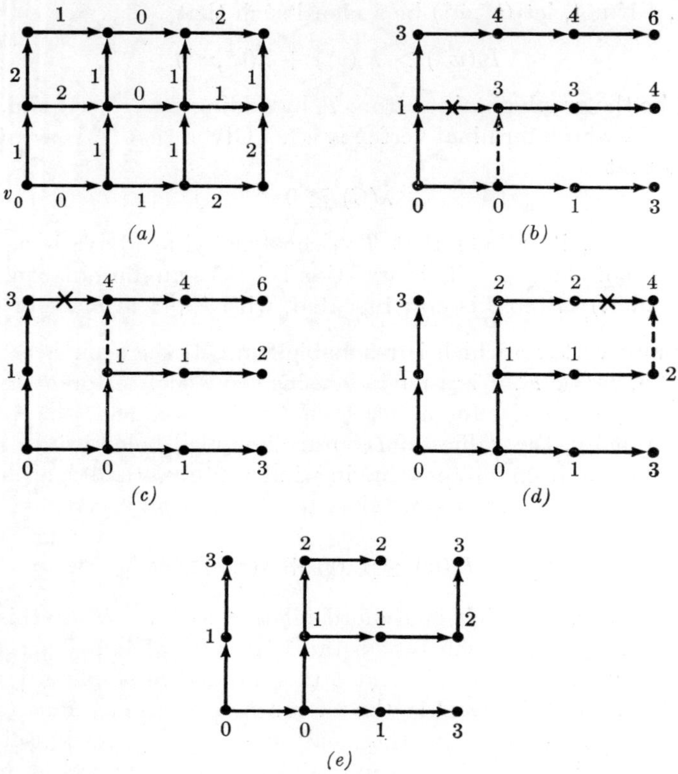

Figure 3-32

Exercise 3-29 Apply the tree-modification method to the directed graph in Fig. 3-33, finding shortest paths from v_1 to all other possible vertices relative to the distance function $\lambda(a) = 1$ for all arcs.

Exercise 3-30 Repeat Exercise 3-29 with the arc lengths indicated in Fig. 3-33.

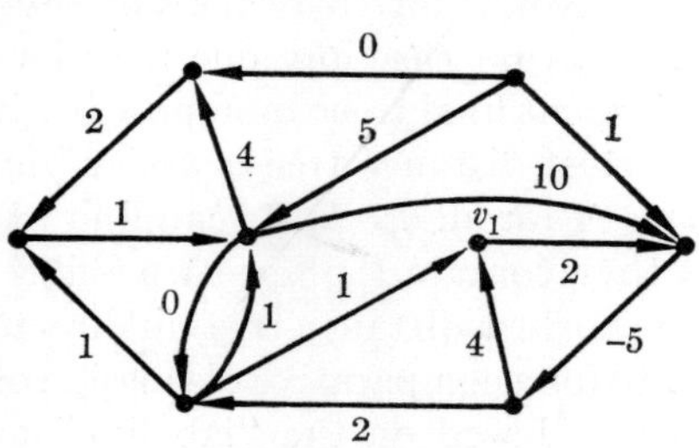

Figure 3-33

The following example illustrates an alternative procedure for finding shortest paths through a directed graph by using the concept of dynamic

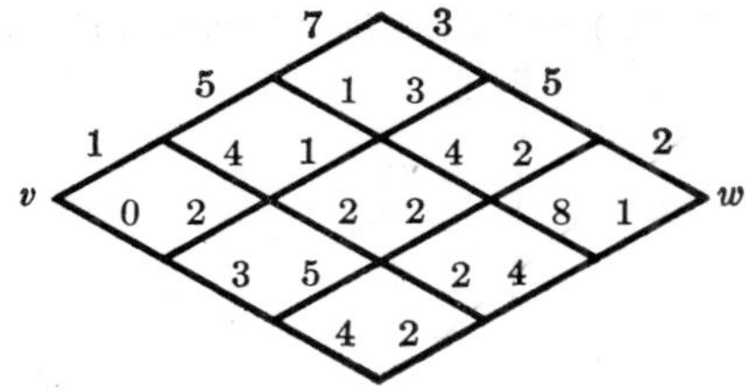

Figure 3-34

programming.[1] Consider the directed graph shown in Fig. 3-34. (All arcs are assumed to be directed to the right, but for simplicity the arrows have been omitted.) The number associated with each arc is to be interpreted as its length, and the problem is to find a shortest path from each vertex x (other than w) to w.

The crucial feature of our problem, from the standpoint of applying dynamic programming, is this: If the sequence of vertices

$$x_1, x_2, \ldots, x_n, w$$

determines a shortest path from x_1 to w, then the sequence

$$x_2, x_3, \ldots, x_n, w$$

necessarily determines a shortest path from x_2 to w. This enables us to consider the vertices in *stages*, considering at the ith stage those vertices which are i arcs removed from w. (Note that for the graph under consideration, all paths from a vertex x to w have the same number, say k, of vertices. Hence the designation of stages is not ambiguous.)

With each vertex x which is one arc removed from w, we associate the (in this case unique) arc $A(x)$ having x as initial vertex. We also associate with x a distance $D(x)$ which is the length of this arc. In our example, these distances are 2 (for the top vertex) and 1 (for the bottom vertex).

In general, having found $A(x)$ and $D(x)$ for all vertices which are k or fewer arcs distant from w, we find $A(x)$ and $D(x)$ for each vertex x which is $k + 1$ arcs removed as follows:

$$D(x) = \min \{\lambda(x,y) + D(y)\}$$

where the minimum is taken over all arcs $a \simeq (x,y)$ having x as initial vertex. (There are at most two such arcs in the example.) In addition, $A(x)$ is taken to be the arc which minimizes the expression. It is readily seen, from the truncation property mentioned earlier and from the fact that the stage 1 distances are clearly minimal, that the stage $k + 1$ distances determined in this manner are minimal for every stage $k + 1$. The bold arcs (again shown without arrows) and numbers in Fig. 3-35 show the values of $A(x)$ and $D(x)$ for the first three stages.

Exercise 3-31 Find $A(x)$ and $D(x)$ for the vertices in stages 4 to 6, Fig. 3-35. In particular, find a shortest path from v to w.

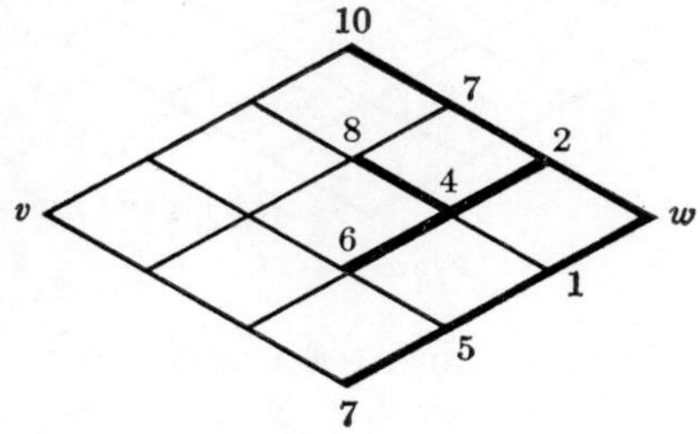

Figure 3-35

REFERENCES

1 Bellman, R. E., and S. E. Dreyfus: "Applied Dynamic Programming," Princeton University Press, Princeton, N.J., 1962.

2 Camion, P.: Chemins et circuits hamiltoniens des graphes complets, *Compt. Rend.*, **249:** 2151–2152 (1959).

3 Dantzig, G., D. Fulkerson, and S. Johnson: Solution of a Large-scale Traveling Salesman Problem, *Operations Res.*, **2:**393–410 (1954).

4 *Eureka*, October, 1961.

5 Ford, L. R., Jr.: "Network Flow Theory," The RAND Corp., P-923, August, 1956.

6 Ford, L. R., Jr., and D. R. Fulkerson: A Simple Algorithm for Finding Maximal Network Flows and an Application to the Hitchcock Problem, The RAND Corp., RM-1604, December, 1955.

7 Gonzalez, R.: Solution of the Traveling Salesman Problem by Dynamic Programming on the Hypercube, *MIT Operations Res. Ctr. Interim Tech. Rept.* 18, 1962.

8 Herz, J. C., T. Gaudin, and P. Rossi: Solution du Problème No. 29, *Rev. Francaise Rech. Operationelle*, **8**(2): 214–218 (1964).

9 Little, J., K. Murty, D. Sweeney, and C. Karel: An Algorithm for the Traveling Salesman Problem, *Operations Res.*, **11:** 972–989 (1963).

10 Pollack, M., and W. Wiebenson: Solutions of the Shortest-route Problem: A Review, *Operations Res.*, **8:** 224–230 (1960).

11 Robbins, H. E.: A Theorem on Graphs, with an Application to a Problem of Traffic Control, *Am. Math. Monthly*, **46**(5): 281–283 (1939).

12 Rédei, L.: Über die Kantenbasen für endliche volständige gerichtete Graphen, *Acta Math. Acad. Sci. Hung.*, **5:** 17–25 (1954).

13 Rosenthal, A.: Solution to Problem E 711, *Am. Math. Monthly*, **53:** 593 (1946).

14 Tutte, W. T.: On Hamiltonian Circuits, *J. London Math. Soc.*, **21:** 98–101 (1946).

15 Ungar, P.: *Am. Math. Monthly*, **57:** 261 (1950).

FOUR

PLANAR AND NONPLANAR GRAPHS, COLORING THEOREMS

4-1 INTRODUCTION

This chapter has two basic aims. The first is to define and give conditions under which a graph is planar, i.e., can be mapped on the plane. The theorem of Kuratowski which excludes the occurrence of two special graphs as subgraphs of a planar graph yields one well-known characterization of planar graphs. Another interesting necessary condition for a graph to be planar is that it be isomorphic to one whose edges are straight. The second aim is to study chromatic graphs with the objective of giving some coloring theorems. These theorems take this form: Given a planar map, i.e., a planar graph together with the regions into which its circuits divide the plane, determine whether it can be colored with n colors in such a way that no two regions with a common boundary of one or more edges have the same color. Or, given n colors, find the conditions on a map for which n is the minimum coloring number. Our emphasis is on the existence rather than on the actual construction of coloring schemes. The concept of dual graph gives another characterization of a planar graph and plays a useful role in studying the coloring problem.

4-2 PLANAR GRAPHS

Recall that a planar graph is one which is isomorphic with a geometric graph in a plane, i.e., it can be drawn in the plane in such a way that its edges intersect only at their end points. A graph can be mapped from a plane on a sphere by placing the sphere at some point of the plane and considering the point of contact as the south pole and then using the north pole as the center of projection of straight lines which join the north pole to every point of the graph in the plane. The intersection of these lines with the sphere yields the desired projection, known as a stereographic projection, which is in 1–1 correspondence with the original graph.

This process can be reversed so that a graph on a sphere can be mapped into the plane, provided that the north pole is chosen to be a point not on the graph. (An interesting difficulty arises if the north pole is selected as a vertex; for then that vertex is mapped at infinity, yielding a needless complication.) Note that any region of a graph drawn in the plane can be made the outside region by means of two mappings. First the graph is mapped on the sphere in such a way that the south pole makes contact with a point inside the desired region. The map is then projected back on the plane by using the former north pole as the point of contact with the plane and the former south pole as the center of the projection lines.

Exercise 4-1 Indicate how a pair of projections can be applied to the graph in Fig. 4-1 in order to place the triangular region inside the rectangle.

In coloring problems one is frequently given a problem on a sphere. The foregoing stereographic projection shows that it is adequate to study coloring problems in the plane if the infinite region is included in the set of regions to be colored.

We now turn to the question of characterizing planar graphs and give basic material which leads to the proof of the theorem of Kuratowski. A simple graph that has the smallest number of vertices and is yet nonplanar is the complete graph on five vertices which we call a type 1 Kuratowski graph (see Lemma 4-4), shown in Fig. 4-2. It is obvious that if any graph contains this five-vertex graph (or indeed any nonplanar graph) as a subgraph, the graph is of necessity nonplanar. An example of

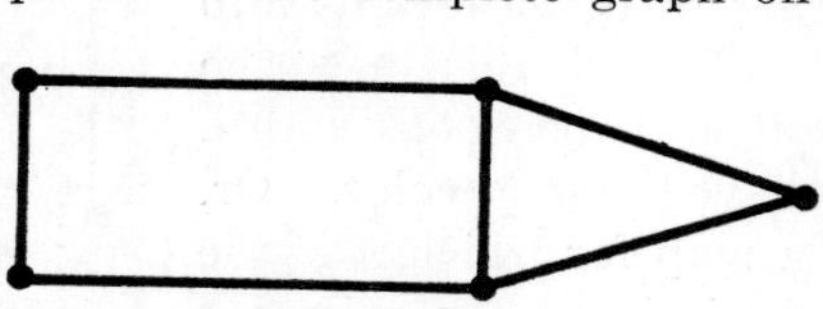

Figure 4-1

a nonplanar graph which does not contain this complete graph is the "three utility–three house" graph shown in Fig. 4-3. We shall call this graph a type 2 Kuratowski graph. The name "utility graph" arises from the problem of connecting each of n houses to each of n utilities by means of conduits which do not pass over one another (i.e., which determine a planar graph). This is not possible for $n \geq 3$, as Lemma 4-4 shows. These two (minimal) possibilities—the complete graph with five vertices and the utility graph—cover the most general situation for nonplanarity.

Definition A vertex v is a *point of articulation* of a connected graph G if the graph $H = G - v$, obtained by deleting v and all edges incident with v, is disconnected. A graph is said to be *separable* if it has at least one articulation point.

Lemma 4-1 A necessary and sufficient condition for a vertex v of a connected graph to be a point of articulation is that v lie on all the chains connecting some pair of vertices.

Proof: If v is a point of articulation, then its removal from the connected graph yields at least two components, and hence any pair of vertices each taken from a different component must have all chains connecting them pass through v. On the other hand, if every chain connecting some pair of vertices passes through v, then the removal of v would disconnect the graph, and hence v is a point of articulation.

Lemma 4-2 (Theorem of Menger-Dirac.[14]) If v_a, v_b are two vertices of a graph of $n > 2$ vertices without a point of articulation and if $U = \{v_a, v_1, \ldots, v_n, v_b\}$ is a chain connecting them, then there are two chains U_1 and U_2 connecting v_a and v_b which have no other vertex in common and along each of them from v_a to v_b the vertices of U (if any) occur in increasing order of subscripts.

Proof: The theorem is proved by induction. It is true if the chain consists of a pair of vertices connected with a single edge. We assume it

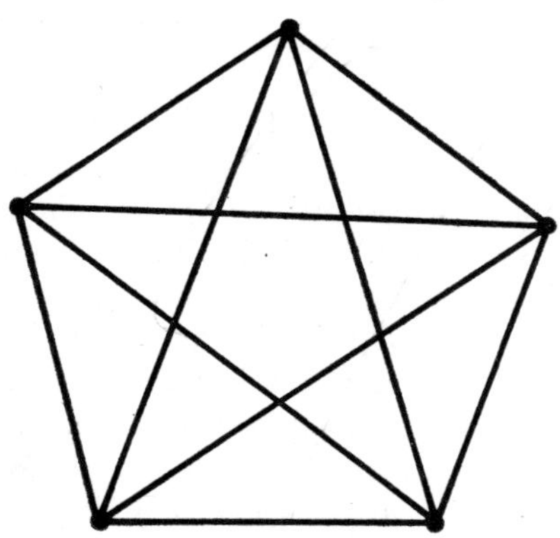

Figure 4-2

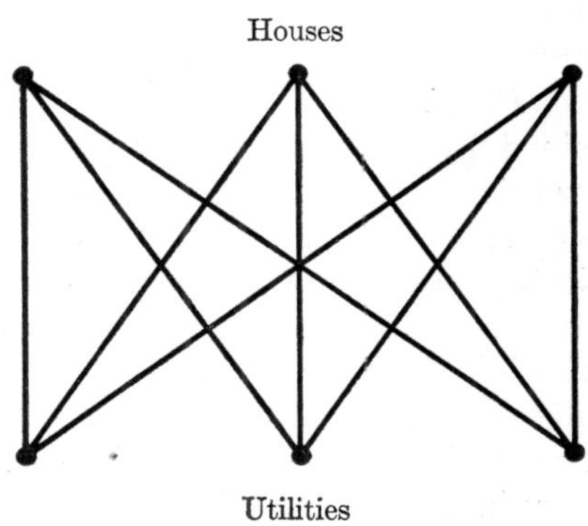

Figure 4-3

is true for any pair of vertices if the length of the chain connecting them is at most m and show that it holds for a chain of length $m + 1$. Define $v_0 = v_a$, $v_{m+1} = v_b$ and suppose that U is the chain joining v_0 with v_{m+1}. Let the vertices of U have the order $v_0, v_1, \ldots, v_m, v_{m+1}$. By the induction assumption v_0 and v_m are connected by two chains $\bar{U}_1$ and $\bar{U}_2$ which satisfy the two properties required in the theorem. Note that the portion of U itself which joins v_0 with v_m may be one of the two chains $\bar{U}_1$, $\bar{U}_2$. Since v_m is not a point of articulation, it does not separate v_0 and v_{m+1} and hence there is a chain connecting v_0 and v_{m+1} which does not pass through v_m. This chain has a part W which connects a last one of the vertices $v_0, v_1, \ldots, v_{m-1}$ to v_{m+1}.

Let v^* be the first vertex (other than v_{m+1}) which is incident with at least one of the chains $\bar{U}_1$, $\bar{U}_2$, or U and which is encountered when U^* is traversed from v_{m+1} toward v_0. Let $W = U^*(v^*,v_{m+1})$, that is, the subchain of U^* joining v^* and v_{m+1}. If v^* is incident with $\bar{U}_1$, then

$$U_1 = \bar{U}_1(v_0,v^*) \cup W$$

and

$$U_2 = \bar{U}_2 \cup U(v_m,v_{m+1})$$

satisfy the conditions of the theorem. An analogous situation holds if v^* is incident with $\bar{U}_2$. Finally, if v^* is incident only with U, then $v^* = v_k$ for some k such that $0 \leq k \leq m - 1$. Let v_p be the vertex with largest subscript not exceeding $k - 1$ which is incident with $\bar{U}_1$ or $\bar{U}_2$. If v_p is incident with $\bar{U}_1$, then

$$U_1 = \bar{U}_1(v_0,v_p) \cup U(v_p,v_q) \cup W$$

and

$$U_2 = \bar{U}_2 \cup U(v_m,v_{m+1})$$

are the desired chains. (If $v_0 = v_p$, the first subchain associated with U_1 is to be ignored.) An analogous pair of chains exist if v_p is incident with $\bar{U}_2$. This completes the proof.

Lemma 4-3 A connected planar graph with n vertices, m edges, and r regions (including the outside or unbounded region) satisfies Euler's formula $n - m + r = 2$.

Proof: Note that $n - m + r$ is unchanged if a vertex of the graph is joined by an edge with another vertex (not intersecting other edges); for this operation produces a new region. (The addition of an edge must produce a circuit, because the graph is connected.) Similarly, this quantity remains the same if a new vertex is introduced and connected to two vertices of the graph by means of two edges; for this also adds a region. The left side also remains unchanged if, instead of adding vertices and edges as described, one removes them according to this procedure. By means of these operations of addition and removal, the figure can finally be reduced to a triangle. This is a graph with $n = 3$, $m = 3$, and $r = 2$, since the region outside the triangle is also counted, from which the result follows.

Lemma 4-4 The Kuratowski subgraphs are nonplanar.

First proof: If the utility graph were planar, we would have, on substituting in Euler's formula, $r = 5$. Each region must have at least four edges for boundary; for if any region had three edges, then two of the three vertices would be either houses or utilities and they would be adjacent, which is a contradiction. Since each edge is the boundary of two regions and is thus counted twice, we must have $4r \leq 2m$ or $m \geq 2r$, that is, $9 \geq 10$. From this absurdity we conclude that the utility graph is nonplanar.

In the case of the complete graph each region has at least three edges for boundary, from which $3r \leq 2m$; and because $r = 7$ and $m = 10$, we have the contradiction that $21 \leq 20$.

Second proof: Another method of proving Lemma 4-4 makes use of the Jordan curve theorem. This theorem states that a simple closed curve (the homeomorph of a circle) divides the plane into two regions whose common boundary is the curve. A corollary of this theorem is that a simple curve joining two points in these two regions intersects the boundary. To prove that the utility graph is nonplanar, connect two of the utilities to two of the houses as in Fig. 4-4. This yields a Jordan curve. The third utility is either inside or outside the region enclosed by the curve. Suppose it is inside. (If it is outside and it is connected to the houses, then another utility must be inside.) We connect it to the two houses. Then no matter which region the third house belongs to, it will be separated by a Jordan curve from a utility and cannot be joined to it. Consequently the graph is nonplanar.

To prove that the complete graph is nonplanar, consider four points joined two at a time. This gives Fig. 4-5. As can be seen from the figure, for v_1 to be joined with v_3 and v_2 with v_4 without intersection, one edge must be inside and the other outside the rectangle. No matter where the fifth vertex is placed, it would be separated by a Jordan curve from one of the other vertices and we again conclude that the graph is nonplanar.

The planarity or lack of planarity of a graph is clearly not affected if an edge is divided into two edges by the insertion of a new vertex of

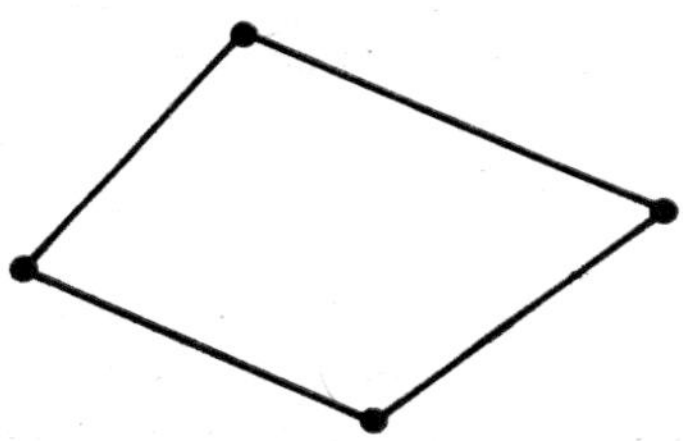

Figure 4-4

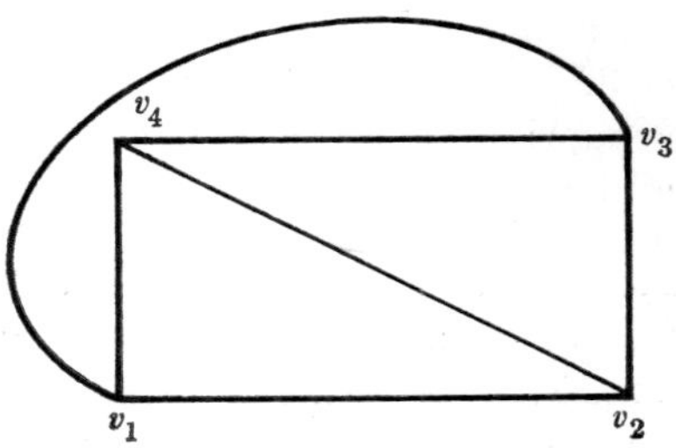

Figure 4-5

degree 2 or if two edges incident with a vertex of degree 2 are merged into one edge by removal of that vertex. This suggests the following definition: Graphs G and G' are *isomorphic to within vertices of degree 2* if they are isomorphic or if they can be transformed into isomorphic graphs by repeated application of the above transformations.

Theorem 4-5 (Kuratowski.) A graph is planar if and only if it does not contain any subgraph which is isomorphic, to within vertices of degree 2, to either of the Kuratowski subgraphs.

Exercise 4-2 Show that the existence of a subgraph which is isomorphic to within vertices of degree 2 to a Kuratowski graph is equivalent to the existence of five or six points together with a set of vertex-disjoint chains (except for their end points) which—when the chains are regarded as single edges—constitute a Kuratowski graph.

The proof we shall present of Theorem 4-5 is a substantial refinement due to Berge[3] of Kuratowski's original proof.[30]

Proof: We have already demonstrated that a planar graph does not contain either of the Kuratowski subgraphs or a subgraph which is isomorphic to one of these to within vertices of degree 2. The proof in the other direction, i.e., that a nonplanar graph contains one or both of the Kuratowski subgraphs, proceeds by induction on the number of edges. We use the equivalent contrapositive statement that if a graph does not contain either of the Kuratowski subgraphs, it must be planar. It is obviously true for a graph with one, two, and three edges. Let the contrapositive be true for less than m edges. We now show by contradiction that it is true for m edges. Hence we suppose that G has m edges, is nonplanar, and contains neither of the Kuratowski subgraphs but that every subgraph with $m - 1$ edges is planar. The contradiction will be that G contains a Kuratowski subgraph.

1. G must be connected; for otherwise all its components have fewer than m edges and are planar, and hence G would be planar.

2. The graph G is inarticulate. Note that a point of articulation may be mapped into the boundary of the infinite region by placing a sphere in a region having this point on its boundary and inverting the graph with respect to the sphere. The removal of this point would leave components planar, because they have fewer than m edges, and thus its addition would yield a planar graph.

3. We show that the existence of an elementary circuit on suppressing an edge, i.e., the removal of an edge $[a,b]$, leaves a circuit S passing through a and b. To see this, observe that the subgraph $G' = G - [a,b]$ must be either (1) articulate or (2) inarticulate. In case 1 there is a point of articulation c through which passes every chain between a and b in G'. However, this leads to a contradiction, for the removal of c leaves two components C_a and C_b. Let C'_a and C'_b be obtained from C_a and C_b by adding the edges $[a,c]$ and $[b,c]$, respectively. By the original statement

(not the contrapositive), neither C'_a nor C'_b contains a Kuratowski subgraph because G does not. By stereographic projection $[a,c]$ and $[b,c]$ can be made into edges of the infinite region. In this region, if a and b are joined, the result is still a planar graph, since the edge is completely within the one outside region. This graph is planar and contains G. Since this is a contradiction to the assumption that G is nonplanar, we conclude that G' cannot be articulate.

Thus, case 2 holds and G' is inarticulate and, by Lemma 4-2, there are two chains joining a and b and forming a circuit, i.e., they have no point in common; therefore, G' has a circuit. This circuit will now be used to contradict the nonplanarity assumption on G.

4. Let S be a simple circuit passing through a and b in G' containing in its interior the maximum number of regions. Let us introduce an arbitrary orientation on S. The parts of G' inside S are called the interior graph, and those outside are called the exterior graph.

The exterior graph cannot contain more than one vertex of the half circuit $S[a,b]$; for otherwise one can construct a circuit passing through a and b which contains in its interior a larger number of regions. This is also true of the half circuit $S[b,a]$. The exterior graph, however, must have one or two points in common with S; for if it had none, the graph G would be disconnected. On the other hand, there exist at least one connected subgraph I of the interior graph and one connected subgraph E of the exterior graph each of which meets both $S]a,b[$ and $S]b,a[$, where $S]a,b[$ denotes the graph obtained from $S[a,b]$ by deleting the vertices a and b. To see this, note that G' is planar and introducing $[a,b]$ into it must make it a nonplanar graph. But for G to be nonplanar, $[a,b]$ must either intersect the subgraph consisting of $S]a,b[$ and $S]b,a[$ plus the interior graph or intersect the subgraph consisting of $S]a,b[$ and $S]b,a[$ plus the exterior graph.

Furthermore, S must contain four vertices which are alternately points of contact with I and with E. To prove this, suppose the contrary and let e_1 and f_1 be two consecutive points of contact of a connected subgraph I_1 of the interior graph which meets $S]a,b[$ and $S]b,a[$. These two points can be joined by a path P lying outside S and not meeting any of the existing edges, since by hypothesis there exists no connected subgraph E of the exterior graph which meets $S]a,b[$ and $S]b,a[$ and which meets $S]e_1,f_1[$ and $S]f_1,e_1[$. Every connected subgraph of the interior graph, including I_1, which meets only $S[f_1,e_1]$ can be transferred to the exterior inside the region limited by P and S. This leaves at least one connected subgraph I_2 of the interior graph which meets $S]a,b[$ and $S]b,a[$ and has two points of contact e_2 and f_2 on $S[e_1,f_1]$, at least one of which is on $S]e_1,f_1[$. We can now replace I_1 by I_2 and continue the transfer procedure. Since the graph is finite, the entire interior graph can be moved to the exterior,

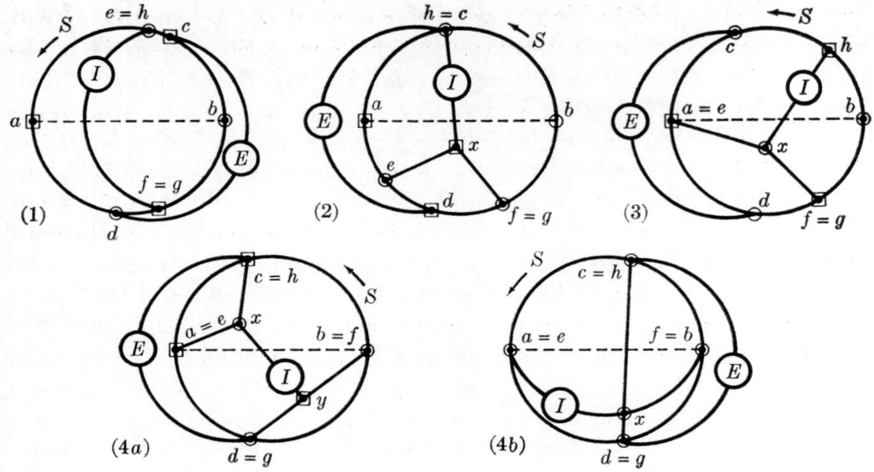

Figure 4-6

which is a contradiction; i.e., the graph must then be planar, since nothing intersects $[a,b]$. Thus there exist connected subgraphs I and E of the interior and exterior graphs, respectively, which meet $S]a,b[$ and $S]b,a[$ and which intersect S at alternate points c, d (of E) and e, f (of I).

The case when both the exterior and interior components have common points of contact at S or coincide with a or b is remedied by introducing additional points to satisfy the requirement of alternation. All the possibilities can be covered by taking four points for the interior component. We suppose that e, f, g, h are the points of contact of I with S such that $e \in S]c,d[$; $f \in S]d,c[$; $g \in S]a,b[$; $h \in S]b,a[$. (See Fig. 4-6.) Obviously e can never be f nor can g be h, but we can have $e = g$ and $e = h$, etc., and so we consider all these cases:

1. If $e \in S]b,a[$ and $f \in S]a,b[$, let $e = h$ and $f = g$. This leads to the graph of type 2, which is contrary to hypothesis.
2. If both e and f are on $S]a,b[$, let $h = c$; for if $h \neq c$, $h \in S]b,a[$ and we are in case 1. But if $h = c$, we again get a graph of type 2.
3. If $e = a$ and $f \neq b$, for example, $f \in S]b,a[$, we obtain a type 2 graph.
4. If $e = a$ and $f = b$, we can assume $g = d$ and $h = c$; for if not, we are in either case 1 or case 3. There are two cases to be considered:

(*a*) If the chains of I joining cd and ef have more than one vertex in common, we get a type 2 graph.

(*b*) If the chains of I joining cd and ef have only one vertex in common, we get a type 1 graph.

As indicated in Fig. 4-6, case 4*b* leads to a type 1 graph and the remaining cases to a type 2 graph and thus, in any case, to a nonplanar graph, a contradiction. This completes the proof. (Note in Fig. 4-6 that circles and squares enclose those vertices which determine the Kuratowski subgraphs. Vertices b and d in Fig. 4-6(2) and (3), respectively, are to be regarded as vertices of degree 2.)

In addition to the theorem of Kuratowski there are other theorems which characterize a planar graph. These are stated without proof.

Theorem 4-6 (MacLane.) A necessary and sufficient condition for a graph to be planar is that it have a circuit basis (see matrix representation of graphs) together with one additional circuit such that this collection of circuits contains each edge of the graph exactly twice.

Theorem 4-7 (Whitney.) A necessary and sufficient condition for a graph to be planar is that it have a dual (see below).

The Dual Graph

Consider a planar graph G with regions R_i ($i = 1, \ldots, n$). Let us associate a point p_i with each region R_i (that is, take a point in the region). If two regions R_i and R_j are adjacent, join p_i and p_j by an edge p_ip_j which intersects the common boundary of R_i and R_j only once and has no point in common with any other boundary of the graph. An edge of G which is not the boundary of a region lies in a region and yields a loop (see Fig. 4-7). This procedure yields a new graph $\bar{G}$ with vertices $p_1, \ldots, p_n$. It is called a *dual graph* of G.

We have illustrated the idea of the dual graph with dashed lines in Fig. 4-7. It is clear from this construction that all graphs dual to G are

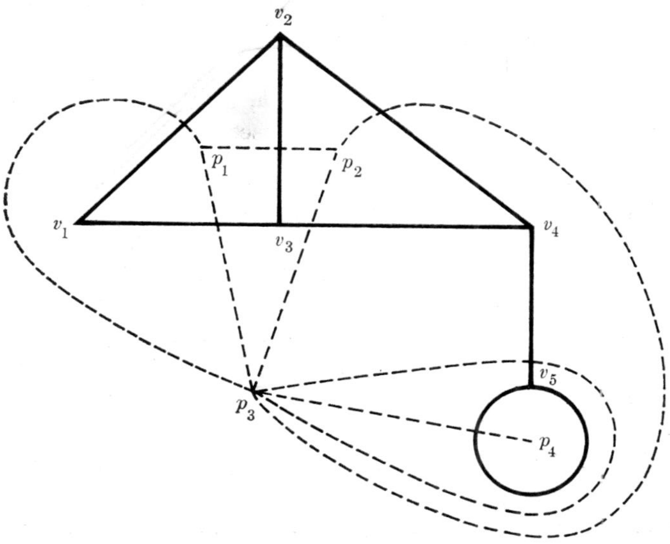

Figure 4-7

topologically equivalent; hence one speaks of *the* dual graph. (Note that $\bar{\bar{G}} \neq G$ if G is not connected, where $\bar{\bar{G}}$ is the dual of $\bar{G}$.)

Exercise 4-3 Start with the dual graph given in Fig. 4-7 and construct its dual. Show that the dual is equivalent to the original graph.

Exercise 4-4 Show that there is no map of five regions in the plane such that every pair of regions is adjacent.

Definition A graph is *trivalent* if it is regular of degree 3.

Theorem 4-8 All regions of the dual of a planar trivalent connected map are three-sided and their number is even.

Proof: It is easy to see that corresponding to each vertex of the graph there is a triangle of the dual. Thus we have for the dual $3r = 2m$, which, when substituted in Euler's formula, gives $r = 2(n - 2)$.

Remark: One can associate a vertex with each edge of a graph. If two vertices are joined by an edge whenever the corresponding edges are adjacent, the resulting graph is known as a *line graph.*

Theorem 4-9 (Krausz.[29]) A necessary and sufficient condition for a graph to be a line graph is the existence of a partition of its edges into complete subgraphs such that no vertex of the graph belongs to more than two subgraphs.

Exercise 4-5 Verify that, for the line graph of the complete graph on n vertices.
(a) Each vertex has degree $2(n - 2)$.
(b) Exactly four vertices are adjacent to both of any two nonadjacent vertices.
(c) Any two adjacent vertices are mutually adjacent to exactly $n - 2$ vertices.
These three necessary conditions are also sufficient (except for $n = 8$) for a graph to be the line graph of a complete graph on n vertices.

Exercise 4-6 Show that exactly $4n - 12$ vertices are adjacent to either one of any pair of adjacent vertices of the line graph in Exercise 4-5.

We shall now show that a necessary condition for a simple graph to be planar is that it can be drawn in a plane with straight lines. (Any simple graph whether planar or not can be drawn with straight lines in three-space.)

Definition A *straight graph* is a planar graph in which every edge is a straight-line segment.

Definition If every region of a planar graph is bounded by a circuit of three edges, the graph is said to be *triangulated.* Fary[21] has proved the following theorem:

Theorem 4-10 Every simple planar graph is isomorphic to a straight graph.

We shall prove this theorem in the sequence of Lemmas 4-11 to 4-15.

Lemma 4-11 If a graph G is isomorphic to a straight graph, then each of its subgraphs is also isomorphic to a straight graph.

Proof: If G is isomorphic to a straight graph $\bar{G}$, then since a subgraph of G is obtained by deleting the vertices and edges not in the subgraph, the corresponding operation in $\bar{G}$ leaves a straight subgraph isomorphic to the desired subgraph in G. (Here $\bar{G}$ does not represent the dual of G.)

Lemma 4-12 Every simple planar graph G is a subgraph of a triangulated graph with the same number of vertices.

Proof: Suppose G consists of at least four vertices. (The case of three or fewer vertices is left as an exercise.) We construct a triangulated graph G' which contains G as a subgraph. If R is a region of G and v_1 and v_2 are two vertices of R without a connecting edge, we connect v_1 and v_2 with an edge lying entirely in R. This yields a new graph which is again simple. This operation of selecting unconnected pairs of vertices on the boundary of a region and connecting them by an edge in the region is continued until every pair of vertices on the boundary of the same region are connected by an edge. Since each time this is done the number of vertices remains the same, the process terminates, yielding a simple graph G'.

Clearly, G' is connected; otherwise, there is a region R adjacent to two components of G and we choose v_1 on the boundary of one of these components and v_2 on the boundary of the other and join them by an edge in R. This contradicts the construction of G'. To show that G' is triangulated, note first that if some region is bounded by only one edge, then the graph consists of this edge and its two vertices. A region cannot be bounded by two edges; for then the graph would contain only three vertices. Both cases contradict our assumption of at least four vertices. Thus the boundary of each region R' consists of at least three edges, e_1, e_2, e_3. If they do not form a circuit, then there are at least four vertices v_1, v_2, v_3, v_4 on the boundary of R'. Now G' contains the edges connecting these vertices, that is, v_1v_2, v_1v_3, v_1v_4, v_2v_3, v_2v_4, v_3v_4 (see Fig. 4-8) which divide the plane into four regions on the boundary of each of which there are only three vertices. But this is a further partitioning of R', which is itself a region in G', and we have a contradiction. Thus e_1, e_2, e_3 form a circuit and every region of G' is bounded by a circuit of three edges.

The next two lemmas are needed for the proof of Lemma 4-15.

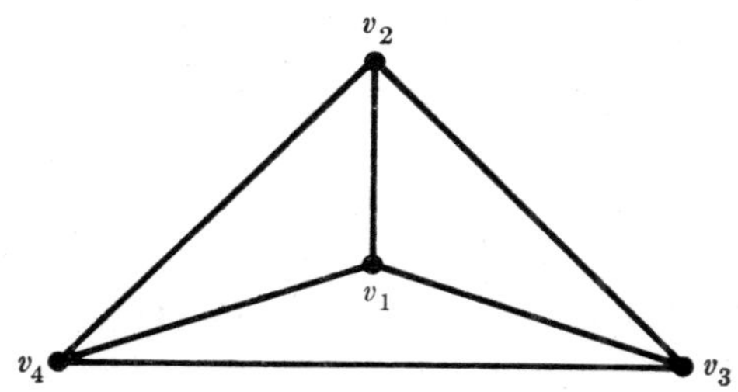

Figure 4-8

Lemma 4-13 Let G be a simple triangulated graph with at least four vertices. If $vv_1, vv_2, \ldots, vv_k$ are edges starting from v in their cyclic (e.g., clockwise) order, then the edges $C_v = (v_1v_2, v_2v_3, \ldots, v_kv_1)$ belong to G and form a circuit which separates v from every other vertex of G.

Proof: It suffices to prove the lemma with the assumption that v does not lie on the boundary of the infinite region (containing the point at infinity); for by means of spherical projection v can be made to lie on the boundary of an interior region. Since the graph is triangulated, v must lie inside some circuits and hence there is an innermost circuit C'_v with vertices $v'_1, v'_2, \ldots, v'_m$ which contains v in its interior. Every consecutive pair of vertices v'_i, v'_{i+1} lies on the boundary of a region which is triangulated and hence contains a third vertex v'. We assert that $v' = v$; for otherwise the circuit $(v'_1v'_2, v'_2v'_3, \ldots, v'_iv', v'v'_{i+1}, \ldots, v'_mv'_1)$ is a circuit having v in its interior and is inside C'_v, a contradiction. Hence the triangular regions adjacent to C'_v have the common vertex v; and since G is simple, C'_v must coincide with C_v and the proof is complete.

If v in Lemma 4-13 (assumed to lie on an interior region R) and its connecting edges are deleted, leaving an empty interior for C_v, and if v_1 is jointed to $v_3, \ldots, v_{k-1}$ by edges not crossing each other but lying inside C_v, perhaps resulting in a not necessarily simple graph $\bar{G}$, then we have the following lemma:

Lemma 4-14 If the graph $\bar{G}$ is not simple, then G has a circuit of three edges which separates two vertices.

Proof: $\bar{G}$ can be nonsimple only if v_1 is connected to some vertex v_i $(3 \leq i \leq k - 1)$ by a new edge and by an edge outside C_v. The circuit (v_1v, vv_i, v_iv_1) in G with v_iv_1 outside C_v separates v_2 and v_k. This proves the lemma.

Lemma 4-15 Every triangulated simple graph is isomorphic to a straight graph.

Proof: The lemma is obviously true for graphs with three vertices. Suppose it is true for graphs with n vertices and let G be a graph with $n + 1$ vertices. Let v be a vertex of G on the boundary of an interior region R and construct $\bar{G}$ as above. The infinite region has the same boundary in $\bar{G}$ as in G.

If $\bar{G}$ is simple, then by the inductive assumption there is a straight graph $\bar{S}$ isomorphic to $\bar{G}$. We must now make sure that $\bar{v}$, the vertex corresponding to v, can be chosen in such a way that it can be connected with straight segments to pertinent vertices and without intersecting other segments. Consider the circuit $\bar{C}_v = (\bar{v}_1\bar{v}_2, \ldots, \bar{v}_k\bar{v}_1)$ of $\bar{S}$ corresponding to C_v in Lemma 4-13, where $\bar{v}_i$ in $\bar{S}$ corresponds to v_i in $\bar{G}$. The straight lines which include $\bar{v}_i\bar{v}_{i+1}$ $(2 \leq i \leq k - 1)$ do not pass through $\bar{v}_1$; for then $\bar{v}_1\bar{v}_i$ and $\bar{v}_1\bar{v}_{i+1}$ would have a common segment. Thus $\bar{v}_1$ belongs to one of the two half planes defined by the straight line $\bar{v}_i\bar{v}_{i+1}$ $(2 \leq i \leq$

$k - 1$) and the intersection of these half planes is a convex region with interior points. Consider the common part K of this region and the interior of $\bar{C}_v$. The segment joining an arbitrary point of K with a boundary vertex of $\bar{C}_v$ lies inside $\bar{C}_v$ except for its end point. If the segments $\bar{v}_1\bar{v}_3, \bar{v}_1\bar{v}_4, \ldots, \bar{v}_1\bar{v}_{k-1}$ were deleted from $\bar{S}$, the interior of $\bar{C}_v$ would be empty. By selecting a point $\bar{v}$ inside K and drawing the segments $\bar{v}\bar{v}_1, \bar{v}\bar{v}_2, \ldots, \bar{v}\bar{v}_k$, we obtain a graph isomorphic to G.

If $\bar{G}$ is not simple, then by Lemma 4-14 there is in G a circuit Δ of three edges which separates two vertices. The circuit Δ, with the outside edges, forms a triangulated graph G_1. It also forms such a graph G_2 with the inside edges. Both graphs have at most n vertices and hence, by the inductive assumption, are isomorphic to straight graphs $\bar{S}_1$ and $\bar{S}_2$, respectively. In $\bar{S}_1$ the interior of the circuit $\bar{\Delta}_1$ corresponding to Δ is empty. On the other hand, $\bar{S}_2$ lies inside $\bar{\Delta}_2$, the circuit corresponding to Δ in $\bar{S}_2$. By an appropriately chosen affine transformation, $\bar{\Delta}_2$ can be mapped into $\bar{\Delta}_1$ so that adjacent regions in G are mapped into adjacent regions. This yields a straight graph isomorphic to G and completes the proof of the theorem.

Polyhedral Graphs

The reader unfamiliar with the terminology of the next two paragraphs may go on without loss of continuity.

A graph is called *polyhedral* if its vertices and edges may be identified with the vertices and edges of a convex polyhedron in some euclidean space. The polyhedron is called the realization or embedding of its graph. We shall say that a graph is n-polyhedral if it corresponds to an n-dimensional polyhedron. Thus a graph is 1-polyhedral if and only if it is the complete graph on two vertices; 2-polyhedral if and only if it is a circuit; 3-polyhedral if and only if it is a 3-connected graph (not so obvious). It is also known that every k-polyhedral graph is k-connected and it contains a subgraph which is isomorphic to within vertices of degree 2 with the complete graph on $k + 1$ vertices (i.e., one which is obtained by inserting additional vertices in the edges of this complete graph). Any 3-connected subgraph of a 3-polyhedral graph is again 3-polyhedral. This need not be the case for $k \geq 4$. The foregoing conditions are not sufficient for polyhedrality, at least for $k = 4$.[25] The complete graph on n vertices is $(n - 1)$-polyhedral, the polyhedron being called a *simplex*.

An astonishing fact is that every complete graph on n vertices with $n \geq 5$ can be shown to be 4-polyhedral. Open questions are whether for every graph the set of integers k for which the graph is k-polyhedral is consecutive or whether one can always join any two realizations in a given dimension by a continuous family of realizations. New polyhedral graphs may be generated from their corresponding polyhedra by (1) taking the dual polyhedron in which the roles of the faces are interchanged, i.e., with a k-dimensional face of the polyhedron is associated an $(n - 1)$-dimensional face of its dual—corresponding incidences are also preserved, (2) cutting a k-dimensional face of an n-polyhedron by a new $(n - 1)$-dimensional face close to it, (3) joining two n polyhedra along a common face, (4) taking the convex hull of two polyhedra of possibly different dimensions contained in skew linear spaces, e.g., the formation of pyramids, (5) taking the direct product of two polyhedra of possibly different dimensions n and m, for example, formation of prisms. The resulting polyhedron is $(n + m)$-dimensional.

4-3 THE COMPLEMENTARY GRAPH

Consider a simple graph with n vertices. We obtain the *complementary graph* by deleting from the complete (simple) graph with n vertices all those edges that occur in the original graph. A graph is self-complementary if it is isomorphic with its complement. Thus the graph of Fig. 4-9 is self-complementary.[26]

Exercise 4-7 Solve the following problem by means of graph theory. Tom, Dick, and Harry are married to Jane, Mary, and Susan, but not necessarily in that order. Each couple has one child. The children's names are Emily, Alan, and Michael. Group together the names in each family by using the following information: (1) Mary's child and Harry's child are the two stars of the school football team. (2) Tom's son is not Alan. (3) Dick's wife is not Susan.

Hint: There are nine vertices in this graph in groups of three. Each member of a group is connected to exactly one member of the others. Consider the complementary graph, ignoring edges joining two vertices in the same group.

Before we give an interesting theorem on complementary graphs, we consider the Euler relation

$$n - m + r = 2$$

and the fact that

$$3n \leq 2m \qquad n \geq 4 \text{ (a useful relation, not needed here)}$$
$$3r \leq 2m \qquad r \geq 4$$

always hold. If we use the second inequality in Euler's formula, we have

$$n - \tfrac{3}{2}r + r \geq 2$$

or

$$r \leq 2n - 4$$

By substituting in Euler's relation and simplifying, we obtain

$$m \leq 3n - 6$$

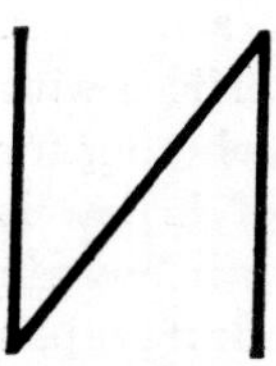

Figure 4-9

which holds for any planar graph. Consequently, a graph

is nonplanar if

$$m > 3n - 6$$

Theorem 4-16 If G is a graph of n vertices and G' is its complement, then: (1) if $n < 8$ at least one of the two is planar, (2) if $n > 8$ at least one of the two is nonplanar, and (3) if $n = 8$ one or the other or both may be planar or nonplanar.

Proof: For case $n \geq 11$, note that $m + \bar{m} = n(n-1)/2$ where $\bar{m}$ is the number of edges of G'. If $\bar{m} \geq m$, then $\bar{m} \geq n(n-1)/4$, otherwise $m \geq n(n-1)/4$. In either case, for $n = 11$ we have $n(n-1)/4 = 27.5$ and $3n - 6 = 27$, from which either m or $\bar{m}$ is greater than $3n - 6$ and the result follows. For $n > 11$ a similar argument applies.

The argument for cases 9 and 10 is elaborate and will not be given here. It can be shown by exhaustion[36a] that for $n = 9$ every planar graph with $3n - 6$ edges has a complement containing one of the two Kuratowski subgraphs. From that it also follows that a graph with $n = 10$ vertices and $3n - 6$ edges has a nonplanar complement, since if we suppress a vertex with its connecting edges we obtain a planar graph on 9 vertices where, by the foregoing, the complement is nonplanar and is contained in the complement of the graph with $n = 10$ vertices under consideration.

For $n = 8$, first consider a graph with four utilities and four houses. This graph is nonplanar and its complement is obviously planar. Again consider the Kuratowski utility subgraph with two isolated vertices. It has 9 edges and its complement has 19 edges, which exceeds 18, the maximum number of edges possible in a planar graph of 8 vertices. Finally consider two concentric squares. Label the vertices of the inner square 1, 2, 3, 4 and those of the outer square 5, 6, 7, 8 where 5 is near 1, 6 near 2, 7 near 3, and 8 near 4. The following pairs give vertices to be joined with straight edges (1,2), (1,4), (1,5), (1,6), (2,3), (2,4), (2,6), (3,4), (3,6), (3,7), (3,8), (4,5), (4,8), (5,6), (5,8), (6,7), (7,8), and join with an exterior arc 6 to 8. This is a planar graph. Its complement can also be shown to be planar.

The cases $n < 8$ are left as exercises to the reader. Note for example that by isolating a vertex in a graph, the complement would have a greater number of edges; this is helpful to remember for cases $n < 8$.

Theorem 4-17 If k and k' are the chromatic numbers of a graph G with n vertices and its complementary graph G', respectively, then[36]

$$2\sqrt{n} \leq k + k' \leq n + 1$$

$$n \leq kk' \leq \left(\frac{n+1}{2}\right)^2$$

Proof: Let n_i vertices of G be colored with the ith color; then

$$\sum_{i=1}^{k} n_i = n \qquad \text{and} \qquad \max_i n_i \geq \frac{n}{k}$$

Two vertices in G with the same color are connected by an edge in G' and hence are colored differently. Thus $k' \geq \max n_i \geq n/k$, or $kk' \geq n$. Now from

$$(k - k')^2 \geq 0$$

we have

$$(k + k')^2 \geq 4kk'$$

This implies that

$$k + k' \geq 2\sqrt{n}$$

We now show by induction on n that $k + k' \leq n + 1$. The theorem is true for $n = 1$. Assume it is true for n vertices and adjoin a vertex v to the complete graph consisting of G and G' taken together. Since v is connected to n other vertices, let m of them be adjoined to G to yield a graph H (with chromatic number h) and $n - m$ to G' to yield a graph H' (with chromatic number h'). Clearly, we may have to use at most one more color in H and H' than we used in G and G'. Thus $h \leq k + 1$, $h' \leq k' + 1$, and $h + h' \leq n + 3$. If in fact $h = k + 1$, $h' = k' + 1$, then if v and its edges in H are deleted, the chromatic number of H will be reduced. In this case $m \geq k$ and $n - m \geq k'$, from which $k + k' \leq n$ and again $h + h' \leq n + 2$, completing the inductive proof.

The theorem is demonstrated if we finally note that from

$$(k + k')^2 \geq 4kk'$$

and from $(k + k') \leq n + 1$ we have

$$kk' \leq \left(\frac{n+1}{2}\right)^2$$

4-4 COLORING THE EDGES OF A GRAPH

Consider the complete graph on n vertices (which is clearly nonplanar for $n \geq 5$) and suppose that some edges of this graph are colored red and the remaining edges blue. What is the smallest number of triangles whose three sides have the same color (monochromatic triangles)? This problem was first solved by Goodman,[23] but here we give a simpler proof due to Sauve.[40]

Theorem 4-18 In a complete graph on n vertices let b be the number of triangles with three blue sides and let r be the corresponding number of red-sided triangles; then the number of monochromatic triangles

$(b + r)$ satisfies

$$b + r \geq \begin{cases} \dfrac{s(s-1)(s-2)}{3} & \text{if } n = 2s \\ \dfrac{2s(s-1)(4s+1)}{3} & \text{if } n = 4s+1 \\ \dfrac{2s(s+1)(4s-1)}{3} & \text{if } n = 4s+3 \end{cases}$$

where s is a nonnegative integer. The bound is sharp; i.e., equality is attained for each n and some (red-blue) coloring.

Proof: For each pair of edges meeting at a vertex, assign the weight 2 if they have the same color and -1 if they do not. The weight at a vertex is the sum of the weights of pairs of edges meeting at the vertex. Also, the weight of a triangle is the sum of the weights of pairs of its edges meeting at each of its vertices. In general, each pair of edges belongs to a single triangle, and hence the sum w of the weights of all the triangles is the sum of the weights of all vertices. The weight of a monochromatic triangle is six, whereas all other triangles have weight zero, from which $b + r = w/6$. The problem now is to determine the smallest value of $w/6$ for all possible colorings. If $n = 2s$, then the weight at each vertex is a minimum when the maximum number of pairs of edges at the vertex have different colors. Now each vertex has $2s - 1$ edges. If there are t edges of one color meeting at a vertex v, then there will be $2s - 1 - t$ edges of the other color also meeting at v and the weight at v is given by

$$2C_2^t + 2C_2^{2s-1-t} - t(2s - 1 - t)$$

where C_2^t is the number of combinations of t vertices taken two at a time, i.e.,

$$C_2^t = \frac{t!}{(t-2)!2!}$$

Similarly,

$$C_2^{2s-1-t} = \frac{(2s-1-t)!}{(2s-t-3)!2!}$$

The reader will readily verify that the above weight is minimum for $t = s$. Thus the total weight at all vertices satisfies

$$w \geq 2s(s-1)(s-2)$$

and

$$r + b = \frac{w}{6} \geq \frac{s(s-1)(s-2)}{3}$$

A similar analysis produces the result for $n = 4s + 1$ with $2s$ edges of each color at each point. If $n = 4s + 3$, then $2s + 1$ cannot be the number of edges of each color at each point, since the total $\frac{1}{2}(4s + 3)$ $(2s + 1)$ is not an integer. Hence $4s + 2$ of the points have $2s + 1$

edges of each color and the remaining point has $2s$ edges of one color and $2s + 2$ of the other.

To prove equality when $n = 2s$, we follow the argument by Lorden.[32] Label the vertices $v_1, \ldots, v_n$ and color with red any edge-connecting pairs of vertices the sum of whose indices is even and color with blue any edge of the opposite case. Vertices that have even indices are connected with red edges, the totality of which is called a red net. Those with odd indices (which exceed the even ones by unity if n is odd) are also connected pairwise with red lines (also forming a red net). Every triangle must have at least one edge in one of the two red nets, and hence there is no triangle that has all three edges colored blue. If $n = 2s$, each of the two red nets has s vertices, and hence $\binom{s}{3}$ red triangles, and there are $2\binom{s}{3}$ red triangles in the graph. Show that if $n = 2s + 1$, there are $\binom{s}{3} + \binom{s+1}{3}$ red triangles. Reference 40 proves equality for odd n.

Theorem 4-19 In any complete graph on n vertices

$$b + r = \binom{n}{3} - \frac{1}{2} \sum_{i=1}^{n} I_i(n - 1 - I_i)$$

where I_i is the number of vertices to which the ith vertex is red-connected.[32]

Proof: Incident with the vertex v_i there are I_i red edges and $n - 1 - I_i$ blue edges, and hence there are $I_i(n - 1 - I_i)$ pairs of edges of mixed color, each pair forming two sides of a triangle of mixed-color edges. Since the third edge is either red or blue, one of the two remaining vertices will have a red-blue pair. Hence the number of mixed color triangles is

$$\frac{1}{2} \sum_{i=1}^{n} I_i(n - 1 - I_i)$$

The total number of triangles $\binom{n}{3}$ is equal to the number of single-color triangles $b + r$ plus the number of mixed-color triangles, and the result follows. Similar questions can be asked[24] for the coloring of a graph using several colors, but there are no analogously simple theorems.

4-5 COLORING REGIONS AND VERTICES: THE FOUR-COLOR PROBLEM

The four-color problem is to prove or disprove that four colors are sufficient to color any planar map so that no two adjacent regions have the

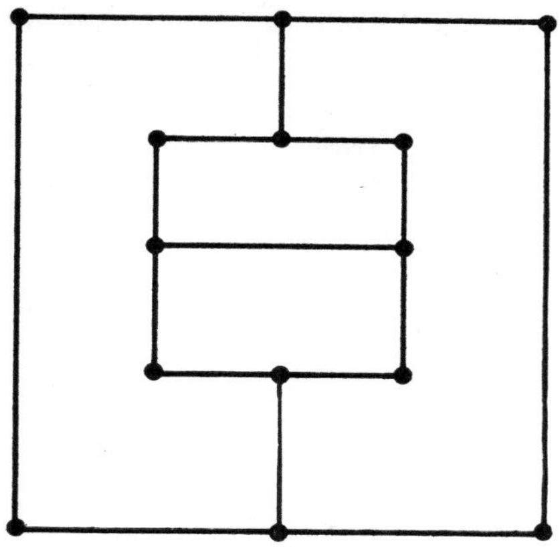

Figure 4-10

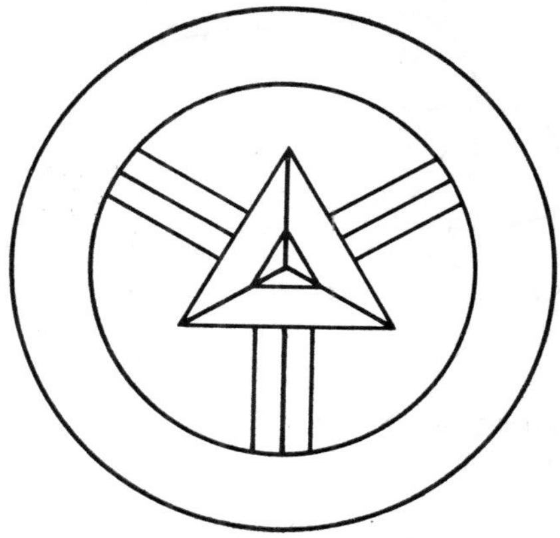

Figure 4-11

same color. (Meeting at a vertex is not considered as adjacency.) Figure 4-10 shows that four colors are necessary. The conjecture was first presented in lectures of Möbius in 1840 and was made better known by De Morgan, who received it through Franci Guthrie around 1850. Cayley in 1878 observed that he had been unable to obtain a rigorous proof. In 1890 Heawood revised a wrong proof of the problem due to Kempe (1879) and was able to prove the sufficiency of five colors. An excellent account of earlier ideas on the problem is given in Ref. 2.

Exercise 4-8 Four-color the regions of the map given in Fig. 4-11.

Remark: If the degree of each vertex of a graph is at most k, it is intuitively obvious that the graph can be $(k + 1)$-colored, because no vertex is joined to more than k other vertices. This suggests the question whether a smaller number of colors would suffice. The complete graph on $k + 1$ vertices has degree k at each vertex but requires $k + 1$ colors.

It turns out that for all graphs of maximum degree k, except the graphs having one component which is a complete subgraph, fewer colors are sufficient. This is the theorem of Brooks[10] applied to a not necessarily planar graph, which we give here without proof.

Theorem 4-20 Let G be a linear graph with n vertices and without loops such that no more than k edges (where $k > 2$) are incident with each vertex. Suppose that no connected component of G is a complete $k + 1$ graph. Then it is possible to color the vertices of G with k colors such that no two vertices have the same color.

The Number of Regional Adjacencies

In a planar map, let us consider the construction of regions such that each region is adjacent to as many other regions as possible. We start

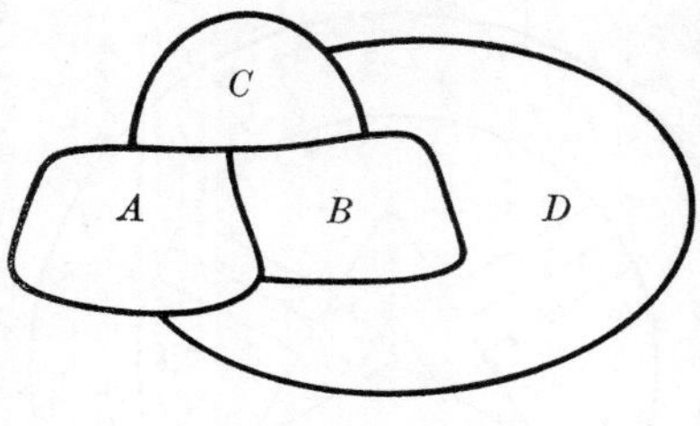

Figure 4-11a

with two adjacent regions A and B. To construct the region C, we connect a point from the boundary of A with another from the boundary of B by an arc. To construct D, we connect the boundary of C with the boundary of A such that B would be adjacent to D, etc. By carrying out this operation n times, we obtain $n + 3$ regions, because we include the outside region also. Each time we draw an arc we obtain three arcs, the new one and a division of the two arc segments it touches into four arc segments (Fig. 4-11a).

Thus the total number of arc segments is $3n + 3$, including the initial three arcs between A and B. Since each arc adjoins two regions, we have a total of $6n + 6$ adjacencies of regions. The average number of regions adjacent to any region is $(6n + 6)/(n + 3) = 6 - 12/(n + 3)$. Note that we could have started the process by constructing A and B and then carrying out $n - 2$ other operations. Our number for $n + 1$ regions (including the outside region) would be $6 - 12/(n + 1)$. Thus in any case for a general map, the average number of regions adjacent to any region is less than 6. Note that this is an average and can, of course, be exceeded for individual regions but compensated for by a small number of adjacencies for other regions.

Another proof that the average of the number of edges bounding the faces of a map is less than six uses Euler's formula excluding the outside region, that is, $n - m + r = 1$, and the relation $3n \leq 2m$, which when combined give $m \leq 3r - 3$. If the regions are numbered $i = 1, \ldots, r$ and if the number of edges of the ith region is e_i, then

$$\sum_{i=1}^{r} e_i < 2m \leq 6r - 6$$

since some but not all edges are on two faces. Hence,

$$\frac{1}{r} \sum_{i=1}^{r} e_i < 6 - \frac{6}{r} < 6$$

It follows that there is at least one vertex having degree 5 or smaller. Yet another proof of this fact appears in the proof of Theorem 4-21.

Sufficiency of Five Colors and Other Theorems

This section is concerned with theorems illustrating in part how far the four-color conjecture has gone toward a proof.

Theorem 4-21 Five colors are sufficient for coloring the regions of a planar map.

Proof: We prove the theorem by induction on the dual; i.e., we color the vertices. Assume it is true for $n - 1$ vertices. The graph has at least one vertex v whose degree ≤ 5. If this were not true, then by using the relationship between the faces and edges $3r \leq 2m$ and $6n \leq 2m$ (that is, we assume the contrary that every vertex has at least degree 6) and substituting in Euler's formula, we obtain $0 = 2m/6 - m + 2m/3 \geq 2$, a contradiction. Now if we remove v from the graph, the resulting graph can be colored with five colors by our induction hypothesis. Let us now consider the worst possibility, i.e., that five vertices $v_1, \ldots, v_5$ are adjacent to v and arranged in clockwise direction. Suppose that the coloring of the graph without v (when v is removed, the edges incident with it are also removed) assigns a different color to each of these vertices (otherwise one of the remaining colors could be assigned to v and the theorem is proved). Let the respective colors be $c_1, \ldots, c_5$. We now show that it is possible to reallocate the colors so that v will receive a color different from the vertices with which it is incident, i.e., at least two of these vertices will be assigned the same color. Consider the subgraph of vertices colored with c_1 and c_3 (the same colors as those of v_1 and v_3). If v_1 and v_3 are not connected (i.e., there is no path between them) in this subgraph, then the vertices colored with c_1 of the component which contains v_1 can be colored with c_3 and those colored with c_3 in that component are now colored with c_1. In this manner both v_1 and v_3 receive the color c_3 and v can be colored with c_1. If, on the other hand, v_1 and v_3 are connected in the subgraph of vertices colored with c_1 and c_3 then, for example, v_2 and v_4 cannot be connected in the subgraph of vertices colored with c_2 and c_4. Otherwise, the path connecting them must meet the path connecting v_1 and v_3 and the vertex where the two paths meet will have assigned to it one color from one subgraph and a second color from the second subgraph. This subgraph which includes v_2 and v_4 can be recolored in the same manner as the disconnected case for v_1 and v_3 above, and hence v_2 and v_4 receive the same color. The other color is then assigned to v.

Theorem 4-22 Two colors are sufficient for coloring the regions generated by the intersection of straight lines in the plane.

Proof: The proof is inductive. A plane is divided by a straight line into two regions, and hence two colors are sufficient. When a second line is drawn, the new map is recolored by reversing the colors on one side of the new line. Suppose the theorem is true for $n - 1$ lines. When the nth line is drawn, if the colors of all regions on one side of the line are reversed, we obtain a total coloring with two colors.

Theorem 4-23 A necessary and sufficient condition for a map to be properly colorable (i.e., regions with a common edge colored differently) with two colors is that every vertex have even degree ≥ 2.

Proof: The faces of a planar graph can be properly 2-colored if and only if the vertices of its dual can be properly 2-colored, i.e., if and only if the dual is bipartite. Using the fact that a tree is bipartite, and considering its chords, one can show that a graph is bipartite if and only if every circuit has an even number of edges. Further, for planar graphs, every circuit will have an even number of edges. Still further, for planar graphs, every circuit will have an even number of edges if and only if every circuit which bounds a face has this property. (The reader should verify this step in detail.) But this is equivalent to the assertion that every vertex of the original graph has even degree.

Exercise 4-9 Prove that the map obtained by drawing any finite number of circles in the plane can be properly colored with two colors.

Exercise 4-10 Use the fact that if every vertex has degree at least 3, then at least one region has at most five sides, to develop a short inductive proof of the sufficiency of six colors for the proper coloring of any planar map.

Exercise 4-11 Show that if every vertex of a map has degree at least 3, then the number of edges and regions satisfy $3r \geq 6 + m$.

Exercise 4-12 Use the result of Exercise 4-11 to show that if every vertex has degree at least 3, and if there are fewer than 12 regions, then at least one region is bounded by 4 or fewer sides.

Exercise 4-13 Prove that four colors are sufficient to color every map which has fewer than 12 regions and for which every vertex is of degree at least 3.

The *flow ratio* of a circuit in a directed graph is the ratio of the number of arcs directed in one direction to the number of arcs directed in the opposite direction, where the denominator must not be more than the numerator.[35] Note that this ratio may be infinite.

Theorem 4-24 A necessary and sufficient condition that the vertices of a graph can be k-colored is that there exist an orientation of the edges of the graph such that the flow ratio of each circuit does not exceed $k - 1$.

Proof: To show necessity, assume that the graph is k-colored and the colors are labeled $0, 1, \ldots, k - 1$. Let each arc be directed from the vertex with the smaller color label to that with the larger color label. Then each circuit has flow ratio $\leq k - 1$. To see this, note that the largest flow ratio is obtained by orienting as many arcs as possible in one direction and as small a number as possible in the opposite direction. This ratio is obtained, for example, by successively coloring the vertices in increasing order and then starting again when all colors are used. In this manner for every set of $k - 1$ arcs oriented in one direction there is one arc oriented in the opposite direction.

To prove sufficiency, suppose the graph is connected and let the flow ratio be $\leq k - 1$ for each circuit. Select a starting vertex v_0 and color it with the color 0 and consider some other vertex v_p. We shall define an auxiliary integer-valued function $g(v_p)$ whose values when reduced mod k yield the desired coloring. As a preliminary operational step for defining $g(v_p)$, we define the gain of a chain of arcs, when traversed from v_0 to v_p, to be the number of arcs traversed in their inherent direction minus $k - 1$ times the number of arcs traversed in the opposite sense.

To show that such a chain with maximum gain exists, note first that if a chain is not simple and hence contains a circuit, the circuit cannot contribute to the gain because of the flow ratio condition and hence it may be deleted from the chain to yield a chain with equal or greater gain. Again, if the resulting chain is not simple, a circuit is deleted, and so on until a simple chain is obtained. Since the number of simple chains is finite, there is one with maximum gain. If with each vertex v_p is associated the maximum gain $g(v_p)$ for a chain oriented from v_0, then we have for two vertices v_{p_1} and v_{p_2} joined by an arc

$$0 \leq |g(v_{p_1}) - g(v_{p_2})| < k$$

for if the absolute value of the difference exceeds $k - 1$, then the maximum gain at one of the vertices can be taken as that at the larger one minus a loss of $k - 1$ along the connecting arc. Thus $g(v_p)$ when reduced mod k yields a k-coloring of the vertices, since the values of g for adjacent vertices differ by less than k and hence cannot produce the same integer mod k. The four-color conjecture can now be stated as follows: The edges of a planar graph can be so oriented that the flow ratio of each circuit is ≤ 3.

Trivalent Maps

To show how the problem of coloring the regions of any map can be reduced to that of coloring the regions of a trivalent map (one which is regular of degree 3), we show how the vertices of the original map whose degree differs from 3 can be reduced to vertices of degree 3. This reduction is accomplished by replacing any vertex of degree $\neq 3$ with a closed polygonal region with as many vertices as there were edges incident with the original vertex. Each of the new vertices has one of these edges incident with it and is of degree 3. The resulting map is trivalent (see Fig. 4-12). When this map has been colored, a coloring of the original map is obtained by shrinking each of the new regions back to its original vertices. Thus if four colors suffice for coloring a trivalent map, then they also suffice for coloring the original map.

We shall say that the regions or edges of a planar graph have been properly colored if they have been assigned colors in such a way that no two adjacent regions or edges, respectively, have the same color.

The following theorem, essentially due to P. Tait and discussed in

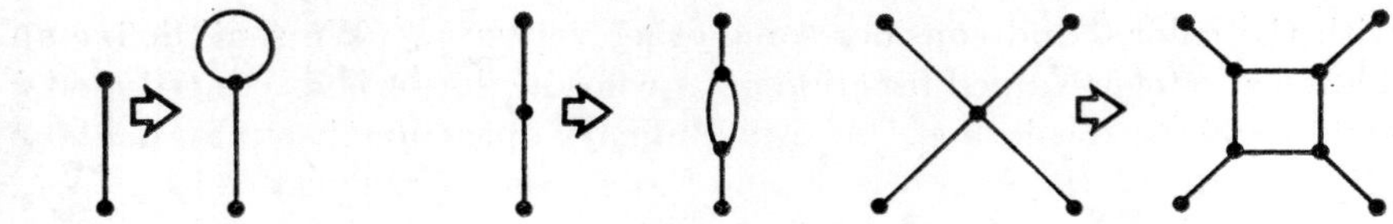

Enlargement of a vertex of degree 1 — Enlargement of a vertex of degree 2 — Enlargement of a vertex of degree 4

Figure 4-12

Ref. 3 and Ref. 34 of this chapter, relates the problem of 4-coloring the regions of a trivalent graph with that of 3-coloring its edges.

Theorem 4-25 If G is a planar, trivalent, 2-connected (i.e., inarticulate) graph, then the regions of G can be properly 4-colored if and only if the edges of G can be properly 3-colored.

Proof: Assume that the regions have been properly 4-colored with colors a, b, c, and d. Assign to each edge of G color 1, 2, or 3, according to the colors of the two regions which it bounds, as prescribed by the following table:

	a	b	c	d
a	–	1	2	3
b	1	–	3	2
c	2	3	–	1
d	3	2	1	–

(Note that the assumption that G is inarticulate is necessary in order to assure that each edge does in fact bound two different regions.) If two adjacent edges had the same color, it would follow that two adjacent regions had the same color. Since this is not the case, it follows that the edges have been properly colored.

Conversely, assume that the edges have been properly colored with colors 1, 2, and 3. The subgraph determined by the edges colored 1 and 2 is bivalent; hence, its regions can be colored with two colors A and B. (Why?) Similarly, the bivalent graph determined by the edges colored 1 and 3 has regions which can be colored with two colors C and D. When the latter two graphs are superimposed, every region of the original graph has one of the four color pairs AC, AD, BC, BD associated with it. If a different color is associated with each pair, the resulting coloring of the regions of G is readily shown to be proper. This completes the proof.

Theorem 4-26, in conjunction with Theorem 4-25, establishes another characterization of the problem of properly 4-coloring the regions of a trivalent graph. This characterization is in terms of numbers associated with the vertices of the graph, and is due to Heawood.[27] See also Ref. 4 of Chap. 1 and Ref. 3 of this chapter.

Theorem 4-26 Let G be a planar, trivalent, 2-connected graph. Then the edges of G can be properly 3-colored if and only if a coefficient $k(v)$, equal to $+1$ or -1, can be associated with each vertex v in such a way that

$$\Sigma k(v) \equiv 0 \pmod{3}$$

where the summation is taken over the vertices occurring in the boundary of any region of G.

Sketch of proof: Assume that the edges of G have been properly colored with colors 1, 2, and 3. Assign to each vertex v the coefficient $+1$ or -1 depending on whether a clockwise movement about v results in encountering the edges incident with v in the color order 1, 2, 3 or 1, 3, 2. Now start with any edge e and traverse the edges which bound some region until e is reached again, determining the color of each successive edge by means of the color of the preceding edge and the coefficient of their common vertex. In order for the color ultimately determined for e to agree with its actual starting color, the mod 3 condition must be satisfied.

Conversely, suppose a coefficient has been associated with each vertex in such a way that the mod 3 condition is satisfied, relative to every region. Assign color 1 to an arbitrary edge and sequentially color the remaining edges as follows: If an uncolored edge e is adjacent to a colored edge f, assign to e the color which is consistent with the color of f and with the coefficient of their common vertex, using the clockwise convention given earlier. The fact that the resulting set of edge colors is proper follows from the mod 3 condition, which essentially guarantees that no color conflicts will arise during the course of this procedure and that the edges may be colored in any order. This completes the sketch of the proof.

Theorems 4-25 and 4-26 can also be viewed in the dual context of *triangular graphs*, i.e., planar graphs such that every region (including the infinite region) is bounded by precisely 3 edges. The dual of a triangular graph is a 2-connected, trivalent, planar graph, and conversely. A proper 3-coloring of the edges corresponds, in the dual context, to a 3-coloring of the edges of a triangular graph such that the three edges bounding any region have distinct colors.

Exercise 4-14 Reinterpret the mod 3 condition of Theorem 4-26 in the dual context of triangular graphs.

Theorem 4-27 (Three-color Theorem.) The regions of a trivalent map can be colored with three colors if and only if every region is bounded by an even number of edges.

Proof: "If" is proved by induction on the number of regions and hinges on the fact proved in Theorem 4-21 that there is at least one vertex of degree no greater than 5. In the dual this implies that there is at least one region bounded by at most five edges, so there is a region bounded by exactly two or four edges. In the two-edge case let R_1 and R_2 be the regions adjacent to this region. If we eliminate one of the two bounding edges, absorb the region, and delete the vertices of degree 2, thus replacing their edges by a single edge incident with their adjacent vertices, by an inductive hypothesis on the number of regions we can 3-color this map. Insertion of the original region requires the third color that is not represented in either R_1 or R_2. The case for a region R bounded by four edges and adjacent to regions R_1, R_2, R_3, R_4 requires the removal of two opposite edges and also the removal of four vertices of degree 2. We assume that the edges removed bounded regions R_1 and R_3. It is easy to show that all resulting regions have an even number of edges and that the resulting map is still trivalent and has two fewer regions. The region R_1 originally had an even number of neighboring regions, one of which was R. When the reduced graph is colored, the two neighbors of R_1 which were also neighbors of R, that is, R_2 and R_4, necessarily have the same color because the neighbors of R_1 must be alternately assigned two colors. Also, R_1 and R_3 have the same color because they are part of the same region in the reduced graph. When the two edges are replaced, R may be assigned the remaining third color.

To prove the "only if" part, suppose that R is bounded by an odd number of edges and hence has an odd number of neighbors. The neighboring regions must be alternately colored with the two colors other than the color assigned to R, but this is impossible.

Exercise 4-15 Dualize the foregoing theorem and give a proof in the dual context.

Chromatic Polynomials

Let $P_n(\lambda)$ be the number of ways in which a map of n regions which covers the sphere can be colored with $\leq\lambda$ colors. The four-color conjecture states that 4 is not a root of any equation $P_n(\lambda) = 0$. Two colorings are essentially different if they cannot be obtained from one another by a permutation of colors. If for the moment we disregard the possible permutation of colors and if m_i (that is, $1, \ldots, n$) is the number of ways of coloring a map with exactly i colors, then there are $m_i\lambda(\lambda - 1) \cdots (\lambda - i + 1)$ ways to color the map with i colors taken from λ colors. Clearly, the total number of ways is an nth degree polynomial known as the chromatic polynomial, giving

$$P_n(\lambda) = m_1\lambda + m_2\lambda(\lambda - 1) + \cdots + m_n\lambda(\lambda - 1) \cdots (\lambda - n + 1)$$

Birkhoff[4] has given an explicit expression for this by finding the values of m_i. It is

$$P_n(\lambda) = \sum_{i=1}^{n} \lambda \sum_{k=0}^{n-i} (-1)^k(i,k)$$

where (i,k) is the number of ways to break down the map of n regions to a submap of i regions by k simple or multiple coalescences (i.e., merging two or more regions, respectively, by removing edges). Thus $(i,k) = 0$ for $k > n - i$, and by definition $(n,0) = 1$, $(i,0) = 0$ for $i < n$ and $(i,n - i)$ is the number of ways to make $n - i$ successive simple coalescences. Thus, letting $n = 3$, we have

$$(2,1) = 3 \qquad (1,1) = 1 \qquad (1,2) = 3$$

and

$$P_3(\lambda) = (3,0)\lambda^3 - (2,1)\lambda^2 + [-(1,1) + (1,2)]\lambda = \lambda(\lambda - 1)(\lambda - 2)$$

Exercise 4-16 Consider a map of five regions formed by a ring of three regions bounding an interior and an exterior region. Show that

$$\begin{array}{llll} (4,1) = 9 & (3,1) = 22 & (3,2) = 51 & \\ (2,1) = 14 & (2,2) = 125 & (2,3) = 150 & \\ (1,1) = 1 & (1,2) = 45 & (1,3) = 176 & (1,4) = 150 \end{array}$$

and $\quad P_5(\lambda) = \lambda^5 - 9\lambda^4 + 29\lambda^3 - 39\lambda^2 + 18\lambda = \lambda(\lambda - 1)(\lambda - 2)(\lambda - 3)^2$

Birkhoff[5] has also shown that for a map on the sphere

$$P_n(\lambda) \geq \lambda(\lambda - 1)(\lambda - 2)(\lambda - 3)^{n-3} \qquad n \geq 3,\ n \neq 4$$

which is obviously true for $\lambda = 1,2,3$ and with $\lambda > 4$ for $n = 3,4$. A map is defined to be maximal if every region in it touches (along an edge) the largest possible number of regions, i.e., there is no other map with the same adjacency of corresponding regions but with an additional regional adjacency. The necessary and sufficient conditions that a map of $m \geq 3$ regions be maximal are that all of its regions be simply connected (i.e., homeomorphic with a circle), its vertices be trivalent, and no two regions touch more than once. In such a map the number of adjacencies is $3n - 6$.

A map is irreducible if all its regions are simply connected, any two adjacent regions form a simply connected region, and any three regions that are adjacent in pairs form a simply connected region about a trivalent vertex.[8] For such a map

$$P_n(\lambda) = \frac{\prod_i P_{n_i}(\lambda)}{\lambda^{\alpha+\beta+\gamma}(\lambda - 1)^{\beta+\gamma}(\lambda - 2)^{\gamma}} \qquad n_i < n$$

where the polynomials $P_{n_i}(\lambda)$ belong to irreducible maps and are $\alpha + \beta + \gamma + 1$ in number and $n = \Sigma n_i - \alpha - 2\beta - 3\gamma$.

A substantial characterization of $P_n(\lambda)$ for different values of λ is given in Refs. 8 and 9.

4-6 GRAPHS AND SURFACES

One often avoids the definition of a surface by appealing to the intuition of the reader in a study in which a rigorous definition would not materially improve understanding. It has been our intention to follow this intuitive appeal generally. But for the student who is curious about how a surface is defined, we shall now give some definitions which require familiarity with ideas from topology, a subject which provides a basis for understanding many modern mathematical concepts.

A 1-1 transformation which together with its inverse is continuous is called a *homeomorphism*. An example of a homeomorphism is the continuous deformation of a pyramid into a sphere, and vice versa. Note that one cannot establish a homeomorphism between a sphere and a torus (a donut-shaped surface) because the latter has a hole.

An n-dimensional topological manifold, or simply an n manifold, is a connected, locally compact topological space with a countable basis, each point of which has a neighborhood homeomorphic to euclidean n space.[1]

When a manifold is compact, it is said to be closed, for example, the n sphere; otherwise, it is open. A closed 2 manifold that is homeomorphic to a polyhedron is called a closed surface. It is known that every closed 2 manifold is homeomorphic to some polyhedron. An example of a closed surface is the 2 sphere, whereas the plane is an open surface.

A geometric n simplex σ^n, $n > 0$, is the set of points $X = \{x_i\}, i = 1, \ldots, n$, defined in terms of $n + 1$ linearly independent points $P_0, \ldots, P_n$ by $x_i = \sum_{j=0}^{n} a_j p_{ji}$, where $\sum_{j=0}^{n} a_j = 1$, $0 < a_j < 1$, $j = 0, \ldots, n$, and p_{ji} is the ith coordinate ($i = 1, \ldots, n$) of p_j ($j = 0, \ldots, n$). The n simplex is closed if $0 \le a_j \le 1$, $j = 0, \ldots, n$ holds. A geometric simplicial complex K is a finite set of disjoint n simplexes, $n = 0, \ldots, p$ in n-dimensional euclidean space such that if a simplex belongs to the complex, so do all its faces and no two distinct simplices have their faces all the same. The dimension of the complex is p. A polyhedron is the point-set union of all the simplexes of K and hence is the set of points which belong to some simplex of K.

A surface S is unorientable if for some point around which is associated a direction of rotation it is possible to move the point on the surface while keeping the same orientation about the point and return to the initial position with a reversed sense of rotation. If there is no such point, the surface is called orientable.

Figure 4-13

If the right end of the rectangle represented in Fig. 4-13 is given a half turn and then joined to the left end, we obtain a band which differs from a ring by the presence of the half turn. It is known as the *Möbius strip*. Note that it has only one side in that if one starts at any point on this surface and traces a curve around the surface back to the point, one covers "both sides" without crossing an edge of the surface, and it can be seen to be unorientable. The sphere and the plane, on the other hand, are examples of two-sided orientable surfaces.

A *map* is a graph together with a surface which contains the graph in such a way that edges intersect only at their end points.

Euler's formula actually applies to a polyhedron in three dimensions. In this case we have $V - E + F = 2$, where V is the number of vertices, E is the number of edges, and F is the number of faces. The proof is obtained by cutting out one of the faces and stretching the polyhedron on a plane without any intersecting edges. The proof proceeds exactly as in the planar graph case except that the outside region is not counted. Instead of it one includes the missing face which has been cut out. The following theorem, which applies to polyhedra, is given because of a similarity of the argument in the proof to that used in graph theory.

Theorem 4-28 Every closed polyhedron has at least two faces with the same number of edges.

Proof: Let e_i be the number of edges of the ith face where we order the labeling such that[31] $e_i \leq e_i + 1$, $i = 1, 2, \ldots, r - 1$, where r is the number of faces. If no two faces have the same number of edges, then $e_{i+1} - e_i \geq 1$ for all i. Thus

$$e_r - e_1 = \sum_{i=1}^{r-1} (e_{i+1} - e_i) \geq r - 1$$

Since $e_1 \geq 3$, we have $e_r \geq r + 2$, that is, the rth face is adjacent to at least $r + 2$ faces, a contradiction.

Remark: One can prove the stronger result that if k is the smallest number of edges of a face, then there are at least k indices i for which $e_i = e_{i+1}$.

Remark: Euler's formula can be generalized to a polyhedron in n dimensions, where we use F_i, $i = 0, 1, \ldots, n - 1$, to denote the number of ith dimensional faces; for example, F_0 is the number of vertices. In that case Euler's formula is

$$\sum_{i=0}^{n-1} (-1)^i F_i = 1 - (-1)^n$$

where the right side is 0 or 2 depending on whether n is even or odd, respectively.

Returning to three dimensions, we note that Euler's formula for surfaces with holes is

$$V - E + F = 2 - 2p$$

where p is the number of independent holes and is referred to as the genus of the surface. The genus of the surface is the largest number of simple closed curves on the surface which do not disconnect the surface. The sphere is of genus 0, since any simple closed curve on its surface disconnects it. The torus (a donut) is of genus 1. Any surface of genus p is equivalent to a sphere with p handles (like the handle on a tea cup). For a nonorientable surface the right side of Euler's formula reads $2 - q$, where q is the genus.

To derive the formula, one may think of the sphere as a polyhedron with handles. A handle may be cut along one joint, leaving a polyhedral-shaped hole in the surface left behind, and may then be pushed into the other joint, creating a similar hole. The reader can now show that $V - E + F = 2$, which holds for any subdivision of the entire sphere, must be reduced by $2p$, which is the number of patches on the surface.

We now derive Heawood's formula which gives for $p \geq 1$

$$\left[\frac{7 + \sqrt{1 + 48p}}{2}\right]$$

as a sufficient number of colors for a map on a surface of genus p. The brackets indicate this is the greatest integer not exceeding the number being taken. Note that for a nonorientable surface the result is $[(7 + \sqrt{1 + 24q})/2]$. Both foregoing formulas give a lower bound to the number of colors when each is reduced by 2.[39]

The same procedure of obtaining a trivalent map from a planar map can also be followed for a map on a surface of genus p. We shall be concerned only with connected maps, because it is not difficult to see that the coloring of a disconnected map is a special case of this.

Now in a trivalent map we have $3V = 2E = aF$, where a is the average number of edges for a given face. By substituting in Euler's formula for a surface of genus p, we obtain

$$a = 6 + \frac{12(p - 1)}{F}$$

We add 1 to each side, put $a + 1 = F$, solve the quadratic in F, and take the positive root α to obtain Heawood's formula. To prove that the greatest integer contained in α gives the sufficient number of colors for a surface of genus p, we note that if $F \leq \alpha$, then this number α is sufficient since F colors are sufficient, F being the number of regions. If $F > \alpha$, then along the curve we always have $a + 1 < \alpha$, as the reader will verify from Fig. 4-14.

From the fact that a is the average number of sides of a face we conclude that there is a face R with at most a sides and thus it has at

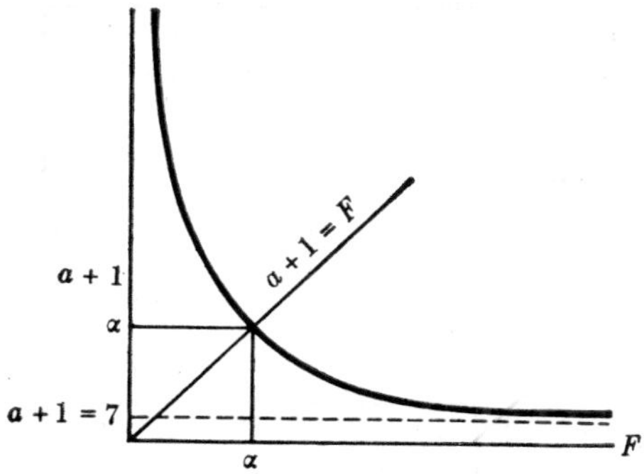

Figure 4-14

most $[\alpha] - 1$ sides, since $a + 1 \leq [\alpha]$. If an edge of this face is removed so that the face is combined with one of its adjacent faces R', then we can show that $[\alpha]$ colors are sufficient for the given graph if $[\alpha]$ colors are sufficient for the reduced graph. For suppose that the reduced graph has been colored with $[\alpha]$ colors; then since R' and the other regions of the reduced graph which bounded R are at most $[\alpha] - 1$ in number, R can be colored with the remaining available color. If the reduced graph has more than $[\alpha]$ regions, we repeat this argument, i.e., we show that to color it with $[\alpha]$ colors would follow if we could color the reduced graph obtained by merging another region having at most $[\alpha] - 1$ sides with one of its neighbors. (Again, such a region must exist, since the average number of sides a still satisfies $a + 1 < [\alpha]$ because the number of regions F is still $> [\alpha]$.) Thus the coloring of the original graph with $[\alpha]$ colors eventually is made to rest on the possibility of coloring a reduced graph of $[\alpha]$ regions with $[\alpha]$ colors, which is clearly always possible. Note that each of the merged regions each having at most $[\alpha] - 1$ sides can be reinstated and colored without conflict with its neighbors. This completes the proof of the sufficiency of $[\alpha]$ colors.

Necessity has been proved for various values of p and requires the construction of special maps (see Ref. 2).

The necessity of seven colors on the torus is illustrated by the map of Fig. 4-15. Heawood's formula shows that seven colors are also sufficient. Here opposite edges must be identified to obtain a torus.

Exercise 4-17 Show that six colors are necessary for some maps on a Möbius strip.

An interesting theorem due to J. Edmonds is concerned with the relation between a graph G and the surface maps (S,G) to which it can give rise. The problem is to characterize maps on orientable surface which are the corresponding surface drawings of a given graph.

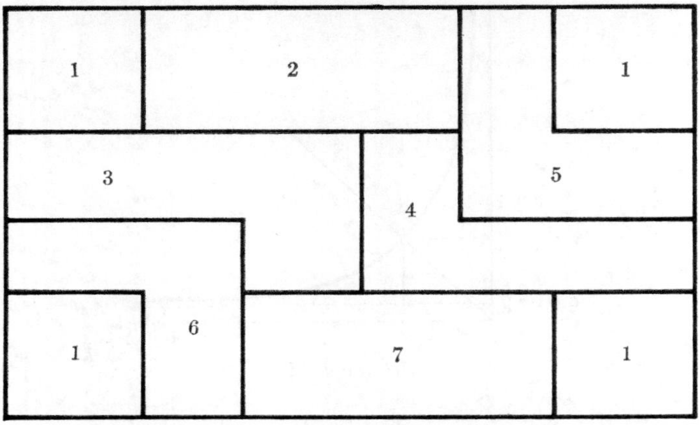

Figure 4-15

It is intuitively clear that if we have a map (S,G) corresponding to a graph G, then the map determines at each vertex a unique cyclic ordering. It turns out that these cyclic orderings of the edges to each vertex are precisely the link between graphs and maps, since we also have the following theorem:

Theorem 4-29 (Edmonds.) For any connected graph G with an arbitrarily specified cyclic ordering of the edges at each vertex there exists a topologically unique oriented surface map (S,G), such that the clockwise edge orderings around each vertex are as specified.

REFERENCES

1 Aleksandrov, P. S.: "Combinatorial Topology," vol. 1, Graylock Press, Rochester, N.Y., 1956.

2 Ball, W. W. R.: "Mathematical Recreations and Essays," The Macmillan Company, New York, 1960.

3 Berge, C.: "Theory of Graphs and Its Applications," John Wiley & Sons, Inc., New York, 1962.

4 Birkhoff, G. D.: A Determinant Formula for the Number of Ways of Colouring a Map, *Ann. Math.*, **14**(2): 42–46 (1912).

5 Birkhoff, G. D.: On the Number of Ways of Coloring a Map, *Proc. Edinburgh Math. Soc.*, **2**(2): 83–91 (1930).

6 Birkhoff, G. D.: The Coloring of Graphs, *Ann. Math.*, **33**(2): 688–718 (1932).

7 Birkhoff, G. D.: The Reducibility of Maps, *Am. J. Math.*, **35:** 115–128 (1913).

8 Birkhoff, G. D.: On the Polynomial Expressions for the Number of Ways of Coloring a Map, *Ann. Scuola Normali Superiore, Pisa, Ser.* 2, **3:** 85–104 (1934).

9 Birkhoff, G. D., and D. C. Lewis: Chromatic Polynomials, *Trans. Am. Math. Soc.*, **60:** 355–451 (1946).

10 Brooks, R. L.: On Colouring the Nodes of a Network, *Proc. Cambridge Phil. Soc.*, **37:** 194–197 (1941).

11 Dynkin, E. B., and W. A. Uspenski: "Multicolor Problems," D. C. Heath and Company, Boston, 1952. (Original in German.)

12 Dirac, G. A.: Note on the Colouring of Graphs, *Math. Z.*, **54:** 347–353 (1951).

13 Dirac, G. A.: A Property of 4-Chromatic Graphs and Some Remarks on Critical Graphs, *J. London Math. Soc.*, **27:** 85–92 (1952).

14 Dirac, G. A.: Some Theorems on Abstract Graphs, *Proc. London Math. Soc.*, *Ser.* 3, **2:** 69–81 (1952).

15 Dirac, G. A.: The Structure of k-Chromatic Graphs, *Fund. Math.*, **40:** 42–45 (1953).

16 Dirac, G. A.: Map Colour Theorems Related to the Heawood Colour Formula, *J. London Math. Soc.*, **31:** 460 (1956).

17 Dirac, G. A., and M. D. Stojakovic: The Four-colour Problem, *Matematicka Biblioteka*, **16:** 1960. MR 22-1946.

18 Erdös, P., and N. G. de Bruijn: A Color Problem for Infinite Graphs and a Problem in the Theory of Relations, *Indagationes Math.*, **13:** 369–373 (1951).

19 Errera, J.: Du Coloriage des Cartes, these, Bruxelles, 1921, *Mathesis*, **36:** 56 (1922).

20 A problem in *Eureka*, October, 1960.

21 Fáry, I.: On Straight Line Representation of Planar Graphs, *Acta Sci. Math.*, **11**(4): 229–233 (1948).

22 Franklin, P.: The Four Color Problem, *Scripta Mathematica*, No. 5, 1941.

23 Goodman, A. W.: On Sets of Acquaintances and Strangers at Any Party, *Am. Math. Monthly*, **66:** 778–783 (1959).

24 Greenwood, R. E., and A. M. Gleason: Combinatorial Relations and Chromatic Graphs, *Can. J. Math.*, **7:** 1–7 (1955).

25 Grünbaum, B., and T. S. Motzkin: On Polyhedral Graphs, *Proc. Symp. Pure Math.*, vol. 7, Convexity, 1963.

26 Harary, F.: A Complementary Problem on Nonplanar Graphs, *Math. Mag.*, **35:** 301–303 (1962).

27 Heawood, P. J.: Map Color Theorem, *Quart. J. Pure Appl. Math.*, no. 24, p. 332, 1890.

28 Kneebone, G. T.: The Three Houses Problem, *Math. Gaz.*, **25:** 78–81 (1941).

29 Krausz, J.: Demonstration Nouvelle d'une Théorème de Whitney sur les Reseaux, *Mat. Fiz. Lapok*, **50:** 75–85 (1943). (In Hungarian.)

30 Kuratowski, G.: Sur le Problème des Courbes Gauches en Topologie, *Fund. Math.*, 15–16 (1930).

31 Linis, V.: *Math. Mag.*, **36**(4): (1963).

32 Lorden, G.: Blue-Empty Chromatic Graphs, *Am. Math. Monthly*, **69**(2): 114–119 (1962).

33 MacLane, S.: A Combinatorial Condition for Planar Graphs, *Fund. Math.*, **28:** 22–32 (1937); *Zbl. Math.*, 15, 375.

34 Marathe, C. R.: On the Dual of a Trivalent Map, *Am. Math. Monthly*, **68**(5): 448–455 (1961).

35 Minty, G. J.: A Theorem on n-Coloring the Points of a Linear Graph, *Am. Math. Monthly*, **69**(7): 623–624 (1962).

36 Nordhaus, E. A., and J. W. Gaddum: On Complémentary Graphs, *Am. Math. Monthly*, **63:** 175–177 (1956).

36a Picard, C.: Graphes Complementaires et Graphes Planaires, *Rev. Franc. Recherche Operationelle*, **8:** 329–343 (1964).

37 Rapaport, E. S.: Cayley Color Groups and Hamilton Lines, *Scripta Math.*, **24:** 51–58 (1959).

38 Ratib, I., and C. E. Winn: Généralisation d'une Réduction d'Errera dans le Problème des Quatre Couleurs, *Intern. Congr. Math.*, Oslo, 1936.

39 Ringel, G.: "Färbungsprobleme auf Flächen und Graphen," VFB Deutscher Verlag der Wissenschaften, Berlin, 1959.

40 Sauvé, Léopold: On Chromatic Graphs, *Am. Math. Monthly*, **68:** 107–111 (1961).

41 Seshu, S., and M. B. Reed: "Linear Graphs and Electrical Networks," Addison-Wesley Publishing Company, Inc., Reading, Mass., 1961.

42 Tietze, H.: Einige Bemerkungen über das Problem der Kartenfarbens auf einseitigen Flächen, *Jabresber., Deutsh. Math. Vereinigung*, **19:** 155–159 (1910).

43 Zeidl, B.: Über 4- und 5-chrome Graphen, *Monatsh. Math.*, **62:** 212–218 (1958).

FIVE

MATRIX REPRESENTATION

5-1 INTRODUCTION

In this chapter we introduce various matrices to represent incidence relations between the vertices and edges of a graph and, more generally, to represent incidence relations between circuits, cut-sets, chains, and the edges associated with them. These algebraic representations apply to both directed and undirected graphs. Whether one manipulates these matrices or not, they are a convenient way to record the structure of the graph. There is a close relationship between the incidence, circuit, and cut-set matrices whereby it is possible to obtain a basis for a certain vector space associated with one in terms of a basis associated with the space for another. This, for example, facilitates the task of determining all the cut-sets of a graph once its circuits are known. Properties of the cut-sets are closely related to the maximal flow through a network, as we shall see in Chap. 7. The notion of a tree, its branches, and chords play an important role in the formulation of basic matrices for circuits and for cut-sets.

In the case of undirected graphs we shall be dealing with incidence matrices whose entries are 0 and 1. The addition of numbers is always reduced mod 2. Thus, $1 + 1 = 0 \pmod 2$, $1 + 0 = 0 + 1 = 1$, $0 + 0 = 0$. One simply adds the elements, divides by 2, and records the remainder as the result of applying the mod 2 operation. This choice of entries enables one to indicate whether a property between two elements holds (with entry 1) or does not (with entry 0). Matrices may be added and multiplied as usual, and then the entries of the resulting matrix are recorded mod 2. For directed graphs we use 1, -1, 0, since an arc, for example, may be incident with a vertex and directed away from the vertex, directed toward it, or not incident with it at all. The mod 2 reductions are no longer adequate in this case, and we must consider matrices with integer-valued entries. In fact, they are more cumbersome to deal with than the mod 2 matrices. However, basically the same kind of theory is obtained for the matrices of a directed graph as is obtained in the undirected case. (It is assumed here that the reader has a working knowledge of elementary matrix theory—the bare essentials.) With the idea of a matrix, one associates the concept of a vector space and related theory revolving about bases and linear combinations.

Definition A real vector space is a set of elements (called vectors) with a rule for addition associating with the vectors x and y a unique vector $x + y$ and another rule for scalar multiplication by real numbers (or the elements of any other field, such as the complex numbers or the integers modulo 2) associating with a vector x a unique vector ax for every scalar a.

In order to have a vector space, the following axioms must be satisfied for all x, y, z in the space and for all scalars a, b:

The vectors must form a commutative group under addition, i.e.,

1. $x + y = y + x$
2. $x + (y + z) = (x + y) + z$
3. $0 + x = x$ for all x, where 0 is the zero vector
4. Each vector x has an inverse y such that $x + y = 0$.

It must also satisfy under scalar multiplication:

5. $a(bx) = (ab)x$
6. $(a + b)x = ax + bx$
7. $a(x + y) = ax + ay$
8. There is a set of vectors which forms a basis, i.e., they are linearly independent and span the space (every vector can be expressed as a linear combination of them). Every basis has the same number of vectors, known as its *rank*.

Two vector spaces V and U are isomorphic if there is a $1 - 1$ correspondence between their elements which preserves addition and scalar multiplication. That is, if v_1 and v_2 are in V and u_1 and u_2 are the corresponding elements in U, then $v_1 + v_2$ corresponds to $u_1 + u_2$. Similarly, kv corresponds to ku, where v corresponds to u and where k is any scalar.[6]

In the case of an undirected graph the elements of the vector space under consideration in our matrix representations are vectors, each of which represents a subset of the edges, whose components are 1 or 0 and indicate whether an edge belongs to the subset or not. In the case of an undirected graph having m edges, the vector space of interest consists of all vectors $(x_1, \ldots, x_m)$, where $x_i = 0$ or 1. The ith component is associated with the ith edge, and each vector uniquely characterizes a subset of edges, namely, those for which $x_i = 1$. In particular, each circuit and cut-set has a vector representation. The sum of two such vectors is again a vector of the same form, and their scalar product (the sum of the products of corresponding components) is 0 or 1 depending on whether the two subsets corresponding to these vectors have an even or odd number of common edges.

The set of all linear combinations of circuit vectors (i.e., vectors which correspond to circuits) determines a subspace consisting precisely of those vectors which correspond to circuits and to edge-disjoint sets of circuits, and similarly for vectors associated with cut-sets.

An important and difficult question has been answered in recent years, namely, when does a given matrix represent a basis for the space associated with the circuits or cut-sets of a graph? This is the problem of the realizability of a matrix. We shall go into this question only briefly, because it involves considerable background material that is outside the scope of this book.

5-2 THE INCIDENCE MATRIX

Let G be a graph having n vertices and m edges. The *incidence matrix* associated with G is the $n \times m$ matrix whose rows and columns cor-

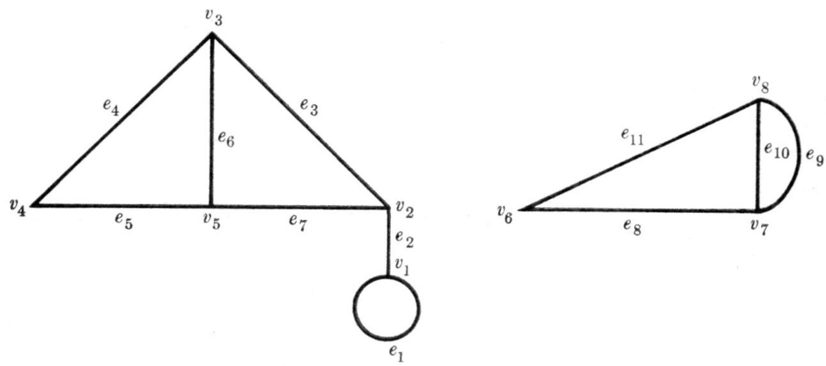

Figure 5-1

respond to the vertices and edges, respectively, where each entry a_{ij} is 1 or 0 depending on whether the jth edge is incident with the ith vertex or not. For a loop, however, every column entry is 0. For example, the 2-component graph shown in Fig. 5-1 has the following incidence matrix:

$$
\begin{array}{cc}
 & \begin{array}{ccccccc|cccc} e_1 & e_2 & e_3 & e_4 & e_5 & e_6 & e_7 & e_8 & e_9 & e_{10} & e_{11} \end{array} \\
A = \begin{array}{c} v_1 \\ v_2 \\ v_3 \\ v_4 \\ v_5 \\ \hline v_6 \\ v_7 \\ v_8 \end{array} &
\left(\begin{array}{ccccccc|cccc}
0 & 1 & 0 & 0 & 0 & 0 & 0 & 0 & 0 & 0 & 0 \\
0 & 1 & 1 & 0 & 0 & 0 & 1 & 0 & 0 & 0 & 0 \\
0 & 0 & 1 & 1 & 0 & 1 & 0 & 0 & 0 & 0 & 0 \\
0 & 0 & 0 & 1 & 1 & 0 & 0 & 0 & 0 & 0 & 0 \\
0 & 0 & 0 & 0 & 1 & 1 & 1 & 0 & 0 & 0 & 0 \\
\hline
0 & 0 & 0 & 0 & 0 & 0 & 0 & 1 & 0 & 0 & 1 \\
0 & 0 & 0 & 0 & 0 & 0 & 0 & 1 & 1 & 1 & 0 \\
0 & 0 & 0 & 0 & 0 & 0 & 0 & 0 & 1 & 1 & 1
\end{array}\right)
\end{array}
$$

Some interesting properties of a graph are observed in its incidence matrix. For example, since an edge of the graph is incident with exactly two vertices, *every column of its incidence matrix contains two unit entries.* A loop, since it is (twice) incident with a single vertex, is the only exception. Hence its column has zero entries. Thus the incidence matrix gives no indication of the existence of loops, because one does not know whether a zero corresponds to a loop or not. Hence it is advisable in the matrix study of graphs to exclude loops, as we shall henceforth.

If the edges and vertices are appropriately indexed, each component of a graph corresponds to a submatrix of a block partitioning of the incidence matrix as follows:

$$
A = \begin{pmatrix}
A_1 & 0 & \cdots & 0 \\
0 & A_2 & \cdots & 0 \\
\cdot & \cdot & \cdots & \cdot \\
0 & 0 & \cdots & A_n
\end{pmatrix}
$$

where A_i is the incidence matrix associated with the ith component. A block-diagonal representation of this type can always be accomplished either by sequential labeling within each component and from one component to the next as illustrated in the example or by rearranging the rows and columns of the incidence matrix, which, of course, produces the same effect. We can now state that two graphs are isomorphic if they have the same incidence matrices to within a permutation of rows and columns. Thus, the incidence matrix provides a complete representation of a graph (if loops are excluded).

Exercise 5-1 Show the equivalence between this definition of isomorphism and that given in Chap. 1.

Theorem 5-1 The rank of the incidence matrix of a p-component graph with n vertices is $n - p$ (where it is understood that arithmetical operations are performed mod 2).

Remark: The rank may be quite different for the incidence matrix considered as a matrix of real numbers, for example.

Proof: Let n_i be the number of vertices belonging to the component P_i ($i = 1, \ldots, p$) which is represented by the submatrix A_i; we shall show that A_i has the rank $n_i - 1$. It is easy to show that the rank of A_i is at most equal to $n_i - 1$. This follows from the fact that the sum of all the rows of A_i is zero mod 2, since each column has two nonzero elements, and hence if the first $n_i - 1$ rows of A_i are added to the last row, the result is a row with zero elements everywhere. Since the rank of a matrix is unchanged by the elementary operation of row addition, the rank of the new matrix which has a row consisting of zeros cannot exceed $n_i - 1$. The sum of any $n_i - 1$ or fewer rows of A_i must have some nonzero element; for if all the elements are zero, then these rows form the submatrix of a component, since none of the vertices corresponding to any of these rows would be connected by an edge to another vertex not in this set, which contradicts the fact that the component P_i is a maximal connected subgraph and proves that the rank of P_i is $n_i - 1$, since every set of $n_i - 1$ rows is independent.

As another method for showing that the rank of A_i is exactly $n_i - 1$, we may start with the first row of A_i and, by permuting its columns, obtain for the main diagonal element in the upper left-hand corner a nonzero element. By the addition of the first row to any row which has a nonzero element in the first column, all elements in the first column except the first become zero. The second row must have another nonzero element which by permuting columns is brought to the (2,2) position and all elements in the second column are similarly reduced to zero. The process is repeated for the third row, etc., until a diagonal matrix is obtained. As indicated above, all the elements of the last row can be made to equal zero. This gives rise to a triangular matrix with n_i rows which with the exception of the last element has unit entries on its principal diagonal. Since these operations do not influence the rank, the rank of A_i must be $n_i - 1$. Thus the rank of A is $n - p$, where p is the number of components.

As indicated previously, the coefficients of the incidence matrix of a directed graph consist of -1, 0, 1. The coefficient is zero if a vertex is not incident with an arc, $+1$ if the arc is oriented away from the vertex, and -1 in the opposite case. Thus the graph of Fig. 5-2 has the incidence matrix

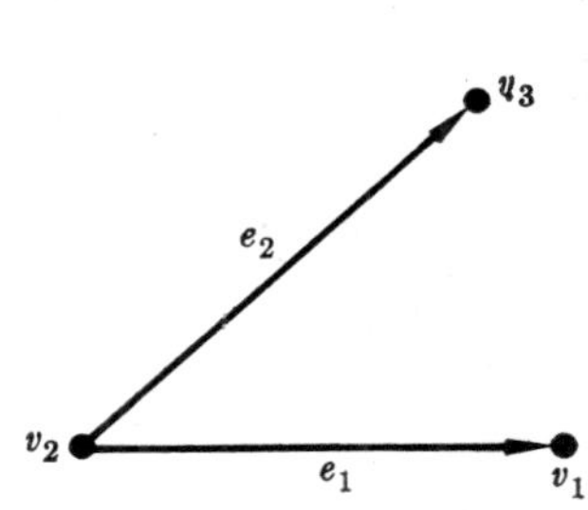

Figure 5-2

$$\begin{array}{c} \\ v_1 \\ v_2 \\ v_3 \end{array} \begin{array}{c} \begin{array}{cc} e_1 & e_2 \end{array} \\ \begin{pmatrix} -1 & 0 \\ 1 & 1 \\ 0 & -1 \end{pmatrix} \end{array}$$

The matrix of a directed graph with n vertices and p components also has rank $n - p$. The term "rank" is associated with a graph that is directed or not, and refers to the rank of the incidence matrix.

Exercise 5-2 Form the incidence matrix of the 2-component directed graph given in Fig. 5-3, after indexing its arcs.

5-3 THE CIRCUIT MATRIX

The circuits C_i of our graph (labeled as in Fig. 5-4) have the following circuit matrix C, each row of which characterizes one of the circuits. (Here it is a coincidence that each edge except e_2 belongs to just two circuits.)

$$C = \begin{array}{c} \\ C_1 \\ C_2 \\ C_3 \\ C_4 \\ C_5 \\ C_6 \end{array} \begin{array}{c} \begin{array}{cccccc:cccc} e_2 & e_3 & e_4 & e_5 & e_6 & e_7 & e_8 & e_9 & e_{10} & e_{11} \end{array} \\ \left(\begin{array}{cccccc:cccc} 0 & 0 & 1 & 1 & 1 & 0 & 0 & 0 & 0 & 0 \\ 0 & 1 & 0 & 0 & 1 & 1 & 0 & 0 & 0 & 0 \\ 0 & 1 & 1 & 1 & 0 & 1 & 0 & 0 & 0 & 0 \\ \hdashline 0 & 0 & 0 & 0 & 0 & 0 & 1 & 0 & 1 & 1 \\ 0 & 0 & 0 & 0 & 0 & 0 & 0 & 1 & 1 & 0 \\ 0 & 0 & 0 & 0 & 0 & 0 & 1 & 1 & 0 & 1 \end{array}\right) \end{array}$$

The reader will observe that the circuit matrix is also block-diagonal, which is a consequence of the fact that every circuit is contained entirely in one component, and the edges of each component are numbered consecutively to avoid the necessity of permutation.

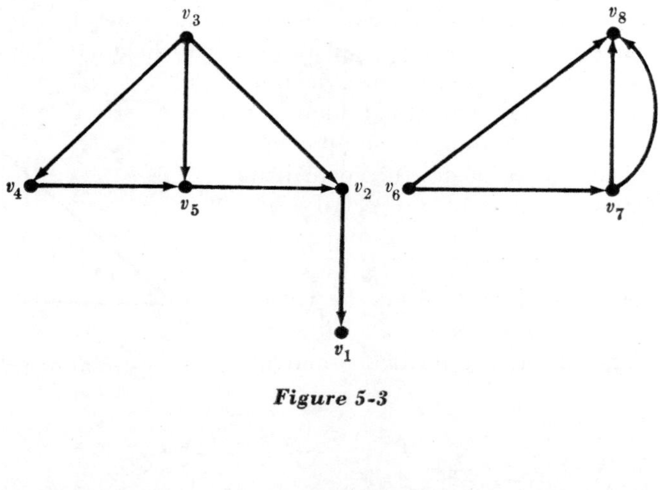

Figure 5-3

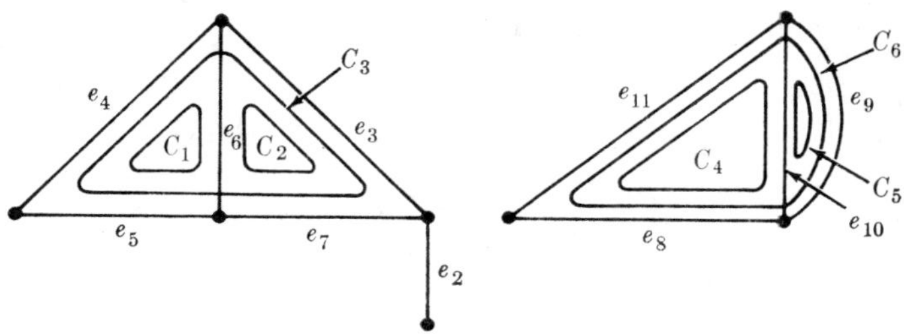

Figure 5-4

Theorem 5-2 The incidence matrix A and the transpose C' of the circuit matrix C, when arranged using the same order of edges, are orthogonal, that is, $AC' = 0$.

Proof: A sketch of the proof of this theorem is as follows: Each vertex incident with a circuit is incident with an even number of edges that are also in the circuit. Thus when a row of A is multiplied by a column of C', we either obtain a sum of elements each of which is zero (if a vertex is not in a circuit) or a sum which is $\equiv 0 \bmod 2$.

Exercise 5-3 Apply Theorem 5-2 to the incidence matrix (without the loop) and the circuit matrix given above.

Remark: If a graph is disconnected, then the incidence, circuit and cut-set (see below) matrices of each component may be treated separately, because none of the properties studied below depend on the fact that there is more than a single component. For this reason we limit our study to connected graphs, that is, $p = 1$ with m edges and n vertices and without loops.

5-4 THE CUT - SET MATRIX

We now write down the proper cut-sets of the graph of Fig. 5-5, thus forming the cut-set matrix, each row of which characterizes one proper

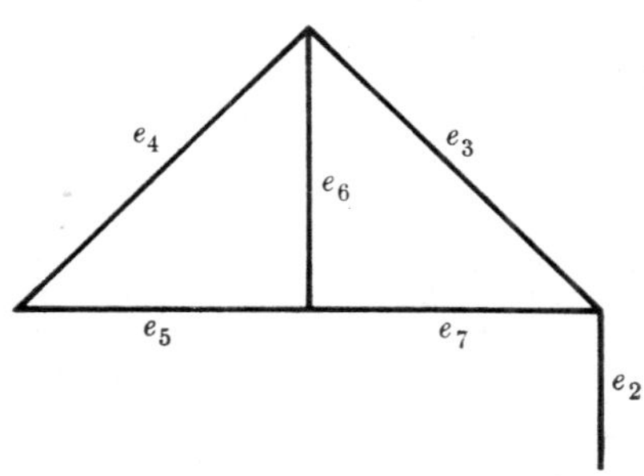

Figure 5-5

cut-set. The proper cut-sets are

$$K_1 = \{e_2\} \qquad K_2 = \{e_3,e_7\} \qquad K_3 = \{e_4,e_5\} \qquad K_4 = \{e_3,e_4,e_6\}$$
$$K_5 = \{e_3,e_5,e_6\} \qquad K_6 = \{e_4,e_6,e_7\} \qquad K_7 = \{e_5,e_6,e_7\}$$

and the cut-set matrix is given by

$$K = \begin{array}{c} \\ K_1 \\ K_2 \\ K_3 \\ K_4 \\ K_5 \\ K_6 \\ K_7 \end{array} \begin{array}{c} \begin{array}{cccccc} e_2 & e_3 & e_4 & e_5 & e_6 & e_7 \end{array} \\ \begin{pmatrix} 1 & 0 & 0 & 0 & 0 & 0 \\ 0 & 1 & 0 & 0 & 0 & 1 \\ 0 & 0 & 1 & 1 & 0 & 0 \\ 0 & 1 & 1 & 0 & 1 & 0 \\ 0 & 1 & 0 & 1 & 1 & 0 \\ 0 & 0 & 1 & 0 & 1 & 1 \\ 0 & 0 & 0 & 1 & 1 & 1 \end{pmatrix} \end{array}$$

We have the following orthogonality relation between the circuit and cut-set matrices arranged with the same order of edges:

Theorem 5-3 The circuit matrix C and the cut-set matrix K of any graph satisfy the relationship $CK' = 0$.

Proof: The theorem follows from the fact that a cut-set always has an even number of edges (which may be zero) in common with every circuit.

Thus since a proper cut-set separates the graph into two parts, it either divides a circuit or it does not. In the latter case the number of common edges is zero; in the former case an even number of edges of the cut-set must belong to the circuit, by Theorem 1-8.

Exercise 5-4 Show that if a graph is connected and inarticulate, then its cut-set matrix contains its incidence matrix.

Remark: For planar graphs the problem of enumerating the cut-sets is equivalent to that of enumerating the circuits in the dual graph.

We now indicate a procedure for deriving the basis matrix for the circuits of a connected graph. We first designate a complete tree of the graph. In the following discussion all trees are understood to be spanning trees. The tree contains $n - 1$ branches and $m - n + 1$ chords. We have from the definition of a tree that to each chord corresponds a circuit of the graph, namely, the chord and the unique chain in the tree joining its end points. Each circuit has at least one edge not in any of the remaining circuits. Thus, the vectors corresponding to these $m - n + 1$ circuits are linearly independent and the rank of the circuit matrix is at least $m - n + 1$. We now show that equality holds and hence that the circuits determined by the chords form a basis for all circuits of the graph.

Here we make use of the following theorem from matrix theory: The rank R of the product of two matrices, the first of order $p \times q$ and the second of order $q \times r$ with ranks k_1 and k_2, respectively, satisfies the inequality

$$R \geq k_1 + k_2 - q$$

If the two matrices are orthogonal, then $R = 0$ and $k_2 \leq q - k_1$. We take the incidence matrix as the first matrix and the circuit matrix as the second. Since the rank of the incidence matrix is $k_1 = n - 1$ and $q = m$ is the common order of the circuit and incidence matrices, it follows that for k_2—the rank of the circuit matrix—we have

$$k_2 \leq m - n + 1$$

However, as we have seen above, $k_2 \geq m - n + 1$ and hence

$$k_2 = m - n + 1$$

We have proved the following theorem:

Theorem 5-4 The rank of the circuit matrix of a connected graph with m edges and n vertices is $m - (n - 1)$.

If the graph has p components, then the rank of the circuit matrix is $m - n + p$. This number is also known as the *cyclomatic number* of the graph.

Exercise 5-5 Use Theorem 5-4, on the rank of the product of two matrices, to prove that the rank of the cut-set matrix is at most $n - 1$.

Remark: There is a 1-1 correspondence between nonsingular submatrices of rank $n - 1$ of the incidence matrix of a connected graph and the trees of the graph. A similar correspondence exists between the submatrices of rank $m - n + 1$ of the circuit matrix and the complements of the trees of the graph.

We shall motivate the first part of the remark. The incidence matrix of a spanning tree is a submatrix of the incidence matrix of the entire graph. It has all the vertices (rows) but only $n - 1$ edges (columns) of the tree. Since a tree is a connected graph, its matrix has the rank $n - 1$. Conversely, a submatrix of rank $n - 1$ corresponds to n rows and $n - 1$ columns. Since the columns are linearly independent, all vertices are connected and we have a tree.

To illustrate the method of generating a basis matrix for the circuits, we again consider the graph given in Fig. 5-6. The bold edges indicate a selected tree whose chords are e_4 and e_7.

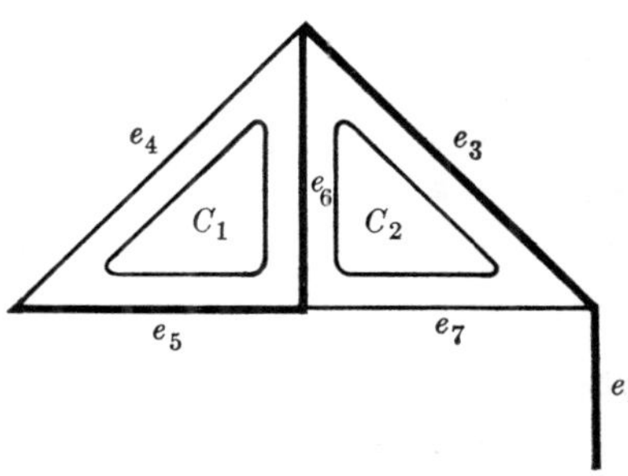

Figure 5-6

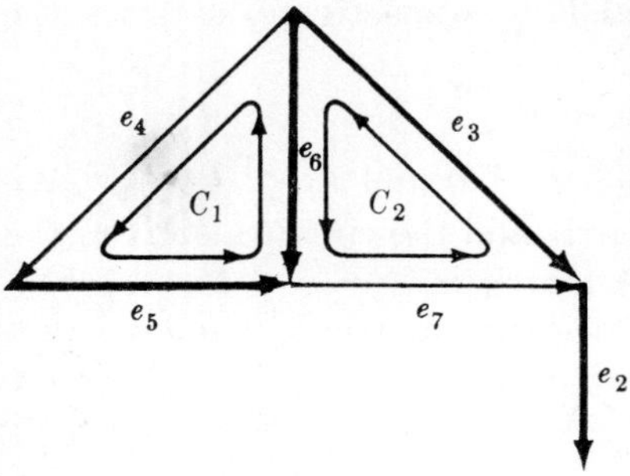

Figure 5-7

The circuit-basis matrix, which is a submatrix of the circuit matrix, includes just those circuits determined by the chords of the selected tree and is given by

$$\bar{C} = \begin{array}{c} \\ C_1 \\ C_2 \end{array} \begin{array}{c} \begin{array}{cc:cccc} e_4 & e_7 & e_2 & e_3 & e_5 & e_6 \end{array} \\ \begin{pmatrix} 1 & 0 & 0 & 0 & 1 & 1 \\ 0 & 1 & 0 & 1 & 0 & 1 \end{pmatrix} \end{array}$$

which has the form

$$\bar{C} = (I|\bar{C}_{12})$$

where I is the identity matrix determined by the incidence of the chords with the circuits and $\bar{C}_{12}$ is the corresponding submatrix for the branches.

In the case of a directed graph we introduce orientations of the circuits as shown in Fig. 5-7, where the orientation of each basis circuit is induced by the inherent direction of its associated chord. In this case the circuit-basis matrix has the form

$$\begin{array}{c} \\ C_1 \\ C_2 \end{array} \begin{array}{c} \begin{array}{cc:cccc} e_4 & e_7 & e_2 & e_3 & e_5 & e_6 \end{array} \\ \begin{pmatrix} 1 & 0 & 0 & 0 & 1 & -1 \\ 0 & 1 & 0 & -1 & 0 & 1 \end{pmatrix} \end{array}$$

Note that -1 occurs where the circuit orientation is opposite to that of the arc direction.

Since the rank of the incidence matrix A is $n - 1$, we can write

$$A = \begin{pmatrix} A_{11} & A_{12} \\ A_{21} & A_{22} \end{pmatrix}$$

where A_{12} is a square nonsingular submatrix of dimension $n - 1$. We have seen that such nonsingular matrices correspond to trees, and hence A_{12} is the submatrix corresponding to a tree which, for example, may be the tree chosen to construct $\bar{C}$ above.

From Theorem 5-2 we have $AC' = 0$, from which it follows that $A\bar{C}' = 0$, where $\bar{C}$ is the circuit-basis matrix. The last relation may be written as

$$\begin{pmatrix} A_{11} & A_{12} \\ A_{21} & A_{22} \end{pmatrix} \begin{pmatrix} I \\ \bar{C}'_{12} \end{pmatrix} = 0$$

where the edges of A must be ordered in the same way as those of $\bar{C}$.

Multiplying out yields

$$A_{11}I + A_{12}\bar{C}'_{12} = 0$$

Since A_{12} is nonsingular, we have

$$\bar{C}'_{12} = -A_{12}^{-1}A_{11} = A_{12}^{-1}A_{11}$$

since $-1 \equiv 1 \bmod 2$. Thus we construct $\bar{C}$ from A.

Exercise 5-6 Use this method to obtain $\bar{C}$ from A for the graph of Fig. 5-6. Let the first two columns correspond to the chords. Note that A_{12} must correspond to the tree.

A basis matrix for the cut-sets is obtained by taking a tree and considering each of its branches at a time. The removal of a branch breaks the tree into two subtrees and partitions the vertices accordingly. The branch and any chords incident with both subtrees comprise a basis cut-set. (In directed graphs, -1 reflects opposition to the branch's direction.)

Exercise 5-7 Use the same tree we used to derive the circuit-basis matrix to show that the cut-sets obtained in this manner are K_1, K_2, K_3, K_6. (See page 106.)

From a circuit-basis matrix $\bar{C}$ it is possible to obtain the cut-set-basis matrix $\bar{K}$, and conversely. By adding rows and permuting columns, $\bar{K}$ can be partitioned as follows

$$\bar{K} = (\bar{K}_{11}|I)$$

where $\bar{K}_{11}$ is the submatrix corresponding to the chords and I is the identity submatrix corresponding to the branches. From the orthogonality relation which exists between the circuit matrix and the transpose of the cut-set matrix we have $\bar{C}\bar{K}' = 0$. Thus

$$(I|\bar{C}_{12})\begin{pmatrix}\bar{K}'_{11} \\ \hline I\end{pmatrix} = 0$$

hence

$$\bar{K}'_{11} + \bar{C}_{12} = 0$$

from which we have $\bar{K}'_{11} = \bar{C}_{12}$, since $-1 \equiv 1 \bmod 2$.

Exercise 5-8 Use this result to obtain $\bar{K}'_{11}$ from $\bar{C}_{12}$ in Exercise 5-6 and verify the equivalence of the answer with that of Exercise 5-7.

Exercise 5-9 Write down the incidence, circuit, and cut-set matrices of the complete graph having five vertices.

5-5 THE VERTEX OR ADJACENCY MATRIX

We can also define a vertex (or adjacency) matrix for both directed and undirected graphs. The element in the (i,j) position of the matrix is equal to the number of edges incident with both vertex i and vertex j (or directed from vertex i to vertex j in the directed case). Thus for the directed graph of Fig. 5-3 (see also Exercise 5-2 on page 104), we have the following vertex matrix:

$$\vec{V} = \begin{array}{c} v_1 \\ v_2 \\ v_3 \\ v_4 \\ v_5 \\ v_6 \\ v_7 \\ v_8 \end{array} \begin{array}{c} \begin{array}{cccccccc} v_1 & v_2 & v_3 & v_4 & v_5 & v_6 & v_7 & v_8 \end{array} \\ \begin{pmatrix} 0 & 0 & 0 & 0 & 0 & 0 & 0 & 0 \\ 1 & 0 & 0 & 0 & 0 & 0 & 0 & 0 \\ 0 & 1 & 0 & 1 & 1 & 0 & 0 & 0 \\ 0 & 0 & 0 & 0 & 1 & 0 & 0 & 0 \\ 0 & 1 & 0 & 0 & 0 & 0 & 0 & 0 \\ 0 & 0 & 0 & 0 & 0 & 0 & 1 & 1 \\ 0 & 0 & 0 & 0 & 0 & 0 & 0 & 2 \\ 0 & 0 & 0 & 0 & 0 & 0 & 0 & 0 \end{pmatrix} \end{array}$$

In general, we have the following theorem regarding the vertex matrix $\vec{V}$ of a graph:

Theorem 5-5 The matrix $\vec{V}^n$ gives the number of arc progressions of length n between any two vertices of a directed graph.

Proof: If a_{ik} is the number of arcs joining v_i to v_k and a_{kj} is the number of arcs joining v_k to v_j, then $a_{ik}a_{kj}$ is the number of different paths each consisting of two arcs joining v_i to v_j and passing through v_k. If this is summed over all values of k, that is, over all the intermediate vertices, one obtains the number of paths of length 2 between v_i and v_j. If we now use a_{ij} to form $a_{ij}a_{jm}$, we have the number of different paths of length 3 between v_i and v_m passing through v_j, and so on. Thus if we assume the theorem true for $\vec{V}^{n-1}$, then the coefficients of $\vec{V}^n = \vec{V}^{n-1}\vec{V}$ give the number of paths of length n between corresponding vertices. This completes the proof.

Remark: Obviously, if for some N, $\vec{V}^n = 0$ for $n \geq N$, then there are no cycles in the graph. If the graph has no cycles, then the elements of $\vec{V}^n$ give the number of simple paths between any two vertices of the graph.

Exercise 5-10 Prove a similar theorem for undirected graphs. Denote their matrix by V.

Remark: We now discuss the use of the vertex matrix as a tool in investigating the properties of a binary relation R in a finite set

$$S = \{a_1, \ . \ . \ . \ ,a_n\}$$

Let $\vec{V} = (v_{ij})$ be the vertex matrix associated with R. If R is reflexive (irreflexive), $\vec{V}$ has 1's (0's) down its main diagonal. If R is symmetric, $\vec{V}$ is a symmetric matrix. If R is asymmetric, then $\vec{V}^2$ will have the same elements along its main diagonal as $\vec{V}$ has, since the only path of length 2 from a_i to a_i is that obtained by traversing a loop at a_i twice (if one exists), unless there is some other element a_j such that a_iRa_j and a_jRa_i. [In the latter case, there will also be an extra path of length 2 from a_j to a_j, and the $\vec{V}^2 = (v_{ij}^{(2)})$ matrix will have elements $v_{ii}^{(2)} = v_{ii} + 1$ and $v_{jj}^{(2)} = v_{jj} + 1$.]

Transitivity is also easily determined. The graph-theoretic statement of this property is the following: If there exists a path of length 2 from a_i to a_j, then there exists an arc from a_i to a_j. Thus, it is only necessary to check whether v_{ij} is positive whenever $v_{ij}^{(2)}$ is positive. If so then R is transitive.

The Index of Primitivity

Note that the vertex matrix in both the directed and the undirected case has nonnegative entries. Zero entries imply that the vertex corresponding to the row of the entry is not connected by an edge to the vertex corresponding to the column. However, these vertices may be connected by a path (or chain) of a certain length. It may be possible that $V^m > 0$ for some integer $m \geq 1$; that is, after V is raised to the power m, all its entries become nonzero. In that case every vertex is accessible to every other vertex in arc progressions of m arcs in length, and the graph corresponding to V^m is complete (in fact it has an arc directed from any vertex to any other vertex) and has a loop at every vertex.

Let us now consider a directed graph D. If D is strongly connected, then its vertex matrix, which we denote by $\vec{V}$, is said to be *irreducible*. In general, a nonnegative square matrix U (from which we form a directed graph according to whether an entry is zero or positive) need not be irreducible. If it is irreducible, i.e., the graph is strongly connected, and if $U^m > 0$ for some integer $m \geq 1$, then U is called *primitive* (and so is its associated directed graph) and the smallest integer m for which this is true (i.e., every vertex is accessible to every other vertex by an arc progression of m edges) is known as the *index of primitivity* of U and is denoted by $\gamma(U)$. We now give the preliminaries in order to show that an irreducible matrix is primitive if and only if the greatest common divisor of the lengths of all its simple cycles is unity (Ref. 73 of Chap. 6).

Definition If H is a subgraph of a graph G, the number of vertices in H is said to be the *order* of H.

Definition If P is an arc progression in a directed graph D, the number of arcs (counting repetitions) in P is said to be the *length* of P.

Lemma 5-6 If $D = (V,A)$ is a primitive graph with index of primitivity m, and if $v,w \in V$, then for all integers $q \geq m$ there exist an arc progression of length q from v to w and an arc progression of length q from w to v.

Proof: Observe that if U is a primitive matrix, then every row and every column of U must contain at least one nonzero element. It then follows that the product $U^{m} \cdot U$ contains only positive elements. In fact, the product $U^{m+n} \cdot U \equiv U^{q}$, where n is any positive integer, contains only positive elements.

Theorem 5-7 A necessary and sufficient condition for a strongly connected graph $D = (V,A)$ with $n \geq 2$ vertices to be primitive is that the GCD (greatest common divisor) of the lengths of all simple cycles in D is unity.

Proof, sufficiency: Suppose $D = (V,A)$ is a strongly connected graph with $n \geq 2$ vertices. Suppose D has r simple cycles with orders p_1, p_2, . . . , p_n and suppose that the GCD of these orders is unity. Choose a closed arc progression P such that P passes through all the vertices of D. We need not be concerned with which vertex we start and terminate; the symbol P will denote any progression equivalent to the one first chosen in the sense that the arcs are encountered in the same order even though the initial (and final) vertices may be different. Let P have length j.

Since D is strongly connected, there exists, for every $v,w \in V$, a path from v to w. With each such ordered pair (v,w) we associate a positive integer $i \leq n$ giving the length of some path from v to w. Now if we wished, we could "travel" from v to w by (1) first traversing P (from v to v) and, at various predetermined vertices, taking "detours" around simple cycles in which these vertices lie and (2) subsequently traversing the path with i arcs. The length of this trip can be expressed by $s = j + \sum_{k=1}^{r} e_{ik}p_k + i$, where e_{ik} is a nonnegative integer for each k. We shall show that, for each i, the e_{ik} can be so chosen that s is constant, i.e., it does not depend on i in that e_{ik} are chosen to help eliminate i.

We recall that if K is a set of integers whose GCD is 1, then 1 can be written as a linear combination of these integers. Hence we can choose positive integers a_1, a_2, . . . , a_r, which may be placed in two complementary sets X and Y such that

$$\sum_{k \in X} a_k p_k - \sum_{k \in Y} a_k p_k = 1$$

Now for each integer i, $1 \leq i \leq n$, and each integer k, $1 \leq k \leq r$, we define (for a sufficiently large integer m)

$$e_{ik} = \begin{cases} (m + 1 - i)a_k & \text{if } k \in X \\ (i - 1)a_k & \text{if } k \in Y \end{cases}$$

$$\text{Then } \sum_{k=1}^{r} e_{ik}p_k + i = \sum_{k\epsilon X} (m + 1 - i)a_kp_k + \sum_{k\epsilon Y} (i - 1)a_kp_k + i$$
$$= m \sum_{k\epsilon X} a_kp_k + \sum_{k\epsilon X} a_kp_k - \sum_{k\epsilon Y} a_kp_k - i \sum_{k\epsilon X} a_kp_k$$
$$+ i \sum_{k\epsilon Y} a_kp_k + i = m \sum_{k\epsilon X} a_kp_k + 1$$

which is a constant.

Hence there exists a fixed integer $s = j + m \sum_{k\epsilon X} a_kp_k + 1$ such that for every ordered pair $v,w \epsilon V$ there exists an arc progression of length s from v to w. Hence there is an integer s such that if U is the vertex matrix of D, then $U^s > 0$. (To see how a given desired path of length s is realized, recall that $s = j + \sum_{k=1}^{r} e_{ik}p_k + i$.)

Necessity: Suppose that $D = (V,A)$ is a primitive graph with $n \geq 2$ vertices and index of primitivity m. By hypothesis (actually, by primitivity) D is strongly connected. Choose a closed arc progression P (from some $v \epsilon V$ to v) such that every *simple cycle* in D is traversed at least once and $t \geq m$, where t is the length of P. Since P can be divided into a succession of simple cycles (generalize Exercise 2-3 to a closed arc progression) t may be written as a linear combination of the orders $p_1, p_2, \ldots, p_r$ of the r simple cycles, *where all the coefficients are positive.* Now by Lemma 5-6 there exists a closed arc progression of length $t + 1$ from v to v. Then (by the lemma) there exist closed arc progressions R (twice P) and S of lengths $2t$ and $2t + 1$, respectively. Notice that both $2t$ and $2t + 1$ can be written as linear combinations of the p_i ($i = 1, 2, \ldots, r$) with strictly positive coefficients. The above generalization of Exercise 2-3 can be applied to the closed arc progression of length $r + 1$, and S, which has length

$$t + t + 1 = 2t + 1$$

can be taken as the desired linear combination.) But to express two consecutive integers as linear combinations of the same set of integers p_i is possible only if the GCD of the p_i is unity. This completes the proof.

An undirected example of nonprimitivity is a bipartite graph, a special case of which is the lattice graph, such as the boundaries of the squares of the chess board, in which some pairs of vertices are joined only by chains having an even number of edges and other pairs are joined only by chains with an odd number of edges. (This fact is true of all bipartite graphs. Prove this as an exercise.) For these graphs all simple cycles have an even number of edges and do not meet the definition of primitivity, since 2 is a common divisor.

The tth power of the matrix $\vec{V}$ of a directed graph D is a matrix with a directed graph having the same vertices and such that the ordered pair (v_i,v_j) is an edge of $\vec{V}^t$ if and only if there is an arc progression in $\vec{V}$ from v_i to v_j of length t.[5]

Exercise 5-11 Show that if $\vec{V}$ is primitive, then so is $\vec{V}^t$ for $t > 0$.

Exercise 5-12 Show that, in a primitive graph, with every vertex v_i there is associated an integer h such that for every vertex v_j there is a path of length h from v_i to v_j. The smallest value of h is called the *reach* of v_i and is denoted by h_i.

Exercise 5-13 Show that, in a primitive graph, if $p \geq h_i$, there is an arc progression of length p from v_i to any v_j. (Note that because the graph is strongly connected, there is an arc from some v_k to v_j and hence a path from v_i to v_k to v_j of length $h_i + 1$, etc., by induction.)

Exercise 5-14 Show that, in a primitive graph D, $\nu(D) = \max [h_1, \ldots ,h_n]$.

Exercise 5-15 If there is a loop at the vertex v_i of a strongly connected graph, then $h_i \leq n - 1$.

The notion of the index of primitivity is important. Consider, for example, a hypothetical chemical reaction with "feedback," i.e., recycling of components, where the vertices of the graph would correspond to various chemical ingredients and where it is desired to know whether a certain product can be obtained by a process consisting of the same number of steps and starting at any of the vertices. Or, in mathematical economics if U is a primitive matrix, the exponent m corresponds to the mth round of transactions such that, if $\gamma(D) = m$, all sectors in an economic input-output model are connected at the mth round and all subsequent rounds.

The question now is: When is a nonnegative irreducible matrix primitive and what is its index of primitivity, or an estimate of it? We shall be briefly interested in the second part of the question.

One of the oldest estimates on the index of primitivity, due to Lowenheim and to Wielandt, is that it cannot exceed $(n - 1)^2 + 1$, where n is the number of vertices and, of course, also the order of the matrix. It is a corollary of the following theorem:

Theorem 5-8 If D is a primitive graph and $\vec{V}$ is its vertex matrix and if s is the length of the shortest simple cycle in D, then the index of primitivity of $\vec{V}$ satisfies

$$\gamma(\vec{V}) \leq n + s(n - 2)$$

Proof: According to Exercise 5-11, D^s (the graph corresponding to $\vec{V}^s$) is primitive, and because D has a simple cycle of length s, D^s has at least s vertices with loops. Hence there is an arc progression of length

$p_i \leq n - s$ from any vertex v_i to some vertex v_k that has a loop in D^s. Again according to Exercise 5-15, there is an arc progression of D^s of length $n - 1$ from v_k to any v_j. Thus the arc progression in D from v_k to v_j has length $(n - 1)s$, and that from v_i to v_j has length $p_i + (n - 1)s$. Hence $h_i \leq p_i + (n - 1)s$ and

$$\gamma(D) = \max [h_1, \ldots, h_n] \leq n - s + (n - 1)s = n + s(n - 2)$$

Corollary 5-9 In a primitive graph, the GCD of the lengths of all simple cycles is unity and hence $s \leq n - 1$, from which it follows that

$$\gamma(\vec{V}) \leq n + (n - 1)(n - 2) = (n - 1)^2 + 1$$

Heap and Lynn[11] show that if the graph of U has at least $k \geq 2$ simple cycles whose lengths are distinct and relatively prime, then the index of primitivity does not exceed

$$(n - 1)^2 + 1 - \tfrac{1}{2}(k - 2)(2n - k - 3)$$

5-6 THE PATH MATRIX

In a connected graph with indexed vertices, a path (properly chain) matrix P is obtained by taking for its rows P_i, the paths from the first to the last vertex, and for its columns the edges of the graph. Hence the entries of the matrix are 1 or 0 depending on whether an edge belongs to a given path or not. For example, the graph of Fig. 5-8 has the matrix of paths between v_1 and v_5 given by

$$P = \begin{matrix} \\ P_1 \\ P_2 \\ P_3 \\ P_4 \end{matrix} \begin{pmatrix} e_1 & e_2 & e_3 & e_4 & e_5 & e_6 \\ 1 & 0 & 0 & 0 & 1 & 0 \\ 1 & 0 & 0 & 1 & 0 & 1 \\ 0 & 1 & 1 & 0 & 1 & 0 \\ 0 & 1 & 1 & 1 & 0 & 1 \end{pmatrix}$$

Theorem 5-10 The product AP' of the incidence matrix and the transpose of the path matrix is a matrix all of whose rows except the first and the last have zero elements. The exceptional two rows have unit elements everywhere.

Proof: An element of AP' is unity if and only if an edge is at the same time in a given path and incident with the first or last vertex. There is exactly one such edge in any path between these two ver-

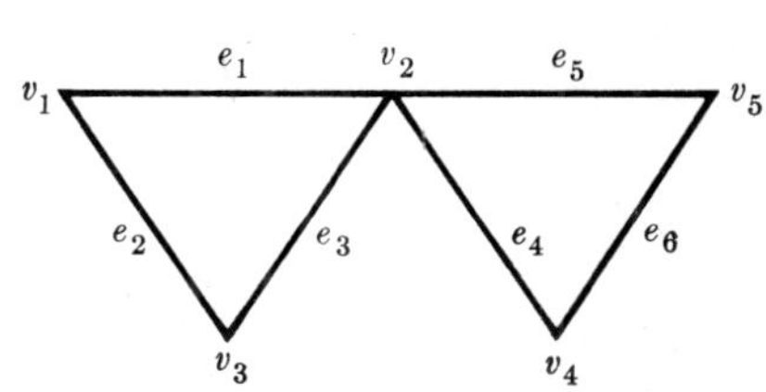

Figure 5-8

tices. Nonterminal vertices along any path have degree 2 or 0, and hence the remaining elements are zero mod 2.[21]

Remark: The rank of the path matrix of a graph with n vertices and m edges is $m - n + 2 - c$, where c is the number of independent circuits in those separable subgraphs between the terminal vertices whose removal from the graph does not remove either of the terminal vertices. From Theorem 5-10 note that the rank of P does not exceed $m - n + 2$. Note also that the chords of the trees in the subgraphs described above do not belong to any path.

5-7 THE REALIZABILITY OF THE CIRCUIT AND CUT-SET MATRICES

We have already examined the problem of constructing and characterizing matrices associated with a graph. The converse problem of constructing a graph corresponding to a given matrix is generally either trivial or very difficult. The first case is readily illustrated by a matrix with exactly two unit elements in every column and zeros elsewhere. A graph can always be constructed with the given matrix as the incidence matrix. Because an edge may be incident with more than two circuits or two cut-sets, the problem of constructing a graph for a matrix which is to serve as a circuit matrix is not so easy.

The question of realizability has been examined by several people. An interesting exposition of the ideas is found in the paper by Ash and Kim.[1] A rigorous and deep theory leading to necessary and sufficient conditions for realizability has been developed by Tutte.[17,18] We shall present the barest essentials for the main theorem.

Consider the column vectors of a matrix. One can say that a subset of these columns is either linearly independent or linearly dependent. Thus the subsets fall into two classes which are not arbitrary since, for example, the following two theorems must hold.[19]

1. Any subset of an independent set is independent.
2. If N_p and N_{p+1} are independent sets of p and $p + 1$ columns, respectively, then N_p together with some column of N_{p+1} forms an independent set of $p + 1$ columns.

A system obeying (1) and (2) is called a matroid. There are theorems not deducible from (1) and (2), that is, there are matroids with no corresponding matrices. Thus every matrix is a matroid, but not conversely.

The above definition of a matroid is a special version of the following general definition. A matroid on a finite set M is a class $\mathbf{M}$ of nonnull subsets of M which satisfies these conditions:

1. No member of $\mathbf{M}$ contains another as a proper subset.
2. If $X, Y \in \mathbf{M}$, $a \in X \cap Y$, and $b \in X - (X \cap Y)$, then there exists $Z \in \mathbf{M}$ such that $b \in Z \subseteq (X \cup Y) - \{a\}$.

The reader should have no difficulty in verifying that the set of all simple circuits or all proper cut-sets of a connected graph forms a matroid on the set of edges of the graph. We shall use R to denote the ring of integers or the ring of residues modulo a prime number.

A chain on a finite set M over R is a mapping f of M into R, that is, if $a \in M$, then $f(a)$ is the coefficient of a in R. One can define the sum $f + g$ of two chains f and g, and the product kf, where $k \in R$, by the relationships $(f + g)(a) = f(a) + g(a)$ and $(kf)(a) = kf(a)$, for $a \in M$, and show that the chains on M over R are the elements of an additive abelian group $A(M,R)$. A chain group on M over R is any subgroup of $A(M,R)$.

A chain f of a chain group N is elementary if it is nonzero and there is no nonzero $g \in N$ such that $|g|$ is a proper subset of $|f|$. Here the set of all $a \in M$ such that $f(a) \neq 0$ is the domain $|f|$ of f.

If R is the ring of integers, we define a primitive chain of N as an elementary chain f of N in which the coefficients $f(a)$ are restricted to the values 0, 1, -1.

A chain group N is called binary if R is the ring of residues mod 2. It is regular if R is the ring of integers and if to each elementary chain there corresponds a primitive chain with the same domain.

If $M(N)$ is the class of domains of elementary chains of N, then $M(N)$ is a matroid on M. It is called the matroid of N. A matroid is binary or regular if it is the matroid of a binary or regular chain group, respectively.

Let $K(G)$ and $C(G)$ denote, respectively, the matroid of proper cut-sets and that of simple circuits of a graph G. A matroid $\mathbf{M}$ is graphic or cographic if there is a graph G such that $\mathbf{M} = K(G)$ or $\mathbf{M} = C(G)$, respectively.

If $\mathbf{M}$ is a matroid on a set E, and if $S \subseteq E$, let $\mathbf{M} \cdot S$ denote the set of minimal nonempty intersections with S of elements of $\mathbf{M}$ (where a minimal nonempty intersection is one not containing another). Then $\mathbf{M} \cdot S$ is a matroid on S, called the *reduction of* $\mathbf{M}$ *to* S. (In a graph G, let $G \cdot S$ denote the subgraph of G determined by the set S of edges. If $\mathbf{M}$ is the cut-set matroid of G, then $\mathbf{M} \cdot S$ is the cut-set matroid of $G \cdot S$.)

Similarly, the set of elements of $\mathbf{M}$ which are subsets of S is a matroid on S, denoted by $\mathbf{M} \times S$ and called the *contraction of* $\mathbf{M}$ *to* S. (In a graph G, let $G \times S$ denote the subgraph obtained by contracting to a point all the edges not in S. If $\mathbf{M}$ is the cut-set matroid of G, then $\mathbf{M} \times S$ is the cut-set matroid of $G \times S$. In circuits, the roles of "dot" and "cross" must be interchanged.)

Finally, if $\mathbf{M}$ is a matroid on E, a matroid of the form $(\mathbf{M} \cdot S) \times T$, where $T \subseteq S \subseteq E$, is called a *minor of* $\mathbf{M}$.

Theorem 5-11 A matroid M is graphic (cographic) if and only if it is regular and has no minor which is the circuit-matroid (cut-set matroid) of a Kuratowski graph.

5-8 RELATION TO COMBINATORIAL TOPOLOGY

In this section we relate the foregoing concepts to certain notions which are basic to combinatorial topology. Some of the results discussed here are repetitions of material presented earlier, but they are shown in a different framework. Also included is the proof of a coloring theorem.

A 1 chain on a directed graph G with n vertices v_i and m edges e_k is a linear combination of the edges, that is, $\sum_{k=1}^{m} a_k e_k$, where a_k are real, although one need not be restricted in this manner but can choose a_k complex or just rational. All 1 chains define a vector space $\mathcal{V}_1$ on G with the edges forming an m basis, where chain addition is defined by adding corresponding coefficients. Similarly, it is possible to define a vector space $\mathcal{V}_0$ of 0 chains by using linear combinations of the vertices with real b_i, that is, they are sums of the form $\sum_{i=1}^{n} b_i v_i$. We define the inner product of two 1 chains or two 0 chains as the sum of the products of corresponding coefficients. If this sum is zero, they are said to be orthogonal.

We can introduce a boundary operator ∂ (and a coboundary operator δ) which transform $\mathcal{V}_1$ into $\mathcal{V}_0$ and $\mathcal{V}_0$ into $\mathcal{V}_1$, respectively. If ∂ is defined on an edge e, then it can also be defined on a 1 chain by requiring it to be linear. Let e be oriented from the end point p_1 to the end point p_2; then we define

$$\partial e = p_1 - p_2 \qquad \text{and} \qquad \partial \sum_{k=1}^{m} a_k e_k = \sum_{k=1}^{m} a_k \, \partial e_k$$

Note that p_1 and p_2 may coincide to form a single vertex of the graph. If v is a vertex of the graph, then the coboundary of v is defined by the "bundle at v"

$$\delta v = \overset{\rightarrow}{\Sigma e} - \overset{\leftarrow}{\Sigma e}$$

where the first sum is taken over the edges directed toward v and the second sum is taken over the edges directed away from v. Similarly, the coboundary of a 0 chain is

$$\delta \sum_{i=1}^{n} b_i v_i = \sum_{i=1}^{n} b_i \, \delta v_i$$

Note that the boundary of a cycle or a linear combination of cycles is zero. The 1 chains having zero boundary form a vector space which is a subspace of $\mathcal{V}_1$. A cocycle is a 1 chain and is the coboundary of some 0 chain. The set of coboundaries is also a vector space which is a subspace of $\mathcal{V}_0$. We have the following interesting theorems.[8]

1. The inner product of the boundary of any 1 chain and any 0 chain is equal to the inner product of the 1 chain and the coboundary of the 0 chain.
2. The inner product of a cycle and a coboundary is zero.
3. A 1 chain which is a cycle and a coboundary is the zero vector of $\mathcal{V}_1$.
4. A 1 chain which is orthogonal to every coboundary is a cycle.
5. A 0 chain which is orthogonal to every cycle is a coboundary.
6. A 1 chain has a unique expression as the sum of a cycle and a coboundary.
7. The dimension of the cycle space is $m - n + 1$ and that of the coboundary space is $n - 1$.

8. Using the integers mod 2, a set of edges is a cut-set if and only if it is a coboundary.

There is an interesting way to introduce incidence matrices by using the boundary operator ∂. Consider the vector $e = (e_1, \ldots, e_m)$ and define $\partial e = (\partial e_1, \ldots, \partial e_m)$; then consider the set of end points $p_1, p_2, \ldots, p_{2m}$ of $e_1, \ldots, e_m$ and form the vector $p = (p_1, p_2, \ldots, p_{2m})$. Let D be a $2m \times m$ matrix whose elements d_{ij} are 0, 1, or -1 depending on whether end point p_i is or is not incident with edge e_j and on whether it is an initial or a terminal point of e_j. Then $\partial e = pD$. If we now associate with each end point p_j the corresponding vertex v_i and form the vector $v = (v_1, \ldots, v_n)$ and the $n \times 2m$ matrix $\mathfrak{D}$ whose i, j element is 0 or 1 depending on whether end point p_j is associated with vertex v_i or not, we have $p = v\mathfrak{D}$. Thus $\partial e = v\mathfrak{D}D$, where $\mathfrak{D}D$ is the incidence matrix A of the graph.

Exercise 5-16 Construct the matrices $\mathfrak{D}$ and D for the graph of Fig. 5-3 and obtain the incidence matrix from the product $\mathfrak{D}D$.

Theorem 5-12 A map on a sphere in which every vertex is of even degree can be colored with two colors.[14]

Lemma 5-13 A closed curve C which does not pass through the vertices of our map crosses edges of regions an even number of times.

Proof of theorem: The lemma (proved below) implies the theorem; for if we use red and blue and start by coloring region R_1 with red, the color of any other region R_i is determined by counting the number of edges that are crossed by an arbitrary curve drawn from a point in R_1 to a point in R_i. We choose red if the number is even and blue if it is odd. From the lemma, any two arbitrary curves from R_1 to R_i yield the same color for R_i. In this manner adjacent regions will have different colors, since a curve from R_1 to one of them yields one color and, when extended across a boundary edge, must yield the other color.

Proof of lemma: We define $f^1(e_i) = 0$ if C crosses edge e_i an even number of times and $f^1(e_i) = 1$ if C crosses e_i an odd number of times (zero is even).

Since the total crossing by C of the boundary of each region is even (each time it crosses into the region it must also cross out), f^1 must be a cocycle and is therefore the boundary of some 0 cochain f^0. Let $v_1, \ldots, v_i$ all be the vertices on each of which f^0 takes on a unit value. Because the total number of edges meeting these vertices is even (an edge incident with two of these vertices is counted twice), the total number of edges which meets just one (but not two) of these vertices is even. These are the only edges for which $\partial f^0 = f^1$ assigns a unit value, and hence f^1 assigns a unit value to an even number of edges. From the definition of

f^1 we have the fact that there are an even number of regional boundary edges which C crosses an odd number of times.

REFERENCES

1 Ash, R. B., and W. H. Kim: On Realizability of a Circuit Matrix, *IRE Trans. Circuit Theory,* **CT-6**(3): 219–223 (1959).

2 Auslander, L., and H. M. Trent: Incidence Matrices and Linear Graphs, *J. Math. Mech.,* **8:** 827–835 (1959). MR 21-757.

3 Auslander, L., and H. M. Trent: On the Realization of a Linear Graph Given Its Algebraic Specification, *J. Acoust. Soc. Am.,* **33:** 1183–1192 (1961).

4 Cederbaum, I.: Applications of Matrix Algebra to Network Theory, *IRE Trans. Circuit Theory,* **CT-6:** 152–157 (1959).

5 Dulmage, A. L., and N. S. Mendelsohn: Gaps in the Exponent Set of Primitive Matrices, *Ill. J. Math.* (To appear.)

6 Gould, R.: Graphs and Vector Spaces, *J. Math. Phys.,* **37:** 193–214 (1958). MR 20-1229.

7 Harary, Frank, and Moses Richardson: A Matrix Algorithm for Solutions and r-Bases of a Finite Irreflexive Relation, *Naval Res. Logistics Quart.,* **6**(4): 307–314 (1959).

8 Harary, F.: Graph Theory and Electric Networks, *IRE Trans. Circuit Theory,* **CT-6:** 95–109 (1959); *IRE Trans. Inform. Theory,* **IT-5:** 95–109 (1959).

9 Harary, F.: A Matrix Criterion for Structural Balance, *Naval Res. Logistics Quart.,* **7**(2): 195–199 (1960).

10 Harary, F.: On the Consistency of Precedence Matrices, *J. Assoc. Computing Mach.,* **7**(3): (1960).

11 Heap, B. R., and M. S. Lynn: The Index of Primitivity of a Non-negative Matrix, *Numerische Math.,* **6**(2): 120–141 (1964).

12 Iri, Masao: A Necessary and Sufficient Condition for a Matrix to Be the Loop or Cut-set Matrix of a Graph and Practical Method for the Topological Synthesis of Networks, *Res. Assoc. Appl. Geometry (Tokyo) Notes* 50.

13 Lowenheim, L.: Potenzen im Relativkalkul und Potenzen allgemeiner endlicher, *Sitz. ber. Berliner Math. Ges.,* **12:** 137–148 (1913).

14 Olum P.: Some Elementary Cohomology Theory, *Pi Mu Epsilon J.,* November, 1951.

15 Parker, F. D.: Matrices, Relations, and Graphs, *Math. Mag.,* **34**(1): 5–9 (1960).

16 Seshu, S., and M. B. Reed: "Linear Graphs and Electrical Networks," Addison-Wesley Publishing Company, Inc., Reading, Mass., 1961.

17 Tutte, W. T.: A Homotopy Theorem for Matroids, I, II, *Trans. Am. Math. Soc.,* **88:** 144–174 (1958).

18 Tutte, W. T.: Matroids and Graphs, *Trans. Am. Math. Soc.,* **90:** 527–552 (1959).

19 Whitney, H.: On the Abstract Properties of Linear Dependence, *Am. J. Math.,* **57:** 507–533 (1935).

20 Wielandt, H.: Unzerlegbare, nicht negative Matrizen, *Math. Z.,* **52:** 642–648 (1950).

21 Wing, O., and W. H. Kim: The Path Matrix and Its Realizability, *IRE Trans. Circuit Theory,* **CT-6**(3): 267–272 (1959).

PART II

APPLICATIONS

Part II contains a variety of representative applications of graph theory. Chapter 6, designed to be a long chapter, contains numerous brief applications illustrating this variety. In it some examples are formulated and solved, some are just formulated, and still others are briefly described to give a flavor of the wide applicability of the subject but must be pursued in the relevant literature for greater familiarity. In contrast, Chap. 7 goes deeper into two particularly significant areas of application. Thus in Chap. 7 linear flows through networks are discussed. General techniques for maximizing flow and minimizing cost are developed from a graph-theoretic viewpoint.

SIX

A VARIETY OF INTERESTING APPLICATIONS

6-1 INTRODUCTION

To study graph theory without being aware of its applications is to miss a major part of its substance. With the intention of emphasizing the importance of this aspect, we have divided the book into two parts, the second of which is devoted to a brief presentation of some interesting applications. This chapter has the aim of presenting a variety of examples to alert the reader to the art of model formulation in science and engineering by using graphs. The next chapter, on network flows, has a more specific objective.

The objective of this chapter is, briefly, to point out various ways in which graphs can be used to formulate and to solve problems. After reading each section, the reader is urged to inquire into the underlying properties of the graph being used in that section. In certain cases, such as the problem of the colored cubes, the use of the simple idea of a graph, when properly chosen, is powerful. For that problem one obtains a lucid explanation of the workings of a puzzle, sold under such names as Tantalizer, which if solved by trial and error is quite unwieldy. In other cases more sophisticated graph-theoretic properties are exemplified. Some of these properties have found their way into industrial and government operation where, for example, PERT is now a basic tool of projected planning for efficient operation to meet prescribed objectives. The utility of the simple concept of a graph emerges throughout.

It is clear that we must be brief in the description of these applications, because most of them can be so expanded that each can occupy a chapter by itself. Therefore, for some sections we have settled for a short summary of some of the ideas in the hope that the interested reader will pursue the subject in the references. We have attempted to group the sections under general headings where they seemed to us to have the greatest relevance, but this grouping is flexible and the reader may be able to find a better framework.

APPLICATIONS TO ECONOMICS AND OPERATIONS RESEARCH

6-2 APPLICATION TO ECONOMICS AND LOGISTICS

One of the most significant large-scale models in mathematical economics and logistics is the input-output model associated with the name of W. Leontief. This model, which we now examine from a graph-theoretic standpoint, constitutes, in fact, a modern realization of the celebrated *Tableau économique* of 1758–1759 due to Francois Quesnay.[75]

In the input-output model a classification grid is imposed on the transactions of an economy in which there are m collections of entities (e.g., establishments, firms, households). Each collection is called an *industry, sector,* or *activity* for entities producing (or consuming) similar goods and services. For some historical period and in some set of valuation conventions, let $x_{ij} \geq 0$ denote the value of purchases made by industry (or activity) i from industry (or activity) j; moreover, let x_{ij} be called the *value of input to industry i from industry j* $(i, j = 1, \ldots, m)$. Thus x_{ij}, in effect, measures the flow of funds from i to j in return for the resource inputs which flow from j to i. Let the sum $\sum_{h=1}^{m} x_{hi} > 0$ be denoted by x_i and be designated as the *value of output of industry i.* Let b_{ij} denote the normalized value of input $x_{ij}/x_i \geq 0$, and let it be designated as the *unit input to industry or activity i from industry or activity j* $(i, j = 1, \ldots, m)$. Let the following (balance) valuation conventions obtain:

$$\sum_{j=1}^{m} x_{ij} \equiv \sum_{h=1}^{m} x_{hi}$$

for all $i = 1, \ldots, m$, that is, the ith row sum corresponds exactly to the ith column sum.

Consider now the matrix $B^* = \|b_{ij}\|$ $(i, j = 1, \ldots, m)$, defined above, and *any* proper principal submatrix B of B^* such that $B = \|b_{ij}\|$, $i, j = 1, \ldots, r$ and $r < m$. Let the system $x(I - B^*) = \theta$, θ being the null vector of order m, be called a *closed input-output model.* Let a system $x(I - B) = w$, where B is *any* proper principal submatrix of order $r < m$ and w is nonnegative, be called an *open-input-output model.* A solution x of an open or a closed input-output model is called *admissible* if, and only if, the vector x is finite and nonnegative but not null. The nonnegative vector w of an open model is construed as a stipulated "bill of goods" or "bill of final demand" for those industries or activities which correspond to rows and columns which appear in B^* but not in B. Admissible solutions x of input-output models (whether open or closed) are construed as activity level vectors which state the level (of value of output) at which each industry or activity is operated. When the inverse of the matrix $(I - B)$ of an open input-output model exists, it may be shown to have the well-known power series form[75]

$$(I - B)^{-1} = \sum_{h=0}^{\infty} B^h$$

We now proceed to give a graph-theoretic formulation of necessary and sufficient conditions for the existence of the inverse of the matrix $(I - B)$ of an open input-output model. For this purpose, we introduce some additional graph-theoretic definitions.[73] Let D be a directed graph and let H be a strongly connected subgraph in D. A strongly connected subgraph H in D will be said to be *maximal in D* if, and only if, every strongly connected subgraph of D is either a subgraph of H or contains no vertex in common with H. A strongly connected subgraph H in D is said to be *closed in D* if, and only if, H is maximal and every vertex of D accessible (by means of a directed path) from any vertex of H is contained in H. Let $A = \|a_{ij}\|$ be a nonnegative square matrix of order n, that is, $a_{ij} \geq 0$ for $i, j = 1, \ldots, n$. *The finite directed graph $D(A)$ of A* is defined to consist of n vertices $\alpha_1, \ldots, \alpha_n$ and the totality of arcs (α_k, α_l) such that the arc (α_i, α_j) exists in $D(A)$ if, and only if, $a_{ij} > 0$ in A. The directed graphs $D(B^*)$ and $D(B)$ associated with the closed and open input-output models introduced above may be shown to have significant technological and accounting-flow interpretations.

A nonnegative square matrix A will be said to be (row) *substochastic* if no row sum of A exceeds unity, that is, $a_{ij} \geq 0$, $r_i \equiv \sum_{j=1}^{n} a_{ij} \leq 1$. If each row sum of A is exactly unity, the matrix is said to be *stochastic*. On the basis of the earlier discussion, it is evident that, in a closed input-output model $x(I - A^*) = \theta$, the matrix A^* is stochastic, whereas, in an open input-output model $x(I - A) = \omega$, the matrix A is substochastic. The following theorem, which we do not prove here, is given in Ref. 75.

Theorem 6-1 Let A be a substochastic matrix. The inverse $(I - A)^{-1}$ exists if and only if there are no closed strongly connected subgraphs in the directed graph $D(A)$ or if, for each strongly connected subgraph H closed in $D(A)$ there exists a vertex of H with associated row sum in A less than unity.

This result may be reformulated in the following fashion:

Theorem 6-2 Let A^* be a stochastic matrix and let A be any proper principal submatrix of A^*. The inverse $(I - A)^{-1}$ exists if and only if no strongly connected subgraph which is closed in $D(A)$ is also closed in $D(A^*)$.

Proof: Assume H is a strongly connected subgraph of $D(A)$. Then H is closed in $D(A)$ with corresponding matrix in A stochastic if and only if H is closed in $D(A^*)$. By Theorem 6-1 no closed, strongly connected subgraph in $D(A)$ is closed in $D(A^*)$.

Theorem 6-2 has the following immediate corollary:

Corollary 6-3 Let A^* be a stochastic matrix. If $D(A^*)$ is a strongly connected graph (that is, A^* is irreducible or indecomposable), then $(I - A)^{-1}$ exists for any proper principal submatrix A of A^*.

Exercise 6-1 Let A be a nonnegative square matrix of order n in which each row sum is positive. Show that, in the directed graph $D(A)$ of A, (a) there exists at least one closed strongly connected subgraph and (b) that every vertex of $D(A)$ is connected to one or more closed strongly connected subgraphs in $D(A)$.

Exercise 6-2 Show by the use of a "bordering" approach that every open input-output model which exhibits at least one admissible solution is equivalent to a "constrained" closed input-output model.

Exercise 6-3 Let A be a nonnegative square matrix of order n. Let x be a positive $(n \times 1)$ column vector and let y be a positive $(1 \times n)$ row vector. Let $\tilde{A}$ denote the following bordered matrix of order $n + 1$: $\begin{Vmatrix} A & x \\ y & 0 \end{Vmatrix}$. The directed graph $D(\tilde{A})$ is strongly connected. Show that, if A in $\tilde{A}$ is distinct from the null matrix, $\tilde{A}$ is primitive and a sharp upper bound for the index of primitivity $\gamma(\tilde{A})$ is 4.

An alternative formulation of an economic "input-output" model, given below, is based on a theorem that is somewhat difficult to prove in limited space. Consider the following matrix whose coefficients give the per unit purchase figures (negative) of material by an industry from other industries and the per unit figures for output (positive) on the main diagonal.[8, 92]

$$\begin{array}{c} \\ v_1 \\ v_2 \\ v_3 \end{array}\begin{array}{c} \begin{array}{ccc} v_1 & v_2 & v_3 \end{array} \\ \begin{pmatrix} 6 & -4 & -2 \\ -3 & 5 & 0 \\ -1 & -1 & 3 \end{pmatrix} \end{array}\begin{array}{c} \\ 0 \\ 2 \\ 1 \end{array} \quad \begin{array}{c}\text{Consumption} \\ \text{(row sums)}\end{array}$$

The determinant of this matrix has the value 38.

Consider the graph associated with this "interindustry" output and purchase model. We introduce a general consumption vertex C from which there is flow (consumption) toward each industry whose amount is indicated by the net output from that industry. These flows are given by the row sums of the matrix. The flow to v_1 is zero, and hence we need not draw that edge of the graph. We then form all possible trees of the graph (Fig. 6-1) and take the product over the edge capacities. The sum of these values taken over all the trees equals the determinant of the above matrix. We have

$$2 + 16 + 0 + 2 + 8 + 0 + 8 + 2 = 38$$

Positive trees correspond to matrices with row sums ≥ 0 and non-

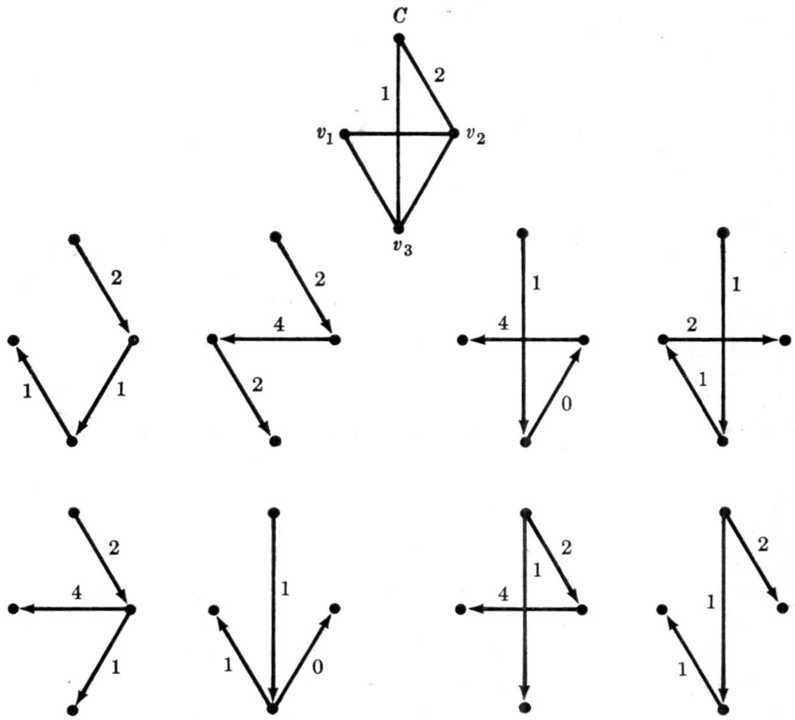

Figure 6-1

positive off-diagonal elements. Thus a single positive tree ensures a positive determinant. Such matrices (e.g., Leontief matrices) with dominant diagonal are of great interest in mathematical economics.

6-3 LINEAR PROGRAMMING AND NETWORK FLOWS

Although network flows are not considered in detail until Chap. 7, we shall now give an informal explanation of the basic concept in order to exhibit one relationship of flow problems with linear programming problems.

Let vertices v_0 and v_n of a directed graph be designated as the origin and destination of a conceptual substance flowing through the arcs. Assume, moreover, that an arc from vertex v_i to vertex v_j has an associated capacity, or upper limit on flow, c_{ij}. Finally, let C_{ij} be the cost per unit of flow in the arc. Our flow problem becomes a linear programming problem requiring the minimization of $\sum_{i,j} C_{ij}x_{ij}$ for a total flow c from v_0 to v_n subject to

$$\begin{aligned}
&\sum_j (x_{0j} - x_{j0}) = c \\
&\sum_j (x_{ij} - x_{ji}) = 0 \qquad \text{for } i = 1, \ldots, n-1 \\
&\sum_j (x_{nj} - x_{jn}) = -c \\
&0 \le x_{ij} \le c_{ij} \qquad \text{for every arc}
\end{aligned}$$

The network flow is sometimes a convenient method for solving this type of linear programming problem, which is known as a transportation problem when it arises in other considerations.

6-4 PERT AND RELATED TECHNIQUES

A directed graph is a natural tool to employ in connection with describing and analyzing complex projects composed of numerous interrelated activities. The overall project might be, for example, the total process of designing, constructing, and testing a piece of equipment or the process of designing and erecting a building, including related considerations such as site acquisition and site preparation. In general, suppose that we are considering some well-defined total project and that we can decompose the set of all actions associated with this project into individual nonoverlapping *activities* $a_1, a_2, \ldots, a_n$. There are, of course, various ways to break a project down into component parts. Determination of an appropriate breakdown depends on considerations which will be evident later

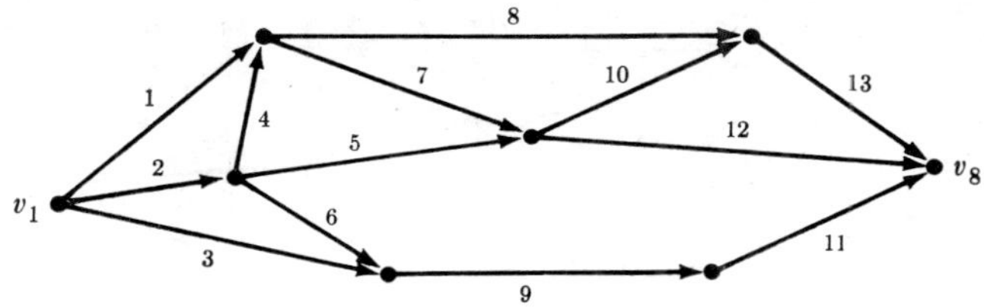

Figure 6-2

in this discussion. (In general, the individual activities must be such that it is possible to acquire the types of quantitative information discussed below and such that all essential precedence relationships, discussed next, can be determined.)

While certain activities are independent of one another, there will in general be certain essential dependencies, with respect to time, which take this form: Activity a_i must be completed before activity a_j can be initiated. If all such time dependencies are given, they can be conveniently summarized by a directed graph such as that of Fig. 6-2. Each arc represents one activity, and each vertex is called an *event* and represents a point in time. In particular, vertex v_1 symbolizes the start of the project as a whole, and v_8 represents its completion. The intermediate vertices serve as the vehicle for indicating the activity precedences. They are chosen, and related to the arcs, in such a way that the following fundamental assertion is valid for every activity (arc) and event (vertex):

If activity a has initial event or vertex v, then a cannot be initiated until every activity which terminates at v has been completed. However, a may be initiated any time thereafter. For example, in Fig. 6-2, activity 10 (as well as 12) can be started only after both activity 5 and activity 7 have been completed. Indirectly, activity 10 also depends on the completion of activities 1, 2, and 4, since 5 and 7 are directly dependent on them. The only activities which can be initiated at the outset are 1, 2, and 3, and the project is considered to be completed only when 11, 12, and 13 (and consequently all activities) have been completed.

The management techniques known as PERT (program evaluation and review technique) and CPM (critical path method), as well as later related methodologies, employ such an activity graph as the fundamental structure on which analyses are based. To illustrate the most basic type of analysis, suppose that each activity a consumes a known amount of time $t(a)$. A group of several activities may, of course, be performed concurrently if no activity in the group is constrained, either directly or indirectly, to start only after another has been completed. (This will be the case if no one activity can be included in a path which leads from v_1 to the initial event of another.) Suppose, specifically, that the quantities

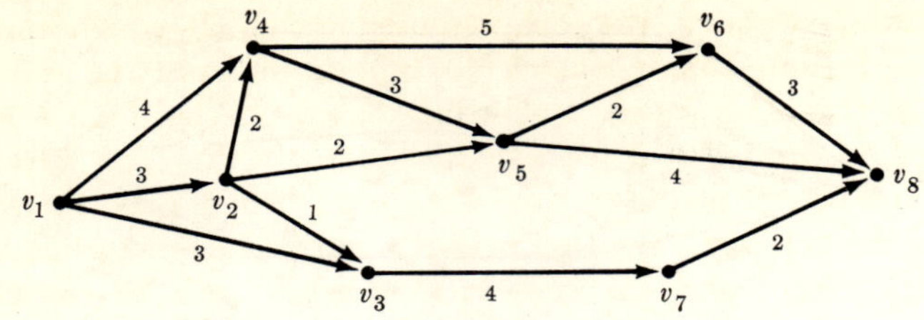

Figure 6-3

associated with arcs in Fig. 6-3 represent activity durations. (For the moment, we are assuming that the duration of each activity is known and constant. In practice, the duration is frequently assumed to vary and to be governed by a probability distribution whose general form is known and whose parameters can be estimated.)

The length (i.e., sum of time durations) of any path from v_1 to v_i clearly represents a lower bound on the elapsed time, measured from the start of the project, before event v_i "occurs" or is "realized" and activities having v_i as initial vertex may be started. This being the case, it is convenient to associate numbers (times) with vertices as follows:

$$T(v_1) = 0$$
$$T(v_i) = \max \{t(P)\} \qquad \text{for } i \neq 1$$

where $t(P)$ denotes the time-length of path P and where the maximum is taken over all paths from v_1 to v_i.

Note that, by its nature, an activity graph is acyclic. (The presence of a cycle would constitute an impossible situation in which no activity of the cycle could be initiated first; for each activity would depend on the completion of another.) This being the case, we can use the method of Chap. 3 to find a spanning tree, rooted at v_1, which simultaneously realizes a longest path from v_1 to each other vertex. (We are assuming that every vertex of the graph is reachable via at least one path.) Figure 6-4 shows an appropriate tree, given the activity durations shown in Fig. 6-3. The corresponding values of $T(v_i)$ are also shown.

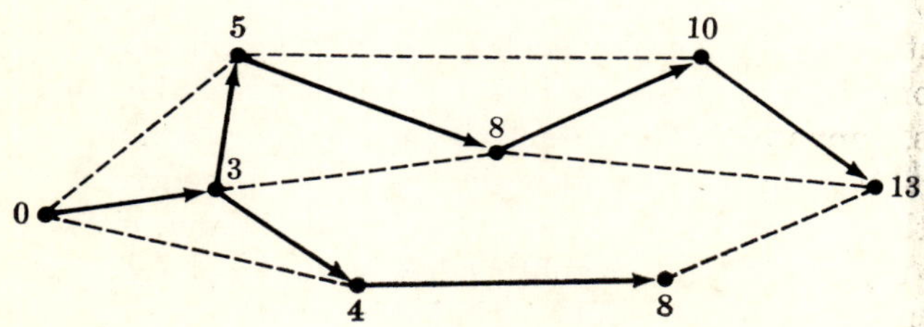

Figure 6-4

As mentioned above, the earliest possible time at which an activity (v_i,v_j) can be initiated is at least $T(v_i)$ time units from the start of the project. On the other hand, a schedule based on the $T(v_i)$'s is in fact feasible. More precisely, if we schedule each activity (v_i,v_j) to start at time $T(v_i)$ and be completed at time $T(v_i) + t_{ij}$ (where t_{ij} is the time associated with the activity), then no activity will be begun earlier than the ground rules permit and the project as a whole will be completed in $T(v_8) = 13$ units, which is the earliest possible completion time.

Exercise 6-4 Prove that if activities are scheduled in this manner, they will not violate any of the precedence relationships.

A (timewise) longest path from the starting event v_1 to the ending event v_n is called a *critical path*. Its length (in our example, 13) is the shortest time in which the entire project can be executed. Moreover, this minimal time will be realized only if each activity of the critical path is initiated as soon as its immediate predecessor is completed. In general, the critical path is not unique. (The reader may note an alternative critical path in our example.) An activity is called a *critical activity* if it is contained in one or more critical paths. Assuming that it is desirable to complete the project in minimal time, there is still a certain amount of latitude in scheduling noncritical activities. Measurement of this latitude is known as the determination of *slack* in the activity graph. This is our next consideration.

Note that each event must be realized (i.e., all activities leading into it must be completed) early enough to execute, successively, all activities in any path from that event to the end event. This suggests associating a second set of numbers (times) with the events, comparable to the $T(v_i)$'s but measured with respect to the end event rather than the starting event. Specifically, we define

$$X(v_n) = 0$$
$$X(v_i) = \max \{t(P)\} \qquad \text{for } i \neq n$$

where $t(P)$ is the length of a path from v_i to v_n and the maximum is taken over all such paths. Note that we can again use the method of Chap. 3 to determine $X(v_i)$. All we need do is temporarily reverse the direction of each activity, find a longest spanning tree rooted at v_n, and then restore the original arc directions. Figure 6-5 shows an appropriate tree for our example, as well as the values of $X(v_i)$.

Since the values of $X(v_i)$ are measured from the end of the project, which occurs $T(v_n)$ time units after the start, it is convenient to relate these times to the start by defining the following quantities:

$$L(v_i) = T(v_n) - X(v_i)$$

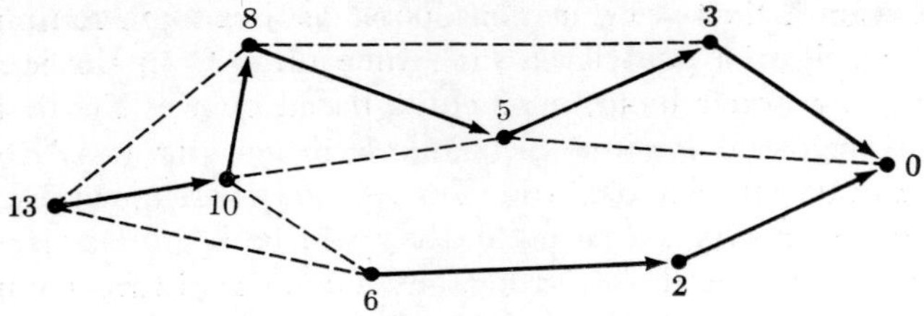

Figure 6-5

Then $L(v_i)$ denotes the latest time at which event v_i can be realized, without lengthening the project as a whole. Figure 6-6 shows the value of $L(v_i)$ for each vertex, alongside the value of $T(v_i)$ determined earlier. The values of $T(v_i)$ and $L(v_i)$ always satisfy the relationship

$$T(v_i) \leq L(v_i)$$

Exercise 6-5 Prove the foregoing assertion, and prove moreover that $T(v_i) = L(v_i)$ if and only if some critical path is incident with v_i.

The values of $T(v_i)$ and $L(v_i)$ serve as the basis for determining the amount of freedom available in scheduling individual activities (without sacrificing minimal project time). For example, the activity from v_2 to v_3 (see Fig. 6-3) requires one unit of time, and we have determined that

$$T(v_2) = 3 \qquad L(v_2) = 3 \qquad T(v_3) = 4 \qquad L(v_3) = 7$$

We could initiate this activity as early as time 3, or as late as time 6, and still "realize" event v_3 by time 7. Naturally, if we "use" some of the available slack by scheduling this activity to start later than time 3, we diminish the latitude available for scheduling subsequent activities, in this case activities (v_3,v_7) and (v_7,v_8). In general, each event v will be realized at a time between $T(v)$ and $L(v)$, depending on how slack is distributed.

So far we have assumed that each activity invariably consumes a fixed amount of time and that this amount is known precisely. If this is not the case (and it never is), one can often reasonably postulate that the

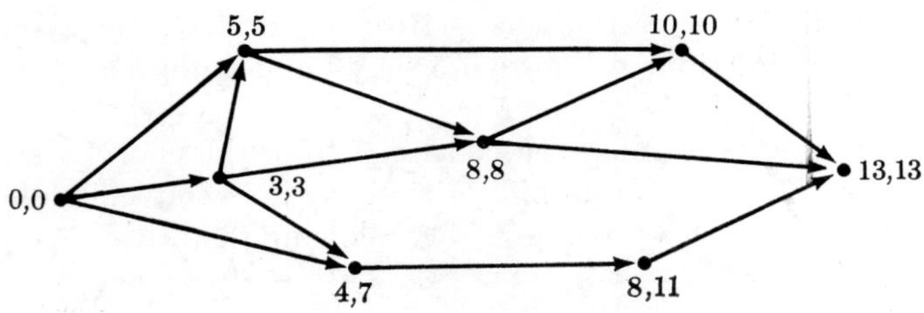

Figure 6-6

duration of each activity can be considered as a random sample from a probability distribution associated with the activity. One then attempts to obtain the best possible estimates of the parameters of this distribution and bases subsequent analyses on them. In the original PERT technique, for example, it was assumed that the activity duration was drawn from a distribution known as a "beta distribution." (In the present context, the precise nature of a beta distribution is immaterial. The interested reader may follow this up by referring to documents describing PERT.) For each activity, one then obtained from appropriate knowledgeable individuals three time estimates: a, m, and b. Here a and b are the estimated shortest and longest times which the activity might reasonably take and m is the time considered most likely. From these the mean time $\bar{t}$ and standard deviation σ are estimated by means of the formulas

$$\bar{t} = \frac{(a + 4m + b)}{6} \qquad \sigma = \frac{(b - a)}{6}$$

With such estimates, one can first of all determine $T(v_i)$, $L(v_i)$, and slacks, based on mean values, in the manner described above. In addition, the variability of the activity durations can be used to estimate such project characteristics as the probability that the project can be completed in a specified total time.

Even if the distribution associated with each activity is considered to be known quite precisely and is assumed to be independent of the distributions of other activities, considerable mathematical complication arises and one must resort to approximation techniques. To illustrate just one complication in a general way, suppose that path P is a critical path and P' is a slightly shorter path from the starting event to the end event, based on mean or expected activity times. It may happen that there is little time variation associated with the activities along P, but a great deal along P'. The distributions associated with P and P' might, for example, be as in Fig. 6-7. In this case there is a substantial probability that P' will in fact be a greater limiting factor than P, so that conclusions based on P as "the" critical path may be rather inaccurate.

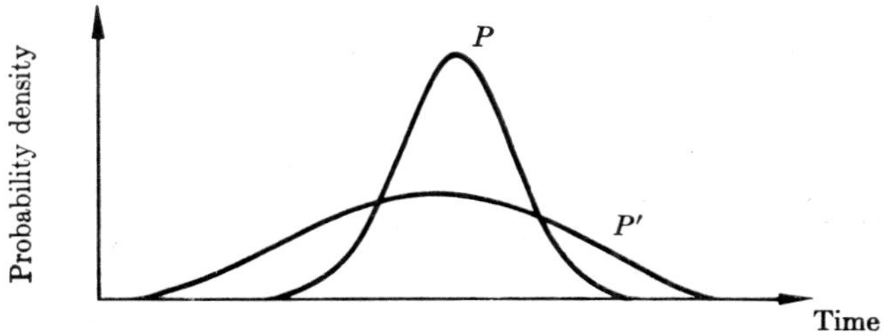

Figure 6-7

In addition to considering project duration, as it is related to activity durations, it is often important to consider additional (or alternative) quantitative characteristics such as the expenditure of manpower and money. Moreover, these characteristics may interact. For example, in some cases one can shorten activity duration by expending more money or allocating additional manpower. A considerable amount of work has been done, and continues to be done, toward solving the variety of problems which result when the objective (i.e., the measure one seeks to optimize) and/or the constraining conditions are varied, or when one's relationship with the project is changed. (For example, if one must irrevocably make all decisions, with respect to allocation of effort to activities, prior to initiation of the project, one approach is required. On the other hand, if one can allocate or reallocate effort during the project, based on performance to date, quite another methodology is appropriate.) Many of these techniques are effectively implemented through the use of digital computers. It has been our objective here to emphasize the basic role played by the activity graph underlying such methodologies. For basic references see papers by Kelley,[53] Kelley and Walker,[54] and Malcolm et al.[61] For an interesting discussion of the underlying assumptions see the paper by MacCrimmon and Ryavec.[60]

Addendum

PERT illustrates that graph theory is a powerful tool for solving a project planning and scheduling problem. From a graph-theoretic point of view PERT is concerned with a time relation defined on a graph. This time relation permits the determination of time schedules, event time distributions, and a longest-time tree (critical path).

The success of PERT stimulated the application of graph theory to other project management problems (Bigelow, 1962). Since original PERT was concerned with a time relation defined on a graph, it is not surprising that other relations such as cost and resource relations were also defined on graphs. Here resources are men, materials, and facilities. In addition to numbers, one can associate with each arc of a graph such functions as time-cost and time-resource trade-off. These functions are interpreted to indicate how the cost or resource usage of an activity varies with activity duration. Defining cost functions for each arc of a project graph permits the computation of cost-time trade-off curves for projects. A number of algorithms have been developed to compute trade-off curves between project duration and project cost. These algorithms can be employed to determine project schedules which minimize project cost for a given project duration.

The project cost curve algorithms of Kelley (1961), Fulkerson (1961), and Grossman and Lerchs (1961) illustrate the use of graph theory in model formulation, computational algorithms, and development of simple proofs. The difficulty of the above three papers is inversely proportional to the amount of graph theory they contain. Kelley's proofs, based on a parametric linear programming method of Gass and Saaty (1955), are forbidding. Fulkerson's method of translating the original parametric linear programming problem into a network flow problem is simpler than Kelley's approach, and the completely graph-theoretic approach of Grossman and Lerchs is nearly intuitively obvious, but nevertheless rigorous. An approach similar to that of Grossman and Lerchs has been applied by Berman (1964) to nonlinear cost functions and by Fey (1964) to multiple project scheduling.

Numerous attempts have been made to deal with resource functions defined on graphs (Lambourn, 1963; Fey, 1964). Here the typical problem is to allocate resources in such a way that schedules will be met and that the amount of resources needed by projects will not exceed the amounts available in any one time period. There remain a number of open problems in this area of resource allocation.

Before leaving PERT, we should point out that its basic assumptions have been questioned (MacCrimmon, 1964). One of the questions arises from the fact that activity durations are sometimes given in terms of a probability distribution and not as real numbers. To face some of the difficulties created by stochastic variables, Fey (1963) and Van Slyke (1963) have implemented a network simulation approach.

REFERENCES FOR ADDENDUM

Berman, E. B.: Resource Allocation in a PERT Network under Continuous Time-Cost Functions, *Management Sci.*, **10:** 734–745 (1964).

Bigelow, C. G.: Bibliography on Project Planning and Control by Network Analysis: 1959–1961, *Operations Res.*, **10:** 728–731 (1962).

Elmaghraby, S. E.: An Algebra for the Analysis of Generalized Activity Networks, *Management Sci.*, **10:** 494–514 (1964).

Fey, C. F.: "Methods of Resource Allocation," I and II, IBM, 1963, 1964.

Fulkerson, D. R.: A Network Flow Computation for Project Cost Curves, *Management Sci.*, **7:** 167–178 (1961).

Gass, S., and T. Saaty: The Computational Algorithm for the Parametric Objective Function, *Naval Res. Logistics Quart.*, **10:** 39–46 (1955).

Goldberg, C. R.: An Algorithm for the Sequential Solution of Schedule Networks, *Operations Res.*, **12:** 499–503 (1964).

Grossman, I. F., and H. Lerchs: An Algorithm for Directed Graphs with Application to the Project Cost Curve and In-process Inventory, paper presented at the Third Annual Conference of the Canadian Operational Research Society, Ottawa, May 4–5, 1961.

Kelley, J. E. Jr.: Critical Path Planning and Scheduling: Mathematical Basis, *Operations Res.*, **9:** 296–320 (1961).

MacCrimmon, K. R., and C. A. Ryavec: An Analytical Study of the PERT Assumptions, *Operations Res.*, **12:** 16–37 (1964).

Van Slyke, R. M.: Monte Carlo Methods and the PERT Problem, *Operations Res.*, **11:** 839–860 (1963).

Wiest, J. D.: Some Properties of Schedules for Large Projects with Limited Resources, *Operations Res.*, **12:** 395–418 (1964).

COMBINATORIAL PROBLEMS

6-5 EXAMPLES OF COMBINATORIAL PROBLEMS IN GRAPH THEORY

In this section we shall briefly allude to computational problems that arise in the subject. Because the methods used can be intricate and outside the scope of the book, we shall be content to deal with one or two interesting problems.

In counting problems one must distinguish between a labeled and an unlabeled or free (topological) graph. Two graphs whose vertices are labeled are the same if and only if for all i and j the same number of edges are incident with the vertices labeled i and j in both graphs. Thus two graphs may be considered distinct even though they may be isomorphic.

One may reverse the process by considering graphs with a given number of unlabeled vertices and a given number of labeled edges. Because labeling distinguishes among graphs in spite of their topological equivalence, it is easier to carry out computations for labeled graphs; for one does not have to determine how many graphs are equivalent and thus reduce the total number. How many graphs have a certain property, e.g., cycles of length 3, is one question that occurs frequently in such investigations.

For example, the number of labeled graphs (which may be disconnected) with n labeled vertices and k unlabeled edges in which a pair of vertices is incident with at most one edge is $\binom{n(n-1)/2}{k}$. Thus we take the edges of the complete graph on n vertices k at a time to obtain the desired number. If we take 4 vertices and 4 edges, there are 15 possible labelings on two topologically different graphs. In fact, Fig. 6-8 gives the number of labelings for topologically unequivalent graphs of four vertices with respectively three, four, five, and six edges.[43]

Many counting problems in graph theory are abstractions of physical problems (especially from statistical mechanics), and their graph-theoretic formulation clarifies the counting process. Some of these concepts are related to special kinds of trees. A graph without articulation points is called a *star*, and hence a connected graph may be regarded as a collection of stars which meet at articulation points. The usual defini-

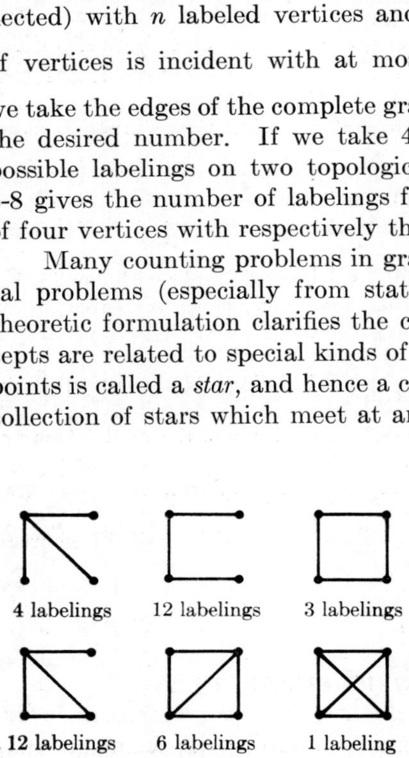

Figure 6-8

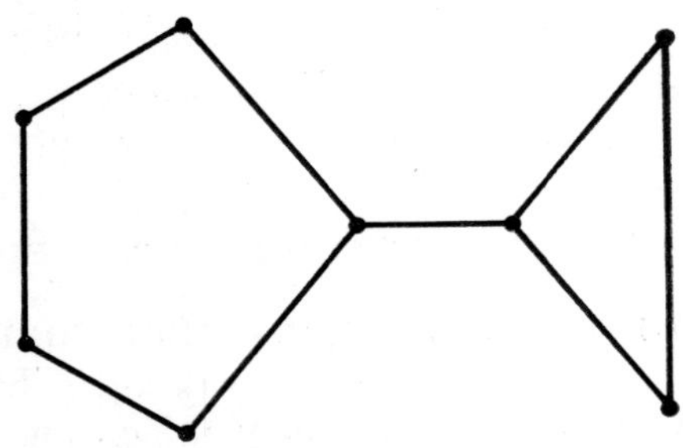

Figure 6-9

tion of a tree indicates that a tree is a graph with articulation points whose constituent stars are single edges. If the constituent stars are polygons, the graph is known as a *Husimi tree.* Figure 6-9 becomes a Husimi tree if the two articulation points are joined by a second chain.

If the constituent stars are more general, the connected graph is known as a *star tree.* If all the stars are the same, it is a *pure* star tree; otherwise, it is called a *mixed* star tree. When the types of stars are not prescribed, we simply have a connected graph. A tree whose edges are designated as either positive or negative is known as a *signed tree.*

Many combinatorial problems relating to graph theory lead to interesting formulas. For example, one can show by elaborate means that the number of graphs with n labeled vertices consisting of k disjoint trees is given by[72]

$$\frac{1}{k!}\sum_{j=0}^{k}\left(-\frac{1}{2}\right)^{j}\binom{k}{j}\binom{n-1}{k+j-1}n^{n-k-j}(k+j)!$$

In the process of the derivation of this quantity it is possible to determine the number of trees of the complete graph with n labeled vertices. This number was first proved by Cayley to be n^{n-2}. We now give an interesting inductive proof of this fact.[72]

Theorem 6-4 The number of trees of the complete graph with n labeled vertices is n^{n-2}.

Proof: To avoid unnecessary manipulations, let us first point out that the following identity is known in analysis:

$$\sum_{j=1}^{n-1}\binom{n}{j}j^{j-1}(n-j)^{n-j-1} = 2n^{n-2}(n-1)$$

The theorem that a complete graph has n^{n-2} trees is clearly true for a graph with a single vertex as $1^{1-2} = 1$. Let us now assume the theorem to be true for the number of trees in a graph with vertices $\leq n - 1$ and prove that it is true for one with n vertices. Let us denote the number of trees of this complete graph by T_n We then divide the n vertices into two sets, one with i elements and the other with $n - i$ elements, where i can assume any of the values $1, 2, \ldots, n - 1$. By the inductive hypothesis the number of trees of the first subgraph is i^{i-2} and that of the second is $(n - i)^{n-i-2}$. Now we examine all ways in which a tree of the first subgraph can be combined with a tree of the second in order to form a tree for the entire graph. Since the connection can be made from any of the i vertices of the first subgraph to any vertex among $n - i$ in the second, the total number is $i(n - i)$. Thus the number of trees of the entire graph obtained from the partition for one value of i is

$$i(n-i)i^{i-2}(n-i)^{n-i-2} = i^{i-1}(n-i)^{n-i-1}$$

However, the i vertices can be chosen from among the n vertices in

$$\frac{n!}{(n-i)!i!} \equiv \binom{n}{i} \text{ ways}$$

and hence if we multiply the above number from a single partition by this quantity and sum over i, we obtain

$$\sum_{i=1}^{n-1} \binom{n}{i} i^{i-1}(n-i)^{n-i-1}$$

There remains the question of duplications. Some trees of the original graph can be obtained in this last sum in more than one way. In fact, since there are $n - 1$ ways to choose the subgraph with i vertices, as the values of i increase toward $n - 1$ those of $n - i$ decrease toward 1, thus interchanging the role of the subgraph with i vertices with that which has $n - i$ vertices, and vice versa. Thus there are $n - 1$ pairs of subgraphs with $(1, n - 1)$, $(2, n - 2)$, . . . , $(i, n - i)$, . . . , $(n - 1, 1)$ vertices, each pair giving rise to two trees of the original graph, since for example, $(1, n - 1)$ can be taken in two ways. Thus the last sum must be divided by $2(n - 1)$ to yield the total number of distinct trees of the original graph. By using our above-cited result from analysis, we obtain n^{n-2} for the number of trees.

Exercise 6-6 Verify the factor of $2(n - 1)$ duplications on the complete graph with four vertices. Note, for example, that the tree (*a*) shown below arises in six ways. Two are accounted for by connecting the subgraphs as indicated in (*b*), where in one case the isolated vertex is the first partition and (*c*) is the second, and where in the other case (*d*) comprises the first partition and the isolated vertex comprises the second. This accounts for two ways. Two more ways are obtained from (*e*). The final two duplications are obtained from (*f*).

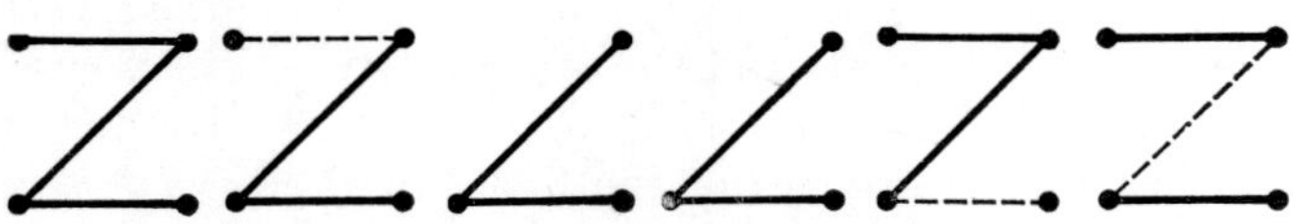

(*a*) (*b*) (*c*) (*d*) (*e*) (*f*)

We now offer a proof due to Trent.[89] Let A be the incidence matrix (less 1 row) of the complete graph on n vertices, with $n - 1$ independent rows. It is known that the number of distinct trees in any graph is given by the determinant of AA', which we denote by $|AA'|$. Here AA' takes the form

$$\begin{pmatrix} n-1 & -1 & \cdots & -1 \\ -1 & n-1 & \cdots & -1 \\ \cdots & \cdots & \cdots & \cdots \\ -1 & -1 & \cdots & n-1 \end{pmatrix}$$

Let T be a second matrix of order $n - 1$ whose elements are given by

$$\begin{aligned} t_{ii} &= 2 & \quad i &< n-1 \\ t_{ij} &= 1 & \quad i &\neq j \\ t_{n-1,n-1} &= 1 \end{aligned}$$

It is easy to show that $|T| = 1$.

Now consider the determinant of the product

$$|T(AA')| = |T|\,|AA'| = \det\begin{pmatrix} n & 0 & \cdots & 0 & 0 \\ 0 & n & \cdots & 0 & 0 \\ \cdots & \cdots & \cdots & \cdots & \cdots \\ 0 & 0 & \cdots & n & 0 \\ 1 & 1 & \cdots & 1 & 1 \end{pmatrix} = n^{n-2}$$

Since $|T| = 1$, it follows that $|AA'| = n^{n-2}$. This completes the proof.

As another illustration let us compute the maximum number of cycles having three arcs that a complete antisymmetric graph of n vertices can have (Ref. 5 of Chap. 1). Consider the vertex matrix associated with this graph. The ith row of the matrix gives the incidence relation for arcs positively incident with the ith vertex, whereas the ith column gives the incidence for arcs negatively incident with this vertex. If r_i denotes the sum of elements in the ith row and c_i denotes the corresponding sum in the ith column, then $r_i + c_i = n - 1$, since the ith vertex is connected by $n - 1$ edges to the remaining $n - 1$ vertices.

The total number of circuits with three edges is $\binom{n}{3}$. However, this is not the number of cycles. A cycle must have all edges cyclically directed. Thus if two arcs are positively incident with a vertex, they cannot be two sides of a cycle, because their orientations are opposite for that cycle.

Since the sum r_i of the elements of the ith row gives the number of arcs directed from the ith vertex, we must exclude from the total number of circuits the quantity $\sum_{i=1}^{n} \binom{r_i}{2}$, that is, the sum of the ith row taken two at a time and summed over all rows. This gives for the number of cycles

$$\binom{n}{3} - \sum_{i=1}^{n} \binom{r_i}{2} = \binom{n}{3} - \frac{1}{2} \sum_{i=1}^{n} (r_i^2 - r_i)$$

Since our graph is complete, the number of its edges is $\binom{n}{2}$ and we must have $\sum_{i=1}^{n} r_i = \binom{n}{2}$, because the total sum of all rows must account for all the edges of the graph.

We now have for the number of cycles

$$\binom{n}{3} + \frac{1}{2}\binom{n}{2} - \frac{1}{2} \sum_{i=1}^{n} r_i^2$$

and the problem is to determine r_i so that this quantity is maximum. The choice of r_i corresponds to a complete graph with special orientation that maximizes the number of cycles. It is sufficient to determine r_i so that $\sum_{i=1}^{n} r_i^2$ is minimum, because this quantity is subtracted from a constant amount in the above expression which is to be maximized. The foregoing argument would also be valid if we used the sum c_i of the column elements and the fact that two arcs having the same orientation toward a vertex with which they are incident cannot be two sides of a cycle. We would then have to find c_i which maximizes

$$\binom{n}{3} + \frac{1}{2}\binom{n}{2} - \frac{1}{2} \sum_{i=1}^{n} c_i^2$$

Thus we would have the maximum number of cycles if c_i which minimizes

$$\sum_{i=1}^{n} c_i^2$$

is found.

Thus the maximum number of cycles is symmetric in r_i and c_i, that is, they must be equal. Since $r_i + c_i = n - 1$, we have, when n is odd, $r_i = (n - 1)/2$.

Exercise 6-7 Substitute this value of r_i and obtain an explicit expression for the maximum number of cycles. Also, find r_i when n is even and complete the computations for this case.

Remark: Another interesting question is to derive a formula for the average number of end points of a tree given at random. Frequently, the computations do not lead to a convenient formula for the result. However, by assuming large values for n, one can obtain an asymptotic formula which is easier for computation. For example, the mean number of end points of a tree taken at random from among the trees whose number was computed above is n/e, where e is the base of the natural logarithm, that is, $e = 2.718 \cdots$. In fact, one can also compute a probability distribution for this mean number.

Formulas have been derived for the number of rooted graphs, i.e., graphs in which a vertex called a root is singled out, and others for computing rooted star trees. Still other results have been concerned with digraphs (directed graphs) and with graphs having strength k; that is, edges of multiplicity up to k can be incident with a pair of vertices.

In a complete graph with n labeled vertices there are $\binom{n}{2}$ edges. The number of graphs with N edges is obtained from

$$\binom{\binom{n}{2}}{N}$$

that is, by taking the $\binom{n}{2}$ edges N at a time. Suppose N edges are chosen at random from among the $\binom{n}{2}$ original ones; what is the probability that the resulting graph is connected? The graph may consist of several components; what is the size of the largest tree, i.e., how many edges does it have? Note that here two points are allowed to be incident with only one edge. However, one may allow multiple (i.e., parallel) edges and ask similar questions.

It can be shown that for large N, the total number of connected graphs is $2^{\binom{n}{2}}$ in the labeled case and $2^{\binom{n}{2}}/n!$ in the unlabeled case.

A birth process or a stationary branching process, also called a multiplicative process, can be represented by a rooted tree, and relevant questions can be studied and answered for the associated tree.[68] Here we have an individual u_0 (corresponding to the root of the tree) who gives birth to w individuals $u_1, u_2, \ldots$, where $w = j$ with probability p_j; u_j in turn gives birth to $u_{j1}, u_{j2}, \ldots$, and so on. The births are mutually independent, and all individuals have the same probability distribution of offspring. The probability P_n that the tree is finite with n vertices has the asymptotic form $P_n \sim A^{-n}n^{-\frac{3}{2}}$ for large n ($n \equiv 1 \bmod q$), where A is a constant and q is the largest integer such that if $p_j \neq 0$, then j is a multiple of q. $P_n = 0$ for other values of n. One can also obtain asymptotic expressions for the probability that $w = j$ and for the distribution of the fraction of vertices with k outgoing segments in trees which are finite and which have n vertices.

Application of Polya's Theorem to Enumeration Problems

Several basic enumeration problems in graph theory can be resolved by appropriate applications of a fundamental combinatorial theorem due to Polya.[71] Among these are the problems of counting the number of nonisomorphic simple graphs having p vertices and q edges, or the number of nonisomorphic simple directed graphs having p vertices and q arcs, and generalizations of these problems when the graphs are not necessarily simple (but when the maximum number of parallel edges or strictly parallel arcs is bounded).

The solution of these and of related counting problems by the use of Polya's theorem is developed fully by Harary.[44] In order to illustrate the basic method in several simple settings, we shall consider the following specific problems:

1. Find the number of nonisomorphic simple graphs having five vertices and q edges, for all q.
2. Find the number of nonisomorphic simple directed graphs having four vertices and q arcs, for all q.
3. Find the number of nonisomorphic graphs having four vertices and q edges, for all q, where at most two edges join any pair of vertices and where loops are prohibited.

Before introducing Polya's theorem and its application to these problems, some preliminaries are necessary. The first of these deals with groups of permutations: A *permutation of degree* k is an operator which, when applied to any ordered set of k elements, rearranges them into a new order. (If each element retains its original position, the permutation is called the *identity permutation.*) Since the nature of the objects being permuted is immaterial, a permutation of degree k can be characterized in terms of the integers 1 to k, which represent the positions of the elements in the ordering. For example, the scheme

Old position:	1	2	3	4	5	6
New position:	3	2	5	6	1	4

characterizes the permutation of degree 6 in which the first element becomes third, the second remains second, the third becomes fifth, etc.

The preceding permutation is also represented, in an obvious way, by the directed graph shown in Fig. 6-10.

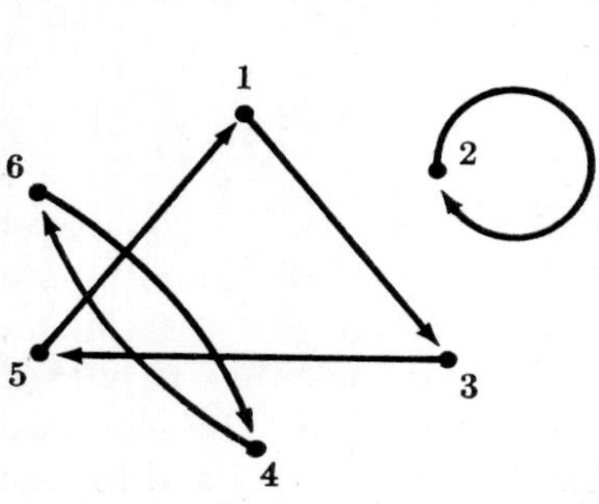

Figure 6-10

More generally, any permutation

of degree k can be represented by an appropriate directed graph whose vertices correspond to the integers 1 to k and which is such that the positive and negative degree of every vertex is 1. We have seen that such a directed graph necessarily decomposes into one or more vertex-disjoint simple cycles (some of which may be loops). In fact, still another useful notation for the above permutation is the so-called *cyclic representation*

$$(1,3,5)(2)(4,6)$$

In general, the cyclic representation is interpreted as follows: The position represented by any integer maps into the position corresponding to the next integer on the right, except that the rightmost position within a grouping maps into the leftmost.

Given a permutation of degree k, the *type* of the permutation is specified by stating how many cycles of length i it contains for $i = 1, 2, \ldots, k$. If n_i denotes the number of cycles of length i, then the permutation's type can be conveniently expressed as the vector $(n_1,n_2, \ldots ,n_k)$. Clearly, the type must satisfy

$$1 \cdot n_1 + 2 \cdot n_2 + \cdots + k \cdot n_k = k$$

(Why?) The specific permutation described above is of type

$$(1,1,1,0,0,0)$$

while the permutation of degree 12 whose cyclic representation is

$$(1,4,2,6)(3)\ (5,7,9,8)(10)\ (11,12)$$

is of type

$$(2,1,0,2,0,0,0,0,0,0,0,0)$$

Another convenient way to represent the type of this permutation is by the expression

$$y_1{}^2 y_2{}^1 y_4{}^2$$

in which subscripts denote cycle lengths and exponents denote corresponding numbers of cycles. (The symbol y has no significance except as a base symbol to which one can attach subscripts and exponents.) Note that if no cycles of length i occur, the factor $y_i{}^0$ may be omitted.

Given a set P consisting of k permutations of degree h, let

$$h_{j_1,j_2, \ldots ,j_k}$$

denote the number of permutations of type $(j_1,j_2, \ldots ,j_k)$. Then the formal series

$$Z(P) = \frac{1}{h} \sum h_{j_1,j_2, \ldots ,j_k} y_1{}^{j_1} y_2{}^{j_2} \cdots y_k{}^{j_k}$$

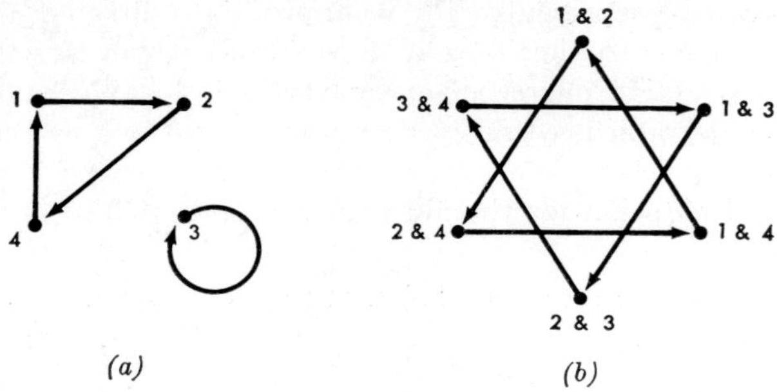

Figure 6-11

where the summation is over all meaningful types, is called the *cycle index* of P.

We shall be concerned with sets P of permutations (of the same degree) which form a group relative to the binary operation of applying two permutations successively. Thus P must contain the identity permutation, the inverse (or reverse) of each of its permutations, and the "product" (i.e., resultant) of any two of its permutations.

The permutations of interest in the present discussion are the permutations of the set of all unordered pairs (or, in the case of a directed graph, ordered pairs) of vertices of a graph which result when the vertices themselves are permuted. For example, if the vertices of a four-vertex graph are permuted as shown in Fig. 6-11*a*, then the unordered pairs of vertices are permuted as shown in Fig. 6-11*b*. In this particular example, note that a permutation of the four vertices whose type is $y_1{}^1y_3{}^1$ induces a permutation of the six unordered pairs of vertices whose type is $y_3{}^2$. Similarly, each of the $n!$ possible permutations of n vertices induces some definite permutation of the $n(n-1)/2$ unordered pairs of vertices [or of the $n(n-1)$ ordered pairs if we wish to study directed graphs].

In what follows, we shall need to know the number of permutations of each type induced by all possible permutations of four vertices, for both unordered and ordered pairs, and by all possible permutations of five vertices, for the unordered case only. A discussion of how this information may be obtained in general is contained in Ref. 44. For the cases of interest, this information is summarized by the following cycle indices:

Four vertices, unordered pairs:

$$\tfrac{1}{24}(y_1{}^6 + 9y_1{}^2y_2{}^2 + 8y_3{}^2 + 6y_2{}^1y_4{}^1)$$

Four vertices, ordered pairs:

$$\tfrac{1}{24}(y_1^{12} + 6y_1^2y_2^5 + 3y_2^6 + 8y_3^4 + 6y_4^3)$$

Five vertices, unordered pairs:

$$\tfrac{1}{120}(y_1^{10} + 10y_1^4y_2^3 + 20y_1^1y_3^3 + 30y_4^2y_2^1 + 15y_2^4y_1^2 + 20y_1^1y_3^1y_6^1 + 24y_5^2)$$

As a further preliminary to describing Polya's theorem, we introduce several auxiliary notions. Consider a set of abstract objects called *figures* and assume that one of several nonnegative integers (we shall use only 0, 1, and 2) is associated with a figure and is referred to as its *content*. (In a more general setting, Polya's theorem permits one to associate a vector of integers with a figure.) If a_k denotes the number of different figures having content k, then the formal series

$$A(x) = \sum_{k=0}^{\infty} a_k x^k$$

is called the *figure-counting series*. (Here x is a dummy variable.)

A *configuration of length s* is a sequence or ordered set of s figures. The content of a configuration is taken to be simply the sum of the contents of its figures. Certain configurations of length s are regarded as being equivalent to one another. Specifically, let P be a group of permutations of degree s, and let h denote the number of permutations in the group. Then two configurations are said to be *equivalent configurations relative to P* if and only if one is transformed into the other by an appropriate permutation in P.

If b_k denotes the number of inequivalent configurations (of length s) having content k, then the formal series

$$B(x) = \sum_{k=0}^{\infty} b_k x^k$$

is called the *configuration-counting series* (relative to P).

Polya's theorem enables one to determine $B(x)$ from the figure-counting series $A(x)$ and the cycle index $Z(P)$ of the permutation group P. Specifically, we have (without proof):

Theorem 6-5 (Polya.) If $A(x)$ and $Z(P)$ denote the figure-counting series and the cycle index of P, respectively, then the configuration-counting series $B(x)$ may be found by substituting $A(x^k)$ for each occurrence of y_k in the cycle index of P.

Now consider once again the problem of counting all nonisomorphic simple graphs having five vertices. Take as figures the 10 unordered pairs of distinct vertices. Adopt the convention that a figure has con-

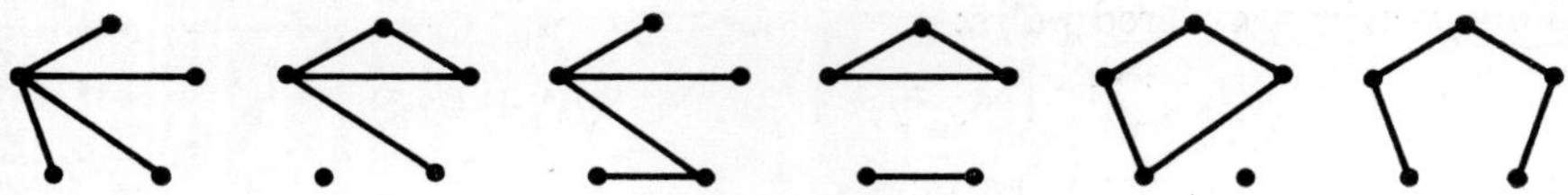

Figure 6-12

tent 1 or 0 depending on whether an edge joins the vertices or not. Thus the figure-counting series assumes the simple form

$$A(x) = 1 + x$$

Next, consider the configurations of length 10 corresponding to sequences formed from the 10 figures. The appropriate permutation group P consists of the set of permutations of the 10 figures (that is, unordered pairs of distinct vertices) induced by the group P^* of all possible permutations of the 5 vertices. (Note that there are 5! such permutations.)

When $1 + x^k$ is substituted for each occurrence of y_k in the cycle index $Z(P)$ given earlier and the resulting expression is simplified, we obtain

$$B(x) = 1 + x + 2x^2 + 4x^3 + 6x^4 + 6x^5 + 6x^6 + 4x^7 + 2x^8 + x^9 + x^{10}$$

We may conclude, for example, that there are 6 distinct graphs having 4 edges, because of the presence of the term $6x^4$. These are exhibited in Fig. 6-12.

To find the number of distinct simple directed graphs having four vertices, we interpret figures as ordered pairs of vertices. The same figure-counting series $A(x) = 1 + x$ is appropriate, since an ordered pair of vertices either is or is not joined by an arc. The cycle index $Z(P)$ for the group P of permutations of the 12 ordered pairs of vertices, induced by all possible permutations of the vertices, was noted earlier. By substituting $A(x^k)$ for y_k in $Z(P)$ and simplifying, we obtain

$$B(x) = 1 + x + 5x^2 + 13x^3 + 26x^4 + 38x^5 + 48x^6 + 38x^7 + 26x^8 + 13x^9 + 5x^{10} + x^{11} + x^{12}$$

From the presence of the term $5x^2$, for example, we conclude that there are five distinct graphs having two arcs. These are shown in Fig. 6-13.

We turn now to the problem of counting the number of distinct graphs having four vertices, no loops, and at most two edges joining

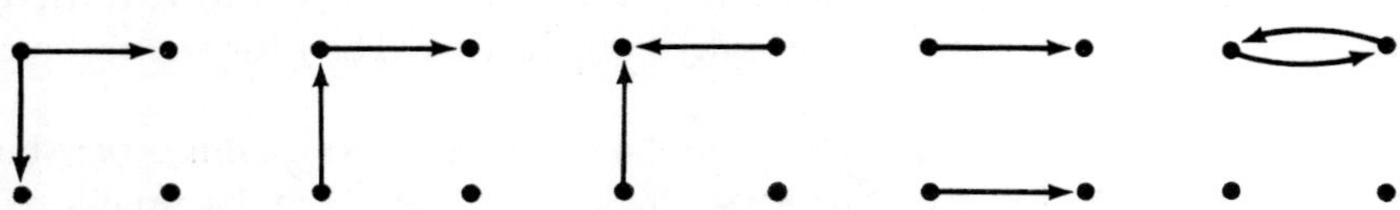

Figure 6-13

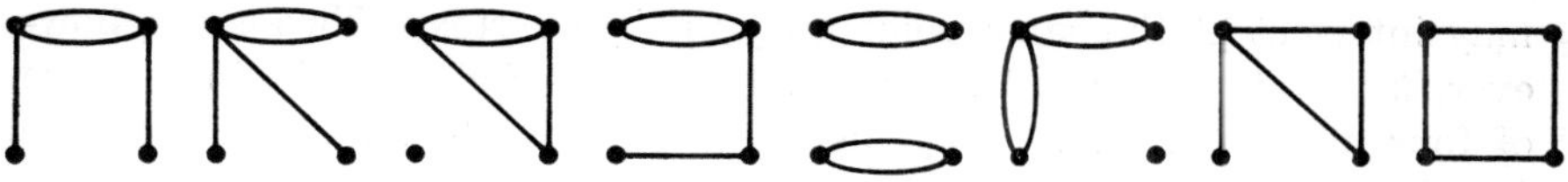

Figure 6-14

any pair of vertices. In this case, a figure is once again interpreted as an unordered pair of vertices. However, the content may now be 0, 1, or 2, depending on the number of edges joining the vertices. Thus the figure-counting series is

$$A(x) = 1 + x + x^2$$

When

$$A(x^k) = 1 + x^k + x^{2k}$$

is substituted for each occurrence of y_k in the appropriate cycle index (given earlier), we obtain

$$B(x) = 1 + x + 3x^2 + 5x^3 + 8x^4 + 9x^5 + 12x^6 + 9x^7 + 8x^8 + 5x^9 + 3x^{10} + x^{11} + x^{12}$$

Thus there are, for example, eight distinct graphs, of the type being considered, having four edges. These are exhibited in Fig. 6-14.

By modifying the definitions of figures, contents, and P, other graph-theoretic enumeration problems can be resolved. See Ref. 44 for a discussion of these variations. Further discussion and a proof of Polya's theorem, in a more general form, may be found, for example, in Ref. 24 of Chap. 1. An interesting discussion of some solved and unsolved enumeration problems may be found in Ref. 43.

6-6 MINIMUM BRICK FACTORY CATASTROPHES[93]

In a brick factory there are m ovens in which bricks are baked. The bricks are then loaded on a small special-rail car at each oven and pushed to any one of n platforms where a loading truck may be available. Since each oven must be connected by rail to each loading platform, the rail lines have a great number of intersections. As the cars go over the intersections they are often derailed, causing spillage of bricks and a traffic jam within the factory. The problem is to construct the rails from the ovens to the destinations with a minimum number of crossings and thus minimize the hazards of derailment.

This problem may be solved within the framework of graph theory, where the rail lines correspond to edges of a graph connecting vertices (corresponding to the ovens) to other vertices (corresponding to the loading platforms). One condition is imposed, and that is that no three edges may intersect at the same point unless it is a vertex. Two edges, however, may intersect at an intermediate point. For example, in the case of four ovens $O_1, \ldots, O_4$ and four platforms $P_1, \ldots, P_4$ we have four intersections marked with x in Fig. 6-15.

Theorem 6-6 The minimum number of interior intersections of edges joining every one of m points to every one of n points in the plane (two edges are assumed to intersect at no more than a single point) is not less than:

$$\begin{array}{ll} (r^2 - r)(s^2 - s) & \text{if } m = 2r,\ n = 2s \\ (r^2 - r)s^2 & \text{if } m = 2r,\ n = 2s + 1 \\ r^2(s^2 - s) & \text{if } m = 2r + 1,\ n = 2s \\ r^2s^2 & \text{if } m = 2r + 1,\ n = 2s + 1 \end{array}$$

Before proving the theorem we shall introduce the notion of a fan and prove a lemma.

Definition A fan at a vertex v consists of all edges incident with v minus their end points. Thus v is also excluded from the fan.

Remark: Note that if we are given two sets of three vertices in the plane and we form the three fans at the vertices of one set, each of which fans has all three vertices of the other set as end points, we obtain the utility graph of Kuratowski, which is nonplanar, and hence the fans must have at least one point of intersection which is not one of the vertices.

Lemma 6-7 Consider a planar graph consisting of three fans at the vertices u_1, u_2, u_3 all having the same end points $v_1, \ldots, v_m$ together with the vertices themselves. The number of interior-point intersections, assuming that no more than two edges intersect at any point, is at least

$$\begin{array}{ll} r^2 - r & \text{if } m = 2r \\ r^2 & \text{if } m = 2r + 1 \end{array}$$

Proof: The proof is inductive. The lemma is true if $r = 1$ by the above remark. Let us assume the lemma to be true for r and show that it is true for $r + 1$, in which case the number of intersections must be at least

$$\begin{array}{ll} r^2 + r & \text{if } m = 2r + 2 \\ (r + 1)^2 & \text{if } m = 2r + 3 \end{array}$$

Now consider the subgraphs C_k $(k = 1, \ldots, m)$ consisting of the

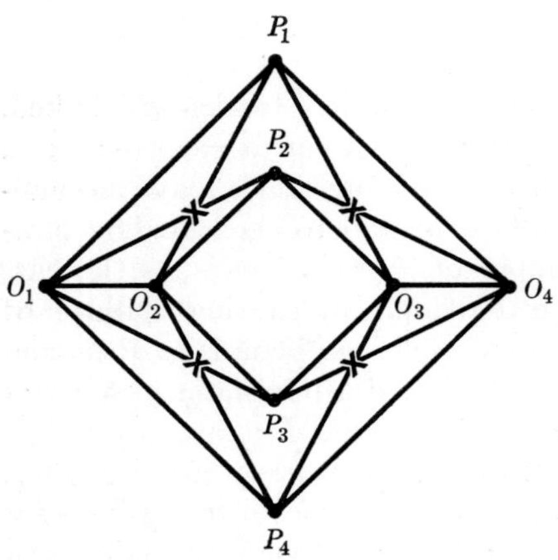

Figure 6-15

vertices v_k $(k = 1, \ldots, m)$, the fans at v_k $(k = 1, \ldots, m)$ determined by the vertices u_1, u_2, u_3, and the vertices u_1, u_2, u_3. If every pair of these subgraphs had a point in common other than the end points of the fans, then the number of intersections would be obtained by considering these subgraphs two at a time. This would give for the intersection number

$$\binom{m}{2} = \frac{m(m-1)}{2} \equiv a$$

But if $m = 2r + 2$, then $a = 2r^2 + 3r + 1 > r^2 + r$; and if $m = 2r + 3$, then $a = 2r^2 + 5r + 3 > (r + 1)^2$. In that case the lemma would be proved. Suppose, then, that some pair of subgraphs C_{k_1} and C_{k_2} have no point in common except u_1, u_2, u_3. Consider the subgraph C' consisting of C_{k_1} and C_{k_2} taken together. By the remark, every other subgraph must have at least one interior intersection point with C'. Since there are $m - 2$ remaining subgraphs, there are at least $m - 2$ intersections with C' and no two intersections are the same, since no three lines can intersect at a nonterminal point. We add to this number what we obtain by applying the induction hypothesis to the minimum possible number of intersections on our graph without v_{k_1} and v_{k_2} and the fans C_{k_1} and C_{k_2}. We have

$$\begin{aligned} (2r) + (r^2 - r) &= r^2 + r &\qquad \text{if } m - 2 = 2r \\ (2r + 1) + r^2 &= (r + 1)^2 &\qquad \text{if } m - 2 = 2r + 1 \end{aligned}$$

Note, for example, that if the number of vertices is $2r$, then there are at least $r^2 - r$ intersections, which we add to the $2r$ intersections with C'. This completes the proof of the lemma.

Proof of the theorem: Again the proof is inductive. The theorem is true for $r = 1, s = 1$ by the remark. We prove that if the theorem is true as stated for m and n, then it is also true for the combinations $(m, n + 1)$, $(m + 1, n)$, and $(m + 1, n + 1)$. Suppose that we have m ovens $O_1, \ldots, O_m$ and $n + 1$ platforms $P_1, \ldots, P_{n+1}$. The graph of this interconnection may be obtained from the graph G of the problem with $O_1, \ldots, O_m$ and $P_1, \ldots, P_n$ by adding the vertex P_{n+1} and its connecting edges to $O_1, \ldots, O_m$. Now G itself may be considered in terms of fans at $P_1, \ldots, P_n$, all with the same end points $O_1, \ldots, O_m$. Let n be even, that is, $n = 2s$; then we shall consider the least number of intersections of the fan at P_{n+1} with the fans at P_1 and at P_2 taken together, with the fans at P_3 and P_4 taken together, and so on. We have s such pairs, and by the lemma if $m = 2r$, there are at least $r^2 - r$ intersections with each pair and hence the total would be at least $(r^2 - r)s$; if $m = 2r + 1$, the number of intersections would be at least r^2s. If n is odd, that is, $n = 2s + 1$, the single remaining fan at P_n can be ignored and the same numbers $(r^2 - r)s$ and r^2s are obtained. Now by the induction hypothesis, G itself has at least $(r^2 - r)(s^2 - s)$ intersections which we now add to the above result, yielding for the smallest possible number of intersections

$$(r^2 - r)(s^2 - s) + (r^2 - r)s = (r^2 - r)s^2 \qquad \text{if } m = 2r,\ n = 2s + 1$$

from which we have

$$(r^2 - r)s^2 + (r^2 - r)s = (r^2 - r)[(s + 1)^2 - (s + 1)] \qquad \text{if } m = 2r,\ n = 2s + 2$$

Also, we conclude that the least number of intersections is

$$r^2(s^2 - s) + r^2s = r^2s^2 \qquad \text{if } m = 2r + 1,\ n = 2s + 1$$

from which we have

$$r^2s^2 + r^2s = r^2[(s + 1)^2 - (s + 1)] \qquad \text{if } m = 2r + 1,\ n = 2s + 2$$

thus proving the theorem.

The construction of the connections with the least number of intersections may be obtained as follows: Consider rectangular coordinates in the plane. If $m = 2r$, take on the x axis the points with abscissas

$$-r, -(r - 1), \ldots, -2, -1, 1, 2, \ldots, r$$

and if $m = 2r + 1$, take on this axis the points with abscissas

$$-r, -(r - 1), \ldots, -2, -1, 1, 2, \ldots, r, (r + 1)$$

If $n = 2s$, take on the y axis the points with ordinates

$$-s, -(s - 1), \ldots, -2, -1, 1, 2, \ldots, s$$

and for $n = 2s + 1$ take the points with ordinates

$$-s, -(s - 1), \ldots, -2, -1, 1, 2, \ldots, s, (s + 1)$$

and then join with a straight line segment all the points from the x axis to all the points of the y axis. In this case one can easily count all the intersections.

Remark: One can also prove by induction that the smallest number of regions in the plane obtained by construction of the foregoing connection graph is

$$(r^2 - r)(s^2 - s) + 4rs - 2(r + s) + 2 \qquad \text{if } m = 2r,\ n = 2s$$

$$(r^2 - r)s^2 + 4rs + 2s + 1 \qquad \text{if } m = 2r,\ n = 2s + 1$$

$$r^2s^2 + 4rs + 1 \qquad \text{if } m = 2r + 1,\ n = 2s + 1$$

Exercise 6-8 Construct the connections for the five-oven four-platform problem according to the foregoing procedure.

Exercise 6-9 Repeat Exercise 6-8 for the five-oven and six-platform problem.

6-7 THE MINIMUM NUMBER OF INTERSECTIONS IN COMPLETE GRAPHS

Zarankiewicz's[93] result, discussed above, concerns the minimum number of edge intersections, when drawn in a plane, of the bipartite graph consisting of two sets of vertices with one edge joining each element of one set to each element of the other set. When each set has three vertices, we have one of the two basic nonplanar graphs appearing in Kuratowski's theorem characterizing planar graphs. By generalizing on this graph, Zarankiewicz was able to establish his result and also to indicate a scheme for its realization.

One can pursue[35,45,80] a similar investigation of the n-vertex generalization of the other basic Kuratowski graph: the (simple) complete graph on five vertices. The main results of the investigation[80] will now be described. Let G_n denote the complete graph on n vertices. We wish to determine I_n, the minimum number of edge intersections when G_n is drawn in a plane in such a way that at most two edges intersect at any point other than a point representing one of the original n vertices.

We need only consider $n \geq 5$, since clearly $I_n = 0$ for $n < 5$. In order to obtain an upper bound M_n on I_n, consider the following specific representation of G_n in a plane, called the *alternating linear model*. Choose a horizontal line segment S in the plane, and subdivide S into $n - 1$ segments by introducing points $p_1, \ldots, p_n$ corresponding (from left to right) to the vertices of G_n. Join p_1 to $p_3, p_4, \ldots, p_n$ by means of semicircles lying above S. Then join p_2 to $p_4, p_5, \ldots, p_n$ by means of semicircles lying below S. In general, join p_i to $p_{i+2}, p_{i+3}, \ldots, p_n$ for $i = 1, 2, \ldots, n - 2$, using semicircles lying above (below) S if i is odd (even).

The construction for $n = 6$ and $n = 7$ is shown in Fig. 6-16. Note

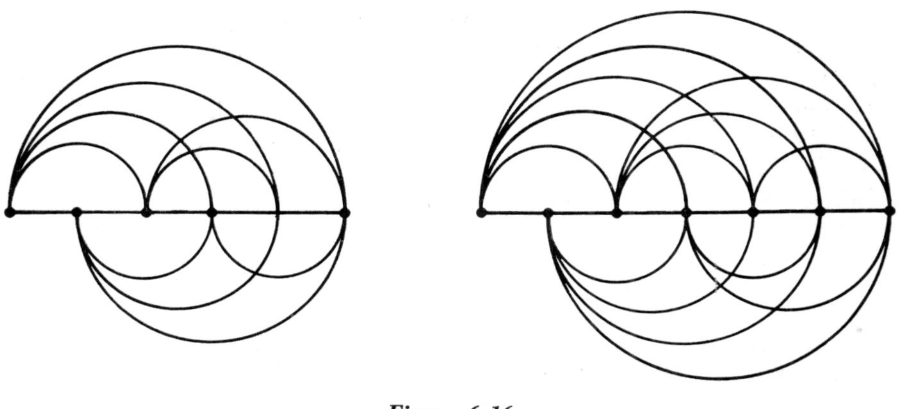

Figure 6-16

that the number of edge intersections is 4 for G_6 and 11 for G_7. In general, if M_n denotes the number of edge intersections in the alternating linear model of G_n, it can be shown (do as an exercise) that

$$M_n = \begin{cases} \dfrac{n(n-2)^2(n-4)}{48} & n \text{ even} \\ \dfrac{(n-1)(n-3)(n^2-4n+1)}{48} & n \text{ odd} \end{cases}$$

Thus M_n as given above is an upper bound on I_n. It is further shown in Ref. 80 that a somewhat smaller upper bound M'_n on I_n is given by the following formulas:

$$M'_n = \begin{cases} \dfrac{n(n-2)^2(n-4)}{64} & n \text{ even} \\ \dfrac{(n-1)^2(n-3)^2}{64} & n \text{ odd} \end{cases}$$

Note that for n even, $M'_n = \frac{3}{4}M_n$. For n odd, M'_n agrees with $\frac{3}{4}M_n$ with respect to the coefficients of n^4 and n^3.

The foregoing result can also be obtained, for example, in the even case by distinguishing $r = n/2$ pairs and forming all G_4's between the pairs. Each such G_4 must be drawn in such a way that the six edges do not intersect one another. Then it can be shown that there are at least $\binom{r}{2}\binom{r-1}{2}$ intersections. On putting $r = n/2$, one has the result.

If there exists a minimal-intersection representation of G_n which has the property that it contains a minimal-intersection representation of

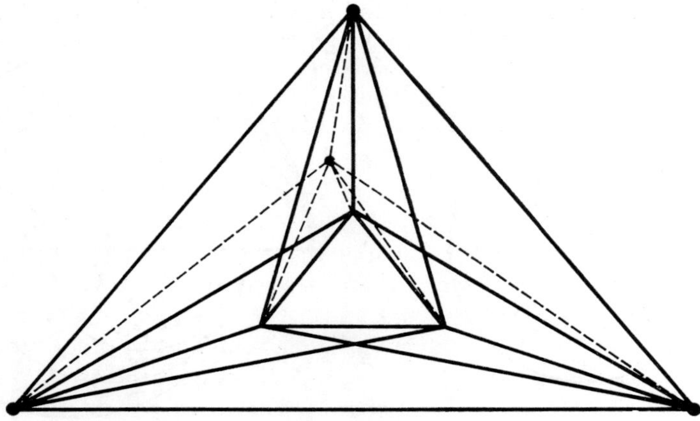

Figure 6-17

G_{n-k} for each even $k < n(k = 2, 4, 6$ are sufficient), it can be shown inductively that M'_n, given above, is precisely I_n.

The values of M'_6 and M'_7 are 3 and 9, respectively. The solid lines in Fig. 6-17 constitute a representation of G_6 which realizes M'_6. By adding the dashed lines, we obtain a representation of G_7 which realizes M'_7. It would be interesting and useful to attempt to find a specific straight-line construction for G_n for all $n \geq 5$.

Exercise 6-10 Draw graphs comparable to Fig. 6-17 for $n = 8, 9$.

Exercise 6-11 To construct a graph with minimum intersections for the case $n = 10$, use two concentric polygons with an equal number of vertices. Join the vertices of the exterior polygon symmetrically on each side in the outer region, those of the interior polygon by straight lines in its interior, and those of the two polygons symmetrically in the enclosed region. Generalize this method.

PUZZLES AND GAMES

6-8 A PROBLEM INVOLVING COLORED CUBES[7]

In most problems considered so far the original formulation has either been directly in terms of a given graph or else has been such that an immediate reformulation in terms of an obvious appropriate graph was possible. In some situations, however, the real problem is that of discovering an appropriate graph, whose structure may bear little superficial resemblance to the original problem. The following example is a simple case in point.

Let C_1 to C_4 denote four congruent cubes and let Y, R, B, and G denote the colors yellow, red, blue, and green, respectively. Suppose that each face of each cube is colored with one of these colors in such a manner that every color is represented on every cube. (Aside from this constraint, the colors are assigned independently to each cube's faces.) Consider the following problem: For a given assignment of colors find (if possible) a way to stack the cubes in a single pile (forming a prism with square base) so that the four squares on each lateral side of the prism have distinct colors.

Remark: Without further constraints, the problem has no solution in some cases. For example, suppose that on each cube all three faces meeting at a corner are red. Then no matter how the cubes are oriented and stacked, the lateral sides of the prism will include eight red squares. A solution requires precisely four.

For any coloring of the cubes, consider the following associated graph having 4 vertices and 12 edges. The vertices correspond to the colors Y, R, B, and G. For each cube C_i there are three edges labeled i. These correspond to the three pairs of opposite faces and join the appropriate

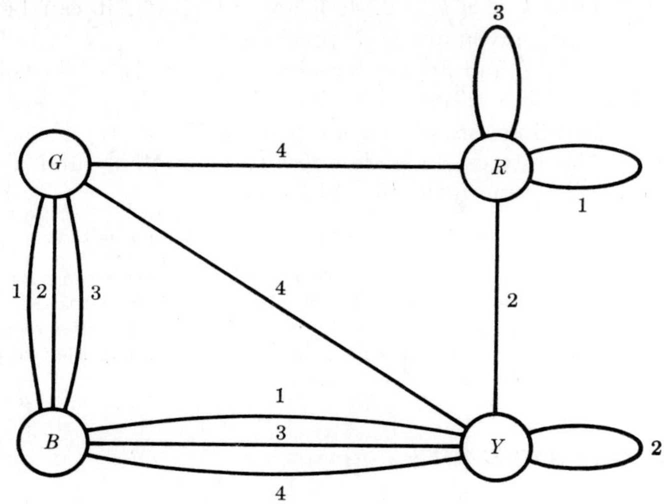

Figure 6-18

vertices (colors). Figure 6-18 illustrates a valid coloring. (Note that a loop results if opposite faces have the same color.) In general, such a graph represents a permissible coloring if every vertex is incident with at least one edge having each of the labels 1, 2, 3, and 4.

Now suppose that a solution to the problem exists. Consider two opposite lateral sides of the resulting prism. The eight corresponding squares represent one pair of opposite sides of each cube, and each color occurs twice. In terms of the graph, there exists a subgraph having four edges, each labeled differently, and such that each vertex is of degree 2. The other pair of opposite lateral sides of the prism determine a second subgraph which has the same properties and is edge-disjoint from the first.

Conversely, it can be shown that if the associated graph has two edge-disjoint subgraphs with these properties, then there exists a corresponding solution to the problem. Figure 6-19 shows two appropriate subgraphs of the graph of Fig. 6-18.

Exercise 6-12 Establish a frame of reference by looking at the prism from a fixed side, referring to the front, right side, and top of each square as x, y, and z, respectively, and referring to the corresponding opposite sides as x', y', and z'. Use the subgraphs in Fig. 6-19 to determine the appropriate color of each side of each cube to solve the problem.

Remark: Note that there are 41,472 possible arrangements of the cubes. The cube on the bottom has three possible arrangements (three essentially different ways of putting it on the table). The remaining

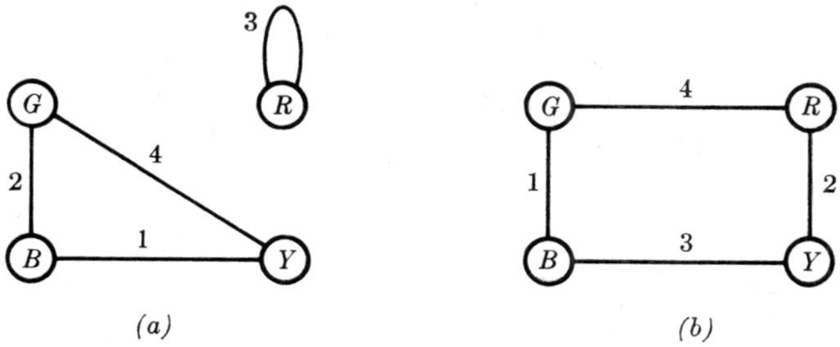

Figure 6-19

cubes each have 24 possible orientations: 6 to choose the side on which it rests and then 4 possible rotations.

6-9 SYSTEM TRANSITION PROBLEMS

Many problems, when formulated abstractly, are of the following general type: A "system" of some sort that at any time can be in only one of a finite number of "states" is given. The set of possible direct (i.e., one-step) transitions is known—either by explicit enumeration or by a suitable general rule. Given a specified initial state, the problem is to determine whether or not one can transform the system into a desired final state by means of an appropriate sequence of direct transitions. (If a cost is assigned to each direct transition, it may further be required to attain the desired state with minimal cost.)

If the states and direct transitions are represented respectively by the vertices and arcs of an appropriate simple directed graph, the problem is one of finding a path joining a specified pair of vertices (states). In many cases, the critical step in the analysis of such problems is the system definition and, more specifically, the selection of a collection of states which adequately characterize the possibilities and in terms of which the direct transitions can be conveniently enumerated.

While various applications of a more serious nature will occur to the reader, for illustrative purposes we shall here consider the puzzle of the missionaries and the cannibals.[83] Three missionaries and three cannibals arrive at bank A of a river and must cross to bank B by using a rowboat which will accommodate only two people. All of the missionaries and one of the cannibals can row. Is it possible to achieve this transfer through a sequence of crossings such that the cannibals never outnumber the missionaries on either bank unless, of course, the number of missionaries is zero? (The missionaries feel rather strongly about this ground rule.)

In attacking this problem, we shall take as the system the collection of cannibals and missionaries on bank A. Letting M, C, and K denote a

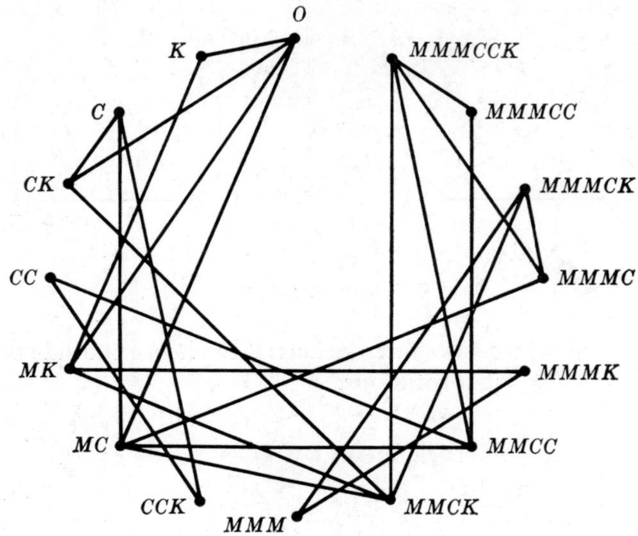

Figure 6-20

missionary, a nonrowing cannibal, and the rowing cannibal, respectively, the system can take on 24 possible states (since M, C, and K can assume 4, 3, and 2 distinct values, respectively). Of these, the following 16 are feasible:

$MMMCCK$	$MMMK$	CCK	CK
$MMMCC$	$MMCC$	MC	C
$MMMCK$	$MMCK$	MK	K
$MMMC$	MMM	CC	O

Here O means that no one is on bank A. (The reader may verify that the other eight states violate the ground rule on one bank or the other.)

Transitions to this system occur through the departure or return of the boat. The graph of Fig. 6-20 enumerates all (25) possible transitions. For simplicity, transitions are shown as edges, since either direction is possible, depending on whether the departure or arrival of the boat is the cause. However, each edge should be regarded as representing two oppositely oriented arcs corresponding to departure (moving clockwise around the circle) and arrival (counterclockwise). The problem, restated, then becomes: Find (if possible) a path from $MMMCCK$ to O whose arcs correspond alternately to departures and arrivals.

Without the last stipulation, the problem is easy. (For example, the sequence of states $MMMCCK$, $MMCK$, CK, O would do.) With it, the problem becomes considerably more difficult. Before proceeding, the reader may wish to spend a few moments with Fig. 6-20, either to find a solution or to become convinced that there is none.

In order to seek a solution systematically, we shall now introduce an auxiliary graph having the same vertices. The objective is to reflect round trips which start from, and return to, bank A, rather than one-way trips. We shall thereby eliminate the need for considering alternation of crossings by building it into the structure of the graph.

Suppose, for example, that the boat is at bank A and the system is in state $MMMC$. From Fig. 6-20 we see that we can go, via state MC, to state $MMCC$ or $MMCK$, so we construct round-trip edges joining $MMMC$ with $MMCC$ and with $MMCK$. Figure 6-21 shows the totality of all such edges (including one-way trips terminating at O). In terms of the graph of Fig. 6-21, the problem becomes: find a chain from $MMMCCK$ to O. We can now readily see that such a chain exists. For example,

$$MMMCCK,\ MMMCK,\ MMMK,\ MMCK,\ MMCC,\ CCK,\ MC,\ O$$

is such a chain. Adding (parenthetically) the intermediate states, the corresponding solution relative to the original graph of Fig. 6-20 is

$$MMMCCK,\ (MMCK),\ MMMCK,\ (MMM),\ MMMK,\ (MK),\ MMCK,\ (MC),\ MMCC,\ (CC),\ CCK,\ (C),\ MC,\ O$$

which is shown in Fig. 6-22. Note that the path is not simple, since two

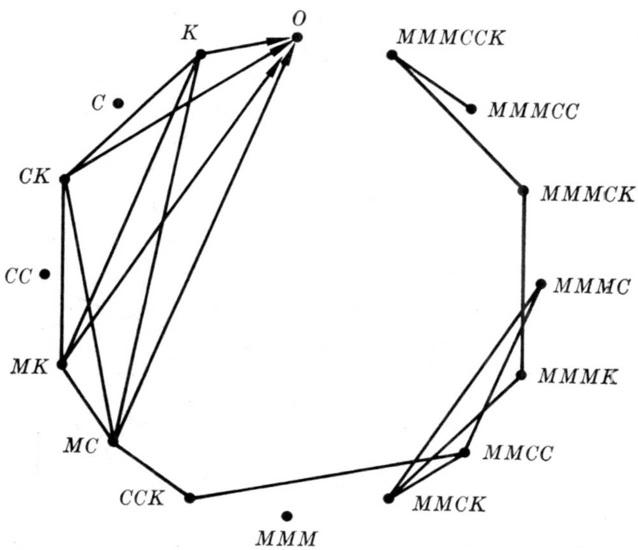

Figure 6-21

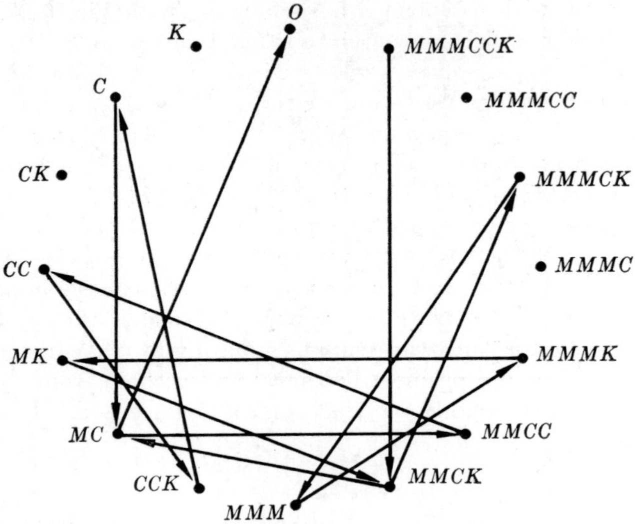

Figure 6-22

arcs enter *MMCK* and *MC*. (In each case, one of the arcs corresponds to a boat departure from bank *A* and the other to a boat arrival.)

Exercise 6-13 Determine whether the problem of the missionaries and the cannibals can be solved in fewer crossings under the less stringent assumption that all cannibals can row. (Note that the more stringent assumption that only one missionary can row makes the problem more interesting.)

Exercise 6-14 Given the directed graph shown in Fig. 6-23, find a path from v to w whose arcs alternate between solid and dashed arcs, the first being solid.

(*a*) Solve the problem by inspection.

(*b*) Solve the problem by applying the method of the text, by constructing an appropriate auxiliary graph each of whose arcs corresponds to a pair of alternating arcs (or a single solid arc terminating at w) in the original graph.

(*c*) Note that the solution is not a simple path in the original graph. (In fact, the path returns to v.)

In some cases, individual feasible transitions are obvious, but it may

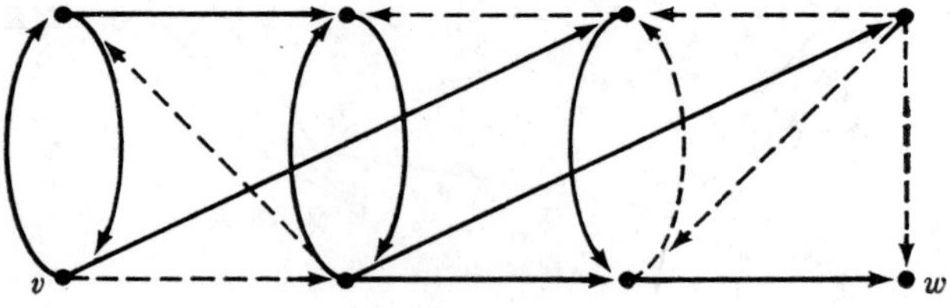

Figure 6-23

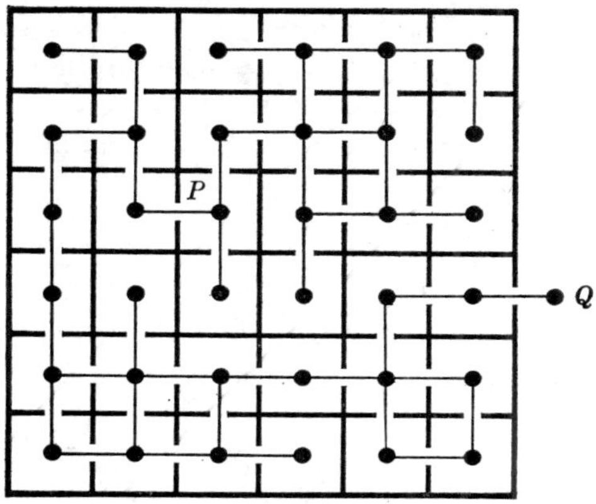

Figure 6-24

not be clear in a particular problem whether one can get from a given initial state to a desired final state. An example of this is the problem of finding one's way through a complex maze or labyrinth, which occurs very frequently in the literature of mathematical recreations. It is, in essence, the problem of finding an edge progression joining specified vertices of an appropriate graph which characterizes the structure of the maze.

Consider, for example, the two-dimensional maze of Fig. 6-24. This maze consists of 36 compartments some of which are joined by "doorways" (indicated by breaks in the lines). Suppose that the problem is to reach point Q, outside the maze, starting from compartment P. Consider the graph whose vertices correspond to the compartments and whose edges indicate those pairs of adjoining compartments which are connected by a passageway. Figure 6-24 shows the associated graph. In terms of this graph, the problem is to determine an edge progression joining P and Q. In principle, the problem as given is very simple. Having constructed the associated graph, we can apply the labeling methods of Chap. 3 to find a tree joining P with all other vertices in the same component. In particular, we shall have found a chain joining P and Q (if the problem has a solution).

We have been overlooking one serious difficulty, however. A straightforward labeling technique requires systematic scanning of the edges and vertices and assumes that we actually know the structure of the entire maze. In practice, if we are *in* the maze, say, at point P, we have at the outset only very local information, namely, we know only which compartments are directly reachable from P. Additional information is learned only gradually, by probing in various directions.

All the numerous existing techniques for finding one's way out of such a maze involve systematization of this process of probing with the objective of avoiding unnecessary repetition of the same route. (Some repetition is unavoidable. The existence of a blind alley cannot be anticipated; it can only be discovered. When a blind alley is found, retracing part of one's route is necessary.) However, by marking the vertices and edges as they are encountered and probed, it is possible to formulate a procedure whereby no edge is traversed twice in the same direction, regardless of the structure of the maze. (It should be noted that the maze may be three-dimensional, in which case the associated graph may not be planar.)

6-10 MATRIX FORMULATION AND SOLUTION OF THE CROSSING PROBLEM[80]

We now illustrate an application of the vertex matrix to the solution of the missionary-cannibal river-crossing problem given in the preceding section. The conditions are that the boat carries no more than two people at a time and that at neither side should there be more cannibals than missionaries who out of habit would eat their outnumbered saintly mentors. Let us consider the simple case of ferrying across the river a group of two missionaries and two cannibals.

Prior to writing down a vertex matrix, we choose vertices corresponding to states at one side. Let us assume that the group starts on the left bank of the river. We now consider all possible states (within our two rules) on the left bank. A state is indicated by a pair of numbers the first of which indicates the number of missionaries and the second the number of cannibals. We have the following possibilities on the left bank:

$$
\begin{array}{lll}
v_1 = (2,2) & v_4 = (1,1) & v_6 = (0,1) \\
v_2 = (2,1) & v_5 = (0,2) & v_7 = (0,0) \\
v_3 = (2,0) & &
\end{array}
$$

Note that (1,0) is not allowed, since the corresponding state on the right bank would be (1,2) and the single missionary would be cannibalized. Similarly, the state (1,2) is not permitted on the left bank. We now form a vertex matrix whose elements are 0 or 1 depending on whether transitions from one state on the left bank to another also on the left bank are possible. The transitions, of course, are effected by the departure of the boat. Thus we list the vertices on the left and on top and enter a 0 or 1 depending on whether it is possible to go from the state represented by the vertex on the left of the matrix to another state represented by a vertex on top.

We have the vertex matrix

$$V = \begin{array}{c} \\ v_1 \\ v_2 \\ v_3 \\ v_4 \\ v_5 \\ v_6 \\ v_7 \end{array} \begin{array}{c} \begin{array}{ccccccc} v_1 & v_2 & v_3 & v_4 & v_5 & v_6 & v_7 \end{array} \\ \begin{pmatrix} 0 & 1 & 1 & 1 & 1 & 0 & 0 \\ 0 & 0 & 1 & 1 & 0 & 1 & 0 \\ 0 & 0 & 0 & 0 & 0 & 0 & 1 \\ 0 & 0 & 0 & 0 & 0 & 1 & 1 \\ 0 & 0 & 0 & 0 & 0 & 1 & 1 \\ 0 & 0 & 0 & 0 & 0 & 0 & 1 \\ 0 & 0 & 0 & 0 & 0 & 0 & 0 \end{pmatrix} \end{array}$$

On the right bank of the river one would have an identical set of states which are essentially the states complementary to those on the left bank. Their matrix would be V', the transpose of the above matrix. The reader will readily verify without difficulty that to obtain the matrix of transitions after one round trip, one must take the product VV'. In general, for the transition matrix of m round trips we have $(VV')^m$, and since the object is to move the group to the right bank, we must multiply this by V. This gives $(VV')^mV$, and the problem is to find the number of round trips m such that the element in the v_1, v_7 position of this product is 1, that is, we have the transition $(2,2) \to (0,0)$ on the left bank. Thus the group has moved to the right bank. Note that in the successive multiplications VV', $VV'V$, $VV'VV'$, etc., elements whose value is more than 1 will appear, indicating the number of ways in which the corresponding transition can be made. Since the object is to determine at each stage one possible transition, all nonzero elements are replaced by unity in these products. It turns out that our problem can be solved with $m = 2$ round trips and one last trip forward [i.e., a unit element appears for the first time in the v_1, v_7 position of $(VV')^2V$]. The computations are as follows:

$$V' = \begin{pmatrix} 0 & 0 & 0 & 0 & 0 & 0 & 0 \\ 1 & 0 & 0 & 0 & 0 & 0 & 0 \\ 1 & 1 & 0 & 0 & 0 & 0 & 0 \\ 1 & 1 & 0 & 0 & 0 & 0 & 0 \\ 1 & 0 & 0 & 0 & 0 & 0 & 0 \\ 0 & 1 & 0 & 1 & 1 & 0 & 0 \\ 0 & 0 & 1 & 1 & 1 & 1 & 0 \end{pmatrix} \qquad VV' = \begin{pmatrix} 1 & 1 & 0 & 0 & 0 & 0 & 0 \\ 1 & 1 & 0 & 1 & 1 & 0 & 0 \\ 0 & 0 & 1 & 1 & 1 & 1 & 0 \\ 0 & 1 & 1 & 1 & 1 & 1 & 0 \\ 0 & 1 & 1 & 1 & 1 & 1 & 0 \\ 0 & 0 & 1 & 1 & 1 & 1 & 0 \\ 0 & 0 & 0 & 0 & 0 & 0 & 0 \end{pmatrix}$$

$$VV'V = \begin{pmatrix} 0 & 1 & 1 & 1 & 1 & 1 & 0 \\ 0 & 1 & 1 & 1 & 1 & 1 & 0 \\ 0 & 0 & 0 & 0 & 0 & 1 & 1 \\ 0 & 0 & 1 & 1 & 0 & 1 & 1 \\ 0 & 0 & 1 & 1 & 0 & 1 & 1 \\ 0 & 0 & 0 & 0 & 0 & 1 & 1 \\ 0 & 0 & 0 & 0 & 0 & 0 & 0 \end{pmatrix} \qquad (VV')^2 = \begin{pmatrix} 1 & 1 & 0 & 1 & 1 & 0 & 0 \\ 1 & 1 & 0 & 1 & 1 & 0 & 0 \\ 0 & 1 & 1 & 1 & 1 & 1 & 0 \\ 1 & 1 & 1 & 1 & 1 & 1 & 0 \\ 1 & 1 & 1 & 1 & 1 & 1 & 0 \\ 0 & 1 & 1 & 1 & 1 & 1 & 0 \\ 0 & 0 & 0 & 0 & 0 & 0 & 0 \end{pmatrix}$$

$$(VV')^2V = \begin{pmatrix} 0 & 1 & 1 & 1 & 1 & 1 & 1 \\ 0 & 1 & 1 & 1 & 1 & 1 & 1 \\ 0 & 0 & 1 & 1 & 0 & 1 & 1 \\ 0 & 1 & 1 & 1 & 1 & 1 & 1 \\ 0 & 1 & 1 & 1 & 1 & 1 & 1 \\ 0 & 0 & 1 & 1 & 0 & 1 & 1 \\ 0 & 0 & 0 & 0 & 0 & 0 & 0 \end{pmatrix}$$

The problem now is to read off the solution from the matrices. We look at the last matrix, that is, $(VV')^2V$ in the (v_1,v_7) position where there is a unit element, and we ask which possible nonzero element of the first row of $(VV')^2$ could have given rise to this unit element in $(VV')^2V$ on multiplication by the seventh column of V. One possibility is the element in the (1,4) position of $(VV')^2$ (first row, fourth column), since there is also a unit element in the (4,7) position of V. Another possibility could be the element in the (5,7) position of V; but we choose the first. Thus the last transition is $v_4 \to v_7$. We next ask where the unit element in the (1,4) position of $(VV')^2$ could have come from. By examining the first row of $(VV')V$ and the fourth column of V', we find that it came from the fact that there is a nonzero element in the position (6,4) of V' [since the corresponding element from $(VV')V$ is also nonzero]. Thus the next-to-last transition is $v_6 \to v_4$. Again we ask for the origin of the unit element in the position (6,1) of $(VV')V$, and we find that it comes from the fact that the (2,1) element of VV' and the (2,6) element of V are unity. Thus the third transition from the end is $v_2 \to v_6$. We similarly find that the fourth and fifth transitions from the end are respectively $v_3 \to v_2$ and $v_1 \to v_3$. Thus the transitions are given by

$$v_1 \to v_3 \qquad v_3 \to v_2 \qquad v_2 \to v_6 \qquad v_6 \to v_4 \qquad v_4 \to v_7$$

or simply v_1, v_3, v_2, v_6, v_4, and v_7.

We could indicate alternative transitions if we wished. To interpret the solution in words, we note that, since v_3 is (2,0), both cannibals must first cross and one must return (because of v_2), the two missionaries must cross in order to have v_6 on the left bank, one missionary must return so that the next state on the left bank will be v_4, and finally both the missionary and the cannibal must cross, giving the final state v_7.

Exercise 6-15 The states of the problem of three missionaries and three cannibals where all missionaries and only one cannibal can row can be represented by (m,r,c), where $0 \leq m \leq 3$, $0 \leq r \leq 1$, $0 \leq c \leq 2$, m refers to the missionaries, r refers to the rowing cannibal, and c refers to the remaining two cannibals. There are 16 possible states for this problem. Write down its transition matrix.

To determine whether a solution exists for a given problem without going through all the manipulation needed to find the solution, the following method can be applied:

The solution must take the form $(VV')^mV$. Suppose the problem begins at vertex v_1 and the ultimate goal is vertex v_k. Then if a solution exists, there is an m such that $(VV')^mV$ is nonzero in the $(1,k)$ element. As shown above, to yield a 1 in the $(1,k)$ element of $(VV')^mV$, there must be at least one nonzero element in row 1 of $(VV')^m$ which corresponds to a nonzero element in column k of V. So the problem reduces to determining which of the elements of row 1 of $(VV')^m$ ever becomes nonzero for any power m. This can be tested easily even for very large matrices. Suppose the set $\{v_a,v_b, \ldots v_c\}$ consists of the vertices corresponding to 1's in the first row of VV'. These are the vertices which can be reached in one round trip from vertex v_1. Add to this set all the vertices which have 1's in row a, row b, . . . , row c of VV'. This new set consists of the vertices which can be reached from v_1 in two round trips. Repeat this addition for any new vertices in the set until it has been done for all vertices of the set. This final set consists of all the vertices which can ultimately be reached from vertex v_1 in any number of round trips. Now, if this set contains any vertices which have nonzero elements in column k of V, the problem has a solution; otherwise, it has not.

For an example, consider the four-missionary and four-cannibal problem, where vertex = (number of missionaries, number of cannibals):

$v_1 = (4,4)$	$v_5 = (4,0)$	$v_8 = (1,1)$	$v_{11} = (0,2)$
$v_2 = (4,3)$	$v_6 = (3,3)$	$v_9 = (0,4)$	$v_{12} = (0,1)$
$v_3 = (4,2)$	$v_7 = (2,2)$	$v_{10} = (0,3)$	$v_{13} = (0,0)$
$v_4 = (4,1)$			

The problem is to get from v_1 to v_{13}.

$$
V = \begin{array}{c} \\ v_1 \\ v_2 \\ v_3 \\ v_4 \\ v_5 \\ v_6 \\ v_7 \\ v_8 \\ v_9 \\ v_{10} \\ v_{11} \\ v_{12} \\ v_{13} \end{array}
\begin{array}{c}
\begin{array}{ccccccccccccc} v_1 & v_2 & v_3 & v_4 & v_5 & v_6 & v_7 & v_8 & v_9 & v_{10} & v_{11} & v_{12} & v_{13} \end{array} \\
\begin{pmatrix}
0 & 1 & 1 & 0 & 0 & 1 & 0 & 0 & 0 & 0 & 0 & 0 & 0 \\
0 & 0 & 1 & 1 & 0 & 1 & 0 & 0 & 0 & 0 & 0 & 0 & 0 \\
0 & 0 & 0 & 1 & 1 & 0 & 1 & 0 & 0 & 0 & 0 & 0 & 0 \\
0 & 0 & 0 & 0 & 1 & 0 & 0 & 0 & 0 & 0 & 0 & 0 & 0 \\
0 & 0 & 0 & 0 & 0 & 0 & 0 & 0 & 0 & 0 & 0 & 0 & 0 \\
0 & 0 & 0 & 0 & 0 & 0 & 1 & 0 & 0 & 0 & 0 & 0 & 0 \\
0 & 0 & 0 & 0 & 0 & 0 & 0 & 1 & 0 & 0 & 1 & 0 & 0 \\
0 & 0 & 0 & 0 & 0 & 0 & 0 & 0 & 0 & 0 & 0 & 1 & 1 \\
0 & 0 & 0 & 0 & 0 & 0 & 0 & 0 & 0 & 1 & 1 & 0 & 0 \\
0 & 0 & 0 & 0 & 0 & 0 & 0 & 0 & 0 & 0 & 1 & 1 & 0 \\
0 & 0 & 0 & 0 & 0 & 0 & 0 & 0 & 0 & 0 & 0 & 1 & 1 \\
0 & 0 & 0 & 0 & 0 & 0 & 0 & 0 & 0 & 0 & 0 & 0 & 1 \\
0 & 0 & 0 & 0 & 0 & 0 & 0 & 0 & 0 & 0 & 0 & 0 & 0
\end{pmatrix}
\end{array}
$$

In column 13, v_8, v_{11}, and v_{12} are nonzero. Therefore, if this problem has a solution, $(VV')^m$ must have a nonzero element in the first row at v_8, v_{11}, or v_{12}.

$$
VV' = \begin{array}{c} \\ v_1 \\ v_2 \\ v_3 \\ v_4 \\ v_5 \\ v_6 \\ v_7 \\ v_8 \\ v_9 \\ v_{10} \\ v_{11} \\ v_{12} \\ v_{13} \end{array}
\begin{array}{c}
\begin{array}{ccccccccccccc} v_1 & v_2 & v_3 & v_4 & v_5 & v_6 & v_7 & v_8 & v_9 & v_{10} & v_{11} & v_{12} & v_{13} \end{array} \\
\begin{pmatrix}
1 & 1 & 0 & 0 & 0 & 0 & 0 & 0 & 0 & 0 & 0 & 0 & 0 \\
1 & 1 & 1 & 0 & 0 & 0 & 0 & 0 & 0 & 0 & 0 & 0 & 0 \\
0 & 1 & 1 & 1 & 0 & 1 & 0 & 0 & 0 & 0 & 0 & 0 & 0 \\
0 & 0 & 1 & 1 & 0 & 0 & 0 & 0 & 0 & 0 & 0 & 0 & 0 \\
0 & 0 & 0 & 0 & 0 & 0 & 0 & 0 & 0 & 0 & 0 & 0 & 0 \\
0 & 0 & 1 & 0 & 0 & 1 & 0 & 0 & 0 & 0 & 0 & 0 & 0 \\
0 & 0 & 0 & 0 & 0 & 0 & 1 & 0 & 1 & 1 & 0 & 0 & 0 \\
0 & 0 & 0 & 0 & 0 & 0 & 0 & 1 & 0 & 1 & 1 & 1 & 0 \\
0 & 0 & 0 & 0 & 0 & 0 & 1 & 0 & 1 & 1 & 0 & 0 & 0 \\
0 & 0 & 0 & 0 & 0 & 0 & 1 & 1 & 1 & 1 & 1 & 0 & 0 \\
0 & 0 & 0 & 0 & 0 & 0 & 0 & 1 & 0 & 1 & 1 & 1 & 0 \\
0 & 0 & 0 & 0 & 0 & 0 & 0 & 1 & 0 & 0 & 1 & 1 & 0 \\
0 & 0 & 0 & 0 & 0 & 0 & 0 & 0 & 0 & 0 & 0 & 0 & 0
\end{pmatrix}
\end{array}
$$

The set of vertices which can be reached from v_1 consists of $\{v_1,v_2\}$. v_2 can reach v_1, v_2, and v_3; so add v_3 to the set, giving $\{v_1,v_2,v_3\}$. v_3 can reach v_2, v_3, v_4, and v_6; so add v_4 and v_6, giving $\{v_1,v_2,v_3,v_4,v_6\}$. v_4 reaches v_3 and v_4, which adds nothing new to the set. v_6 reaches v_3 and v_6, which also adds nothing new to the set. We have now exhausted the possibilities. Therefore, v_1 can reach only v_1, v_2, v_3, v_4, and v_6. It cannot reach v_5, v_7, v_8, v_9, v_{10}, v_{11}, v_{12}, or v_{13} in any number of round trips. But in column 13 of V only v_8, v_{11}, and v_{12} are nonzero. Since none of these are included in the set, the problem is unsolvable.

Remark: A transition (boolean) matrix may also be used to represent a directed graph associated with postulates introduced on a set of statements $E_1, E_2, \ldots, E_n$. The postulates give relations of the form $E_i \rightarrow E_j$. Note that we always have $E_i \rightarrow E_i$. By taking the second power of this matrix, we obtain a matrix in which each unit entry indicates either a postulate or a proposition which can be proved in two steps, for example, $E_6 \rightarrow E_7$, $E_7 \rightarrow E_{10}$, jointly imply $E_6 \rightarrow E_{10}$. Unit elements in the third power of the matrix indicate postulates or propositions which can be proved in two or in three steps. Finally, the $(n-1)$st power displays all propositions. An interesting problem is to find the smallest number of postulates from which a given set of propositions can be deduced. Note that there are at least 2^n sets of equivalent postulates which yield n theorems.[70,73]

Exercise 6-16 Show that the total number of $n \times n$ matrices whose elements are 0 or 1 and all of whose diagonal elements are unity is 2^{n^2-n}.

6-11 TRIANGLE DIVISION PROBLEM

Suppose that one divides a triangle ABC into smaller triangles by drawing n lines parallel to the sides (see Fig. 6-25, where $n = 2$) and then assigns letters A, B, and C to the resulting points as follows: Points on BC are labeled B or C but not A, and similarly for CA and AB. The points interior to the large triangle can be labeled by any of the three letters. It is desired to show that the number of small triangles with three differently labeled vertices is odd.

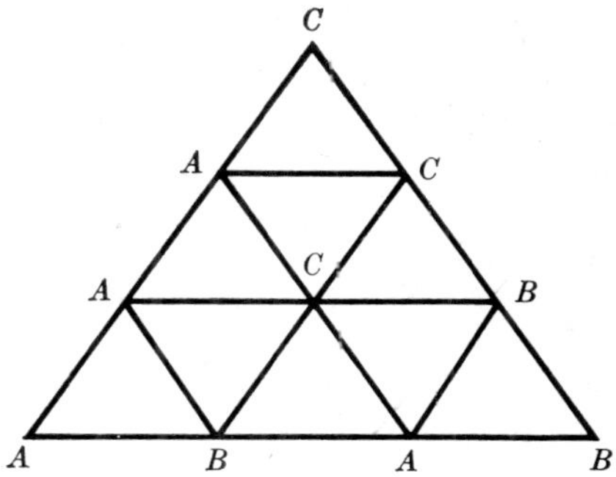

Figure 6-25

The proof begins by assigning the number 0 to a line segment whose end points are alike and the number 1 in the opposite case. The sides of a triangle whose vertices are labeled ABC sum to 3. This sum is 0 or 2 in the opposite case. The sum along the segments on each side of the large triangle must be odd, as can be easily verified. The sum over all small triangles must be odd because the interior segments are counted twice. Hence, there is an odd number of triangles numbered 3.

6-12 TWO-PERSON GAMES

A directed graph is a natural mathematical model to employ when analyzing some types of competitive situations between two individuals or groups of individuals having conflicting objectives. The following brief discussion does not attempt to present the most general context in which graph theory is applicable to such "games," or to precisely identify concepts with those from the formal theory of games.

Consider a situation in which two persons alternately make incremental modifications to a structure (such as a chessboard and its pieces). Assume that there is a standard starting "state" (e.g., the positioning of the chessmen at the outset) and a "rule book" which completely specifies the list of permissible moves, i.e., incremental changes of state, for each player. If there are only a finite number of distinct game states, the rules of the game can be completely characterized by a finite directed graph with two categories of arcs. Each state is considered as a vertex v_i, and there exists an arc of category 1 (or 2) from v_i to v_j if and only if the game can be transformed from state i to state j by a valid move by player 1 (or 2). A complete play of the game consists of an arc progression of

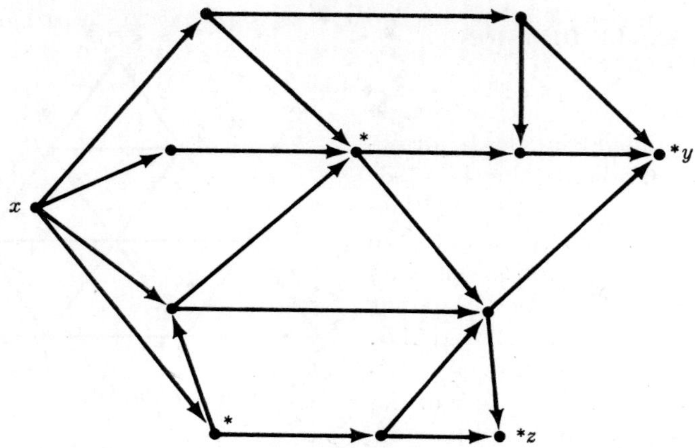

Figure 6-26

arcs of alternating types starting at the appropriate initial state with an arc of the type associated with the starting player.

Suppose that the game is such that no state is ever repeated, i.e., the associated graph has no cycles. Then the number of individual moves in a complete play of the game is bounded. Assume, moreover, that to "win" means to reach one of a specified set of states or vertices for the first time. For example, in the acyclic graph of Fig. 6-26 assume that y and z are the winning states and that x is the starting state. (In this example the set of permissible moves is assumed to be the same for both players, so that it is not necessary to specify two categories of arcs.)

Remark: One must distinguish between a play of the game and the game itself. Thus, for example, a player, in more general games, can be in a "winning state" and lose the game, or for that matter, be in a "losing state" and subsequently win. The particular kind of game illustrated here can be called "two-person perfect-information perfect-recall finite-win-lose alternation game in extensive form" (to give a vague idea of how special is its category). This is perhaps the simplest kind of game involving more than one player.

Note that the graph has a set S of distinguished vertices (marked with asterisks in Fig. 6-26) having the following three properties:

1. No two vertices of S are joined by an arc.
2. Every vertex not in S is joined to at least one vertex in S by an arc.
3. All winning states are in S.

Properties 1 and 2 are sometimes referred to as *internal* and *external stability*, respectively. A set of vertices, such as S, which possesses both properties is called a *nucleus* or *kernel*. Now suppose that player 1 plays first, that a nucleus S is known to player 1, and that the starting state is not in S. Then by property 2 player 1 can move to a state in S. If this is not a winning state, according to property 1, no matter what move player 2 makes it will lead to a state not in S (and hence by property 3 not a winning state). Another judicious move on the part of player 1 will return the game to S. By this procedure, play will ultimately terminate with player 1 the winner.

Thus, in principle at least, if the structure of the game is completely itemized and a nucleus S exists and is located, player 1 has a winning strategy if the initial state is not in S. (If it is in S, player 2 has a winning strategy.) In practice, of course, the structure of nontrivial games is too complex to actually itemize and consider the entire graph. However, if the game rules are such that the graph—though very large—has a highly systematic structure, it is sometimes possible to devise techniques which in effect generate elements of S as required in the vicinity of the current game state.

Remark: For the reader who is familiar with the game of Nim and of the strategies for winning, based on binary manipulations, these manipulations actually amount to determining whether the current state, i.e., the remaining numbers of sticks in the various piles, is in S and finding an appropriate transition into S if it is not.

A Switching Game

The solution of the following game, formulated by C. E. Shannon, is due to Lehman.[58] Two individuals, a "cut" player and a "short" player, play a game on a graph. They single out two vertices called the terminals, and the cut player will make a move by deleting an edge of the graph while the short player in his turn will make a move by designating an edge which cannot be deleted. Thus alternately the cut player deletes an edge and the short player states an edge which cannot be subsequently deleted. The object for the short player is to maintain a chain between the terminals while the cut player attempts to remove all chains. A game in which the cut player can win on going first or second in his deletions is known as a cut game, whereas a game in which the short player can win by going first or second is known as a short game. A game is neutral if whichever player goes first can win. Here we only give the condition for a short game:

Theorem 6-8 A game is a short game if and only if the graph contains two edge-disjoint trees whose vertices are the same and include the two terminals.

The condition is sufficient. If the short player can win by playing second, he can obviously win by going first. His move must always consist of preserving an edge which connects the two components of the trees produced by the deletion of an edge by the cut player. Since both trees have the same number of edges, he can always connect the resulting components by using the other tree. Necessity is harder to prove and will not be shown here.

A version of the switching game, known as "Gale" and, in commercial form, as Bridg-it, is described in Ref. 27 and is also treated in Ref. 58. The graph involved is essentially the graph depicted by the bold arcs in Fig. 6-27. In order to make the description of play more symmetric, the cut player makes moves on the light graph shown in the figure, while the short player makes moves on the bold graph. The former seeks to construct a chain joining w' and w'', while the latter seeks to construct a chain joining v' and v''. One "move" by either player consists of designating one of his edges which was not previously designated (initially, no edge is designated) and which does not cross an edge already designated by his opponent. Except for a minor technicality, the bold and light graphs are duals of each other. (What is the technicality?)

For this particular switching game, O. Gross devised a simple winning strategy, described in Ref. 27. The game is a neutral game, and the strategy is a "pairing strategy," in which the selection of an edge by the

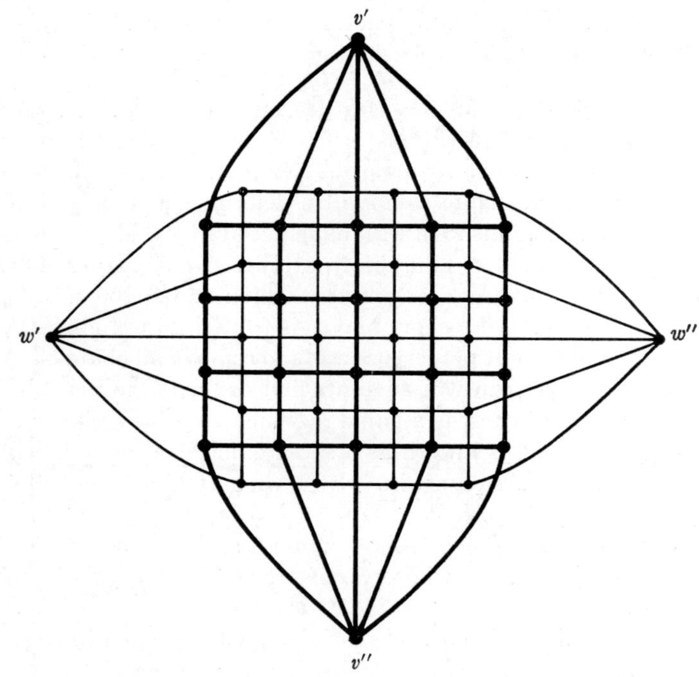

Figure 6-27

second player is always countered by the selection of a specific corresponding edge by the first player (except if when the corresponding edge has already been played).

6-13 APPLICATION TO BOARD GAMES[52a,69]

Suppose we are given a set of squares of a chess board, as illustrated in Fig. 6-28, where from an even-numbered square it is possible to move one square vertically or horizontally and from an odd-numbered square it is possible to move one square diagonally. By associating a vertex with each square, one can form the vertex matrix V of the associated graph.

Exercise 6-17 Determine by examining the element in the seventh row and third column of V^4 the number of ways one can move from 7 to 3 in four moves; then obtain the (7,3) element from $V + V^2 + \cdots + V^5$ to accomplish the same task in less than six moves, and finally from the diagonal elements of V^5 obtain the number of ways a piece can return to its starting position in five moves.

The reader may be familiar with various puzzles having the general objective of precisely covering a given plane "board" with a set of plane "pieces" whose total area equals that of the board to be covered. In the case where each piece has only a finite set of permissible board positions, this problem can be phrased in graph-theoretic terms and solved (at least in principle, and in some cases in fact) in such terms. The following brief discussion outlines this approach to such problems and is based on work by P. Jullien.[52a]

Assume, for the sake of concreteness, that the board, as well as each of the pieces, consists of a rectangle whose dimensions are integers. (Much more general assumptions could be made.) Suppose, moreover, that the board has been subdivided into unit squares and that the only permissible positions for any given pieces are those in which no squares are partially covered.

With this understanding, we can itemize all the permissible board positions of each piece. Having done so, two board positions will be considered *compatible* if and only if they correspond to two different pieces and do not have any overlapping area.

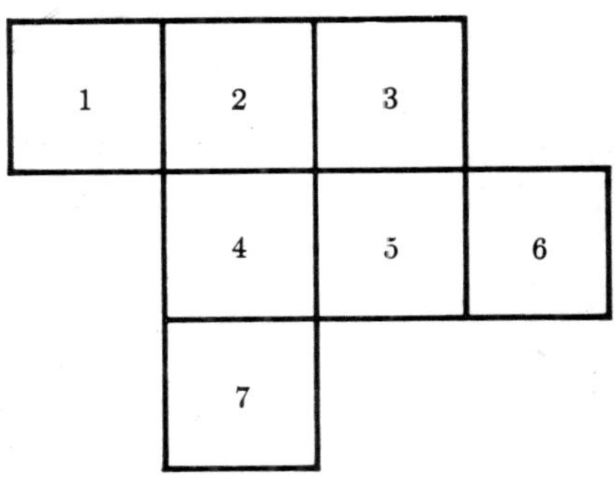

Figure 6-28

Let k denote the number of pieces, and let n_i represent the number of permissible positions of the ith piece. Consider the graph G

having $\sum_{1}^{k} n_i$ vertices, corresponding to all possible positions of all pieces, and such that vertices v and w are adjacent if and only if the corresponding positions are compatible in the above sense. It is readily seen that every solution to the puzzle corresponds to a complete subgraph having k vertices, and conversely. Thus the task of finding all solutions to the puzzle is equivalent to that of finding all complete subgraphs having k vertices.

For the remainder of the discussion, let $A, B, \ldots, Z$ denote the pieces, with A_i denoting a particular position (i.e., vertex) of piece A. Jullien describes a succession of transformations which start with the graph of the puzzle and which end with a graph whose vertices are the various solutions, i.e., the complete subgraphs of the original graph.

As the first step, all vertices corresponding to pieces A and B (where these may be any two of the pieces) are removed, being replaced by vertices of form AB_{ij}, restricted to those ij combinations such that A_i and B_j are compatible. In addition, AB_{ij} is joined with positions of pieces $C, D, \ldots, Z$ whenever those positions are compatible with both A_i and B_j. It may be easily seen that there is a 1-1 correspondence between complete graphs with k vertices in the original graph and complete subgraphs with $k - 1$ vertices in the new graph.

By repetition, one finally arrives at a graph whose vertices are of the form $ABC \ldots Z_{ij \ldots q}$, signifying that $A_i, B_j, \ldots, Z_q$ determine a complete subgraph of the original graph of the puzzle. (Can you fill a $12 \times 12 \times 12$ cube with $2 \times 4 \times 8$ bricks? Prove.)

MATCHINGS

6-14 MAXIMAL MATCHINGS AND RELATED TOPICS

So far, matching is a theoretical subject. It is an excellent example of a mathematical theory that needs an application. Generalizations of the basic problem considered here are exemplified by the section dealing with the interconnection of electric power stations.

Let $G = (V,E)$ be a graph having no loops. A set of edges $M \subset E$ is called a *matching in G* if no two edges of M are adjacent. Thus each vertex of G is incident with at most one edge of M. A vertex is said to be either *covered* or *exposed*, relative to M, depending on whether or not it meets an edge of M. The empty set is a legitimate (though uninteresting) matching relative to which every vertex is exposed.

A matching in G will be called a *maximal matching in G* if no matching has higher cardinality. If every vertex is covered, the matching is said to be *perfect*, and it is sometimes called a 1 factor.[90] A perfect matching, if one exists for G, is clearly maximal. (Note that no perfect matching can exist if $|V|$ is odd.)

Tutte[90] has characterized the set of graphs having perfect matchings. The main objective of this section is to describe a recent algorithm due to Edmonds[20] for finding a maximal matching in an arbitrary graph. In particular, it will determine a perfect matching if one exists. In contrast to other known approaches to solving this problem for general graphs, the maximum effort required to implement this algorithm increases algebraically, rather than exponentially, with the number of vertices of G.

Suppose that G is bipartite, having a vertex partition $\{V_1,V_2\}$ such that every edge joins a vertex in V_1 to a vertex in V_2. Matching problems often arise in the context of bipartite graphs, particularly when the vertices in V_1 and V_2 represent different kinds of entities (e.g., men vs. women, men vs. jobs, jobs vs. machines). In such settings one frequently wishes to "pair" or "match" the two types of entities in such a way that there are as few as possible residual unmatched entities (i.e., exposed vertices). The structure of the original graph serves to itemize all potential or feasible pairings.

Given a matching M in a graph $G = (V,E)$, a simple chain C in G is called an *alternating chain*, relative to M, if (when the chain is traversed from one end to the other) its edges are alternately *matching edges* (edges of M) and *nonmatching edges* (edges of $E - M$). Given a matching M and an alternating chain C, consider the edge set $M' = M \oplus C$ consisting of those edges appearing in M or in C but not in both. Expressed differently, M' is obtained by deleting from M the matching edges in C and adding to M the nonmatching edges in C. Since the number of matching and nonmatching edges in C differs by at most 1, the number of edges in M and M' also differs by at most 1.

To illustrate the preceding remarks, consider the matching M indicated by the bold arcs in Fig. 6-29*a*. The set $M' = M \oplus C$ shown in (*c*) is obtained by taking as C the alternating chain shown in (*b*). If, as in the preceding example, the end points of alternating chain C are exposed, relative to M, then it is readily seen that M' is necessarily a matching and

(*a*) (*b*) (*c*)

Figure 6-29

that $|M'| = |M| + 1$. For this reason, an alternating chain both of whose end points are exposed is called an *augmenting chain*. The existence of an augmenting chain is a necessary, as well as sufficient, condition for M to be nonmaximal. Stated formally, we have the following theorem due to Berge,[3] with proof along the lines of Edmonds:[20]

Theorem 6-9 A matching M in a graph G is maximal if and only if G contains no augmenting chain relative to M.

Proof: To complete the proof, it remains to show that if M is not maximal, then there exists an alternating chain joining two exposed vertices relative to M. (The converse has already been established.)

Assume that matching M' has one more edge than M has, and consider the graph $G' = (V', M \oplus M')$, which includes precisely those edges which occur in M or in M' but not in both. No vertex v of G' has degree greater than 2, since at most one edge in M, and one in M', is incident with v. The components of G', other than isolated vertices, correspond to alternating chains and alternating simple circuits relative to M (and to M'). Each of the latter has the same number of edges from M as it has from M'. Hence there must be at least one component which corresponds to an alternating chain and which has one more edge from M' than from M. The end points of such a chain are necessarily exposed vertices relative to M. This completes the proof.

In view of the preceding theorem, the problem of finding a maximal matching can be solved if one has an acceptable method for finding an augmenting chain whenever one exists. For we could simply start with the "empty" matching as M_0, find an alternating chain C_1 (a single edge, in this case), define $M_1 = M_0 \oplus C_1$, and in general define $M_i = M_{i-1} \oplus C_i$, where C_i is an alternating chain joining two exposed vertices relative to M_{i-1}. Eventually, a stage i for which no augmenting chain C_{i+1} exists will be reached. M_i is then maximal.

Effective algorithms for finding maximal matchings in bipartite graphs have been known for some time; see Refs. 22 and 49 for a description of several of them. The algorithm to be described, which is applicable to any graph, systematically searches for augmenting chains and subsequent enlargements of the matching. In the process it identifies certain regions of the graph (associated with specialized trees called hungarian trees) which can be eliminated from further searching without danger of overlooking any additional augmenting chains. From time to time it may also *shrink* (or *contract*) certain subgraphs (associated with appropriate circuits of odd length) to a single "pseudovertex," thus simplifying subsequent search.

Before describing the algorithm, several auxiliary concepts[20] will be introduced. An *alternating tree* is a tree in which the vertices have been partitioned into two classes, called *inner* and *outer* vertices, in such a way

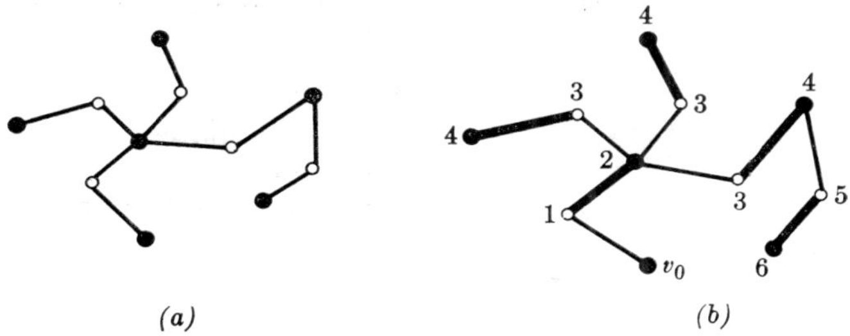

Figure 6-30

that every inner vertex is of degree 2 and every edge joins an inner vertex with an outer vertex. (Note that all "end vertices," i.e., vertices of degree 1, are necessarily outer.) Figure 6-30a is an example of an alternating tree (the bold vertices being the outer ones). An alternating tree necessarily has precisely one more outer vertex than it has inner vertices. To show this, note that every edge of the tree is incident with precisely one inner vertex. Hence if there are m edges, there are $m/2$ inner vertices and $(m + 1) - m/2 = m/2 + 1$ outer vertices.

In order to determine a maximal matching in an alternating tree, select any outer vertex as v_0 and define d_i to be the number of edges in the tree chain joining v_0 with any other vertex v_i, with $d_0 = 0$. (This is illustrated in Fig. 6-30b.) Every inner vertex v_i is adjacent with precisely one outer vertex v_j such that $d_j = d_i + 1$ (that is, v_i lies "between" v_0 and v_j). The set of edges joining all such pairs v_i and v_j is a matching. It is necessarily maximal, since only v_0 is exposed. Corresponding to every outer vertex there is a maximal matching leaving only that vertex exposed, and every maximal matching of the alternating tree is of this type.

Exercise 6-18 Prove that the set of edges defined in the preceding paragraph is necessarily a matching.

Exercise 6-19 Prove that the vertex left exposed by a maximal matching in an alternating tree could not be an inner vertex.

Exercise 6-20 Prove that a maximal matching in an alternating tree has the property that the tree chain joining the exposed vertex with any other vertex is an alternating chain. (Other vertices do not have this property in general.)

Given a matching M in a graph G, a *planted tree* in G is an alternating tree T in G such that all matching edges which are incident with vertices of T are edges of the tree and such that these edges constitute a maximal matching for T. The (unique) exposed vertex of T is called the *root* of planted tree T.

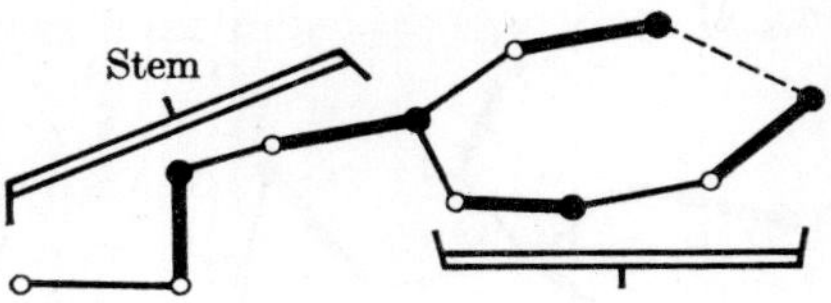

Figure 6-31

Let T be a planted tree in G, relative to matching M, and let v_0 be its root. If there is an edge of G joining an outer vertex of T with an exposed vertex of G other than v_0, then T is called *augmenting* relative to e. Clearly, an alternating chain joining two exposed vertices exists, so M can be enlarged in the manner described earlier.

A *flowered tree* is a planted tree T together with an edge of G joining two outer vertices of T. If C_1 and C_2 denote the alternating chains joining these vertices with v_0, then $C_1 \cup C_2$ is called a *flower*, $C_1 \cap C_2$ is called a *stem*, and $C_1 \oplus C_2$ is called a *blossom*. Figure 6-31 illustrates these concepts. (If the root of the tree is incident with the blossom, there are no edges in the stem.) The vertex where the stem meets the blossom is called the *tip* of the stem.

Let T be a planted tree in G relative to matching M, and suppose that some edge e joins an outer vertex v of T to a covered vertex w not in T. Then the edge f which covers w is not in T, nor is its other end point x. (Why not?) In this case, we can enlarge T by adding edges e and f and vertices w and x (as inner and outer, respectively).

Note that all of the foregoing three concepts (an augmenting tree, a flowered tree, and a tree which can be enlarged) are related to consideration of a planted tree and an edge meeting one of its outer vertices. They differ with respect to the other end point of the edge under consideration (whether it is in the tree as an outer vertex or whether it is not in the tree, either exposed or covered). Naturally, the same planted tree may, with respect to different edges, meet several of these conditions. If it does not meet any of these conditions with respect to any edge incident with any of its outer vertices, we have the concept of a hungarian tree. A planted tree T in G is called a *hungarian tree* if every edge of G which is incident with an outer vertex of T joins it to an inner vertex of T. Thus T is not an augmenting tree or a flowered tree relative to any edge e of G, nor can it be enlarged as in the preceding paragraph.

The following theorem is of central importance for the algorithm for finding a maximal matching in a graph:

Theorem 6-10 A planted tree T in G can be enlarged to either an augmenting tree, a flowered tree, or a hungarian tree.

To see this, we need only consider the definitions and look for an edge e of G relative to which T is augmenting or flowered. If there is no such edge, and if T cannot be enlarged, then T is necessarily hungarian.

We are now in a position to describe the algorithm for finding a maximal matching. Initially, take any matching M in G and any planted tree T relative to M. (We can take as M the empty set and as T any vertex.)

1. If T is augmenting relative to some edge e of G, use the corresponding alternating chain to enlarge M to a matching M' having one more edge. If M' is not perfect, pick a new planted tree T (e.g., any exposed vertex).
2. If T is flowered relative to some edge e, shrink the vertices and edges of the blossom to a single new pseudovertex and remove any loops created.
3. If T can be enlarged by adjoining two new vertices and edges in the manner described earlier, enlarge it.
4. Continue steps 1 to 3 until either a hungarian tree or a maximal matching is found (in the transformed graph in which a number of blossoms may have been shrunk).

In the former case, if the hungarian tree T contains all vertices of the transformed graph G^*, then the current matching M is clearly maximal in G^*. If it does not, remove from G^* the vertices of T and all edges incident with these vertices and repeat steps 1 to 3 for the remaining graph. In this manner we gradually remove one hungarian tree after another until either we have a perfect matching in the last remaining portion of the graph or else the final, remaining portion is hungarian.

The maximal matchings in these several portions are now used to reconstruct a maximal matching for the original graph by reexpanding the previously shrunken blossoms (in reverse order from the order of shrinking) and adding appropriate edges to the matching. If the blossom has $2n + 1$ vertices, then n edges of the blossom are added to the matching in the one possible way which does not involve an edge incident with the tip of the stem (which may already be incident with an edge of the matching).

The algorithm for finding a maximal matching will be illustrated with the graph of Fig. 6-32*a*. Throughout the discussion, the edges and root of the current planted tree will be shown boldly, and the edges in the current matching will be denoted by asterisks.

Initially, we take as M the empty set and choose any exposed vertex, say, F, to start a planted tree T. T is augmenting relative to edge 13, so we add 13 to M and start a new planted tree at C, as shown in Fig. 6-32*b*. Relative to the new T, edges 11 and 13 meet the conditions for enlarging T, as in (*c*). Relative to edge 10, the current planted tree is augmenting. We enlarge M, discard T, and start a new tree at H, as shown in (*d*). By two augmentations (based on edges 6 and 7) we enlarge T, as in (*e*). Now T is augmenting relative to edge 12. We enlarge M and start a new tree at I, as shown in (*f*). Two enlargements (based on edges 8

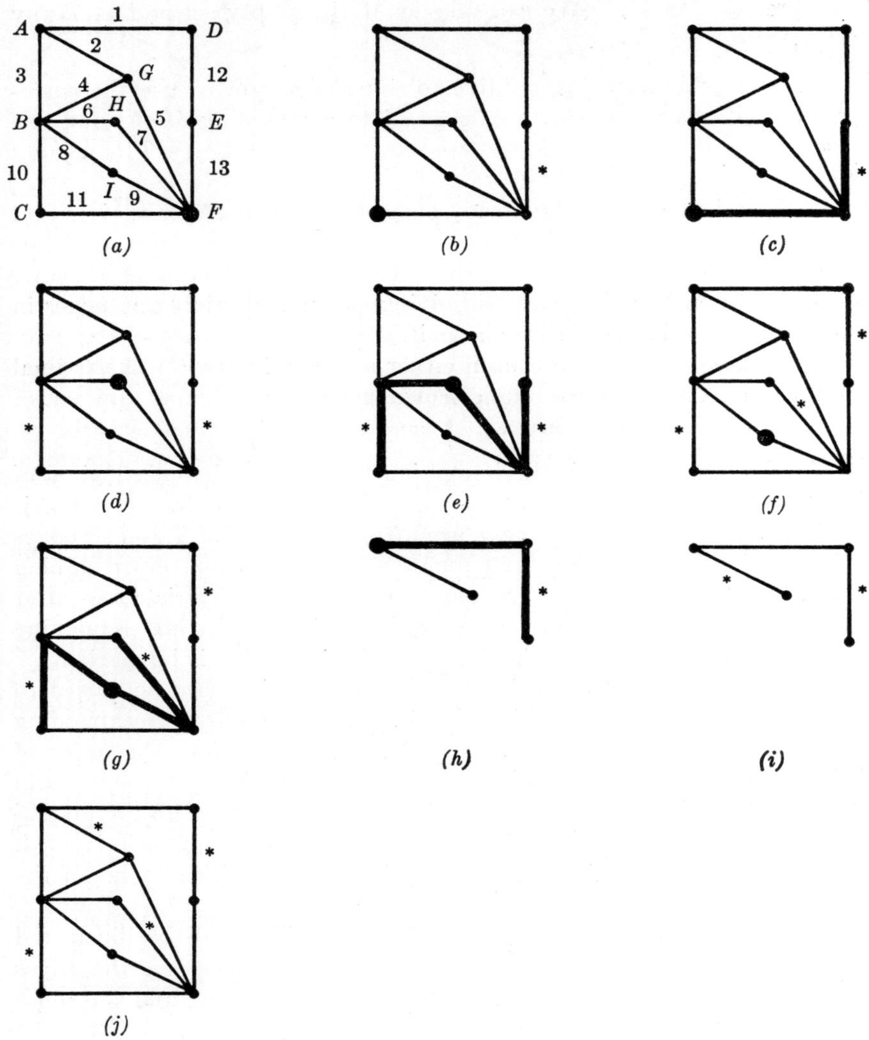

Figure 6-32

and 9) yield the tree shown in (g). This tree is hungarian. Its outer vertices (C, H, and I) meet only inner vertices of the tree. We next remove the tree temporarily from consideration and consider the subgraph G_1 obtained by eliminating the vertices of the hungarian tree and all edges incident with these vertices. This subgraph is shown in (h). Here we have chosen a new root A and enlarged the tree, based on edge 1. Relative to edge 2, this tree is augmenting, and it yields a new matching

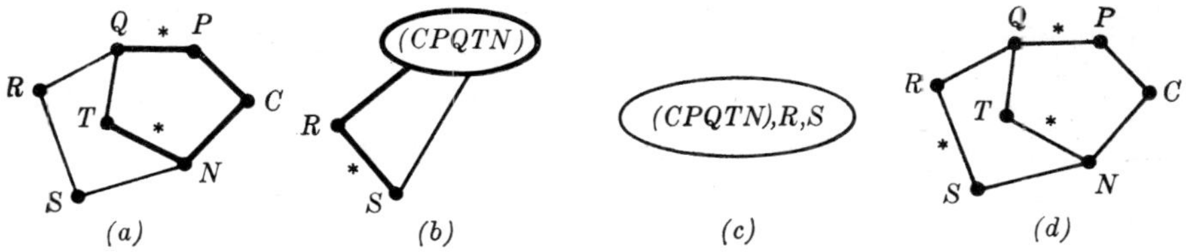

Figure 6-33

in G_1 consisting of the two edges shown in (*i*). This matching is clearly maximal (in fact perfect) in G_1. When combined with the maximal matching in the hungarian tree found earlier, it constitutes (for reasons discussed later) a maximal matching for G, shown in (*j*). The present example did not require shrinking of any blossoms.

In more complex graphs, it is necessary to find and successively remove several hungarian trees T_1, T_2, . . . and hence consider several residual subgraphs G_1, G_2, . . . before finally arriving at a subgraph in which we find either a perfect matching or else a hungarian tree which contains all remaining vertices. By successively restoring the hungarian trees in reverse order and assigning a maximal matching to each as it is added, a maximal matching for the entire graph is obtained.

To illustrate the shrinking and subsequent reexpansion of blossoms, suppose that, in addition to the edges shown in Fig. 6-32*a*, vertex C was incident with the structure shown in Fig. 6-33*a*. Relative to the planted tree and matching edges shown in (*a*), the circuit C, P, Q, T, N is a blossom; and when shrunken, it yields (*b*). Relative to the new planted tree in (*b*), the remaining graph is a blossom, and it can be shrunken to the single pseudo-pseudovertex shown in (*c*). Reexpanding blossoms in reverse order and recalling the matching edges destroyed in the shrinkings yields the maximal matching shown in (*d*).

Assuming that vertex C in our original example was actually the above pseudo-pseudovertex (i.e., assuming that this double shrinking had preceded the considerations shown in Fig. 6-32), a maximal matching for the total graph is as shown in Fig. 6-34.

Exercise 6-21 Find maximal matchings for the graphs shown in Figs. 6-15 and 6-17.

Thus far we have introduced the terminology and preliminary results required to describe the algo-

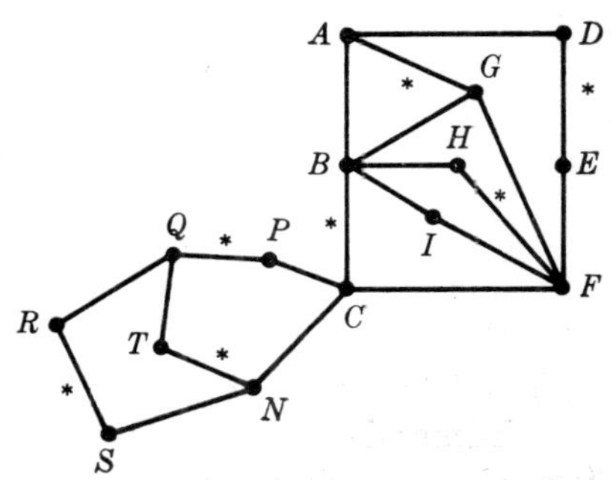

Figure 6-34

rithm and informally illustrated the algorithm by means of a limited example. We shall now restate the algorithm in general terms.

Statement of the algorithm Let $G = (V,E)$ denote the graph for which a maximal matching is to be found and let M denote an initial matching in G. (The empty set will do.) Let S denote a set of edges, initially empty, which will subsequently be enlarged by adding certain edges. If any vertices of G are exposed relative to M, select one of them and regard it as a planted tree T. Consider, in any order, the edges which are not in T but which meet outer vertices of T. If T is flowered relative to an edge, shrink the corresponding blossom to a single pseudovertex, record the blossom involved, and henceforth regard the pseudovertex as an outer vertex of T. If T can be enlarged by adjoining two edges and vertices as described earlier, enlarge it. If T is augmenting relative to some edge, use the associated augmenting chain to enlarge M. (In this case T is no longer a planted tree. Start a new planted tree by selecting—if one exists—an exposed vertex relative to M.)

If a point is reached at which the present planted tree T is hungarian, and if not all vertices of the graph under consideration are in T, temporarily eliminate all vertices of T and all edges which meet these vertices. Denote by G_1 the remaining graph, and repeat the above process for this graph.

Ultimately, after removal of a finite number of hungarian trees $T_1, T_2, \ldots, T_n$ and consideration of the corresponding residual graphs $G_1, G_2, \ldots, G_n$, a point such that either (1) G_n is empty or (2) M includes a perfect matching for G_n is necessarily reached. At this point restore $T_n, T_{n-1}, \ldots, T_1$, in turn, with a maximal matching for each. If any vertices are pseudovertices, reexpand them. (A safe way to do this is in the reverse order from their order of shrinking.) As each vertex is expanded, assign to the corresponding blossom the maximal matching which does not meet the vertex which was the tip of the stem at the time of shrinking. After all blossoms have been expanded, the resulting matching is maximal for the original graph. A complete validation of this algorithm is presented in Ref. 20.

Let $G = (V,E)$ be a graph each of whose vertices v has an associated nonnegative integer $d(v)$. By a *degree-constrained subgraph* of G is meant a subgraph $G' = (V,F)$ such that the degree of v in G' is at most $d(v)$ for every vertex v. By a *maximal* degree-constrained subgraph is meant one having the largest possible number of edges. [The reader will recognize the maximal matching problem as corresponding to the case when $d(v) = 1$ for every vertex v.]

Edmonds has developed a straightforward generalization of the foregoing algorithm to solve the general problem of finding a maximal degree-constrained subgraph for arbitrary nonnegative $d(v)$'s. He has also considered a second important generalization of the maximal matching problem. It is that of finding a matching whose total "weight" is maximal, where each edge of the graph has an associated numerical "weight." (The basic matching problem corresponds to the case when all weights are unity.) It is possible to combine an algorithm[18] for solving this problem with one for finding shortest paths (as in Chap. 3) to solve the *Chinese postman's problem*, first studied by Kwan.[56a] It is to find a closed edge progression which includes every edge of a connected graph at least once, and which has minimal total weight. (Thus the problem is to duplicate enough edges to obtain a unicursal graph, adding minimal weight in the process.)

SOME ENGINEERING APPLICATIONS

6-15 ANALYSIS OF PHYSICAL SYSTEMS

In this section we shall discuss the way in which an appropriate directed graph can be used to assist in deriving essential information about a physical system as a whole, given information about the characteristics of its component parts and the manner in which they are interconnected. The method to be discussed, developed, for example, by Trent,[86] is perhaps most widely applied to electrical circuitry, but it is equally applicable to systems involving other manifestations of energy employing, for example, translational or rotational mechanical devices or fluid devices. In addition to "pure" systems, the technique can be extended to "mixed" systems in which several energy types occur in various system elements and are interrelated through appropriate "coupling" devices.

Consider a collection of m two-terminal devices or system *elements* $E_1, \ldots, E_m$ whose terminals are joined, in some manner, at n junction points $P_1, \ldots, P_n$. These might be, for example, a collection of resistors, capacitors, and inductors, together with one or more voltage generators (in the simplest instance batteries, in other cases time-dependent devices). It is assumed that each individual system element can be adequately characterized by a known equation which relates two fundamental variables, a *through variable* x_i and an *across variable* y_i, which are associated with the element E_i and measured in a specified direction. The choice of variables and the use of the terms "through" and "across" will be clarified shortly.

For example, if x_i and y_i denote current and potential difference, respectively, each passive element (element other than a generator) may have a characterizing or *constitutive equation* of the form

$$y_i = kx_i \qquad \text{resistor}$$

$$y_i = k\frac{d}{dt}x_i \qquad \text{inductor}$$

$$y_i = k\int_{t_0}^{t} x_i\,dt \qquad \text{capacitor}$$

where t corresponds to time. An active element, or generator, is characterized by an equation expressing just one of the basic variables as a (possibly constant) function of time. For example,

$$y_i = f(t)$$

characterizes a voltage generator.

Assume now that an arc a_i is associated with each element E_i and that a vertex v_j is associated with each junction point P_j. If the end points of the arcs are taken to be the appropriate junction points, the resulting directed graph provides a convenient characterization of the structure of the corresponding physical system. The essential characteristic of the through variables, in the present context, is that they satisfy the following postulate at each vertex:

Vertex postulate The algebraic sum of the through variables associated with the arcs incident with any given vertex is zero.

By algebraic sum is meant the following: each through variable is added or subtracted, depending on whether the corresponding arc is positively or negatively incident with the vertex under consideration. In Figure 6-35, for example, the postulate is satisfied at v_1, since

$$(4) - (7) - (-3) = 0$$

and it is readily seen that the postulate is satisfied at the other vertices as well. In the context of electrical networks, this is Kirchhoff's current law. In general, one must choose as one basic variable, namely the through variable, one with appropriate physical dimensions so that the vertex postulate is satisfied.

The across variables are also assumed to satisfy a basic postulate, in this case related to circuits rather than to vertices.

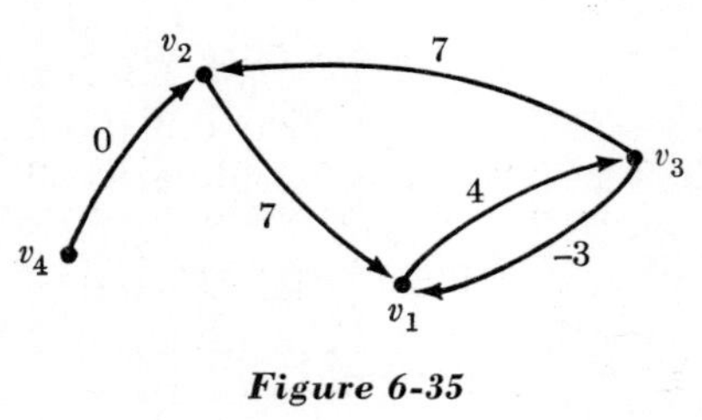

Figure 6-35

Circuit postulate The algebraic sum of across variables associated with the arcs of any simple circuit is zero.

In this case the circuit is assumed to be oriented (in either of the

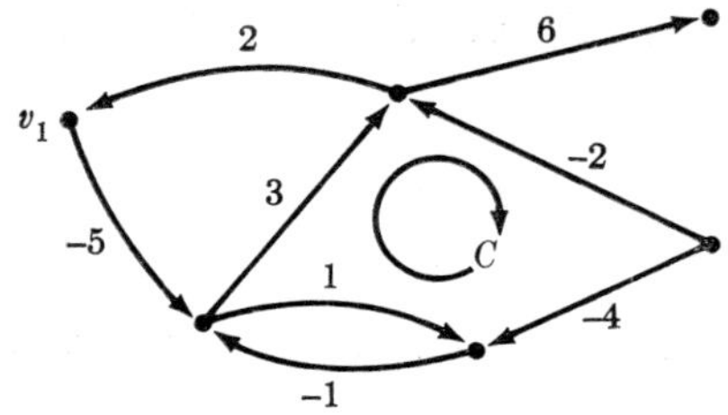

Figure 6-36

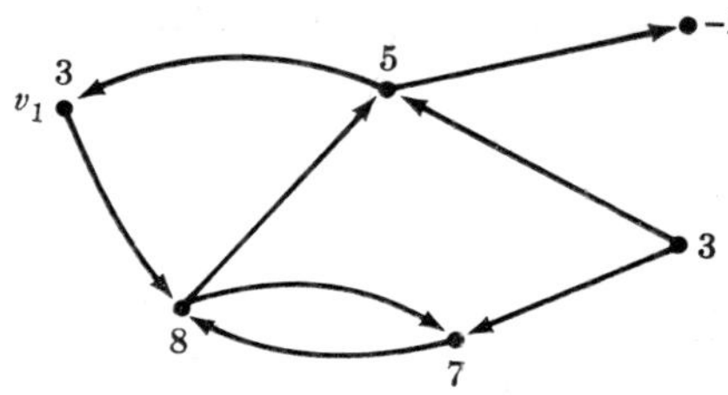

Figure 6-37

possible ways) and each across variable is added or subtracted, depending on whether the corresponding arc has a direction which agrees or disagrees with the induced orientation. In Fig. 6-36, for example, the postulate is satisfied for the oriented simple circuit C, since

$$(3) - (-2) + (-4) - (1) = 0$$

It can also be seen to be satisfied for each of the five other simple circuits of this particular graph. For electrical networks, the reader will recognize Kirchhoff's voltage law.

The circuit law has the following alternative formulation: If v_1 is a fixed vertex and v_i is any other vertex, then the algebraic sum of across variables associated with any chain oriented from v_1 to v_i is independent of the particular chain selected. (It is assumed here that the graph is connected, so that at least one such chain exists.) Using this formulation, we can associate a quantity S_j with each vertex v_j as follows: Assign S_1 arbitrarily, and for other j set $S_j = S_1 - K$, where K is the algebraic sum of across variables along any chain oriented from v_1 to v_j. By setting $S_1 = 3$ in the preceding example, we obtain the values of S_j shown in Fig. 6-37. The across variables determine the values of S_j to within an additive constant, and one may choose any convenient reference vertex. In an electrical context, the values of S_j can be considered as potentials relative to the reference potential established at the reference vertex. The values of the across variables are then seen to be differences in potential.

The process for deriving equations which characterize the system as a whole from the constitutive equations of its elements, together with the system structure, proceeds in two stages. First, the vertex and circuit postulates are exploited to reduce the totality of through or across variables to a (relatively) small set of independent variables in terms of which all the variables can be expressed. The constitutive equations are then brought into play to interrelate through variables with across variables. We now turn to the first stage of this process.

When applied to a particular vertex v_i, the vertex postulate asserts that

$$\sum_{j=1}^{m} a_{ij}x_j = 0$$

where $a_{ij} = +1(-1)$ if the ith vertex is positively (negatively) incident with the jth arc and is zero otherwise. Expressed differently, the vectors

$$A_i = (a_{i1}, \ . \ . \ . \ , a_{im})$$

and

$$X' = (x_1, \ . \ . \ . \ , x_m)$$

are orthogonal. Note that A_i is a row of the graph's incidence matrix A. Since the space spanned by the rows of A is the same as the space spanned by the rows of the cut-set matrix K, it follows that X' is a linear combination of circuit vectors (rows of the circuit matrix C) since A (or K) and C determine orthogonal subspaces which jointly span m-dimensional space.

Referring to material from Chap. 5, $\bar{K}$ and $\bar{C}$ can be written as

$$\bar{K} = (\bar{K}_{11}|I)$$

and

$$\bar{C} = (I|\bar{C}_{12})$$

where the I's are identity matrices, by choosing as the first $m - n + 1$ columns the chords of a spanning tree. By partitioning the vector of through variables in the same way, we have

$$X = \left(\frac{X_c}{X_b}\right)$$

where X_c and X_b correspond to the chords and branches, respectively.

The vertex postulate then implies that

$$(\bar{K}_{11}|I)\left(\frac{X_c}{X_b}\right) = 0$$

Hence

$$X_b = -\bar{K}_{11}X_c = \bar{C}'_{12}X_c$$

and the branch through variables have been expressed in terms of the chord through variables.

In a similar way, the circuit postulate leads to the matrix equations

$$(I|\bar{C}_{12})\left(\frac{Y_c}{Y_b}\right) = 0$$

and

$$Y_c = -\bar{C}_{12}Y_b = \bar{K}'_{11}Y_b$$

expressing the chord across variables in terms of the branch across variables.

Application of these relationships constitutes the first stage of the analysis: We have reduced to a minimum the number of through and across variables which must be explicitly considered. (The particular variables selected depend, of course, on the selection of a spanning tree.)

The constitutive equations of the elements can be conveniently written in matrix form as follows if the across variables are explicit functions of the through variables:

$$Y = \Omega_x X - Y_g$$

where Ω_x is an $m \times m$ diagonal matrix whose ith diagonal element is either a constant or a differentiation or integration operator and Y_g is a column vector which is zero in positions corresponding to passive elements and is an appropriate function $f(t)$ for generators. (The corresponding diagonal entries of Ω_x are 0.)

If the through variables are expressed as explicit functions of the across variables, we obtain a comparable expression

$$X = \Omega_y Y - X_g$$

The former expression can be rewritten as

$$\Omega_x \bar{C}' X_c = Y + Y_g$$

Premultiplying by $\bar{C}$ yields

$$\bar{C}\Omega_x \bar{C}' X_c = \bar{C}Y + \bar{C}Y_g = \bar{C}Y_g \tag{6-1}$$

in which the only unknowns are the chord through variables.

The latter expression can be rewritten as

$$\Omega_y \bar{K}' Y_b = X + X_g$$

and then as

$$\bar{K}\Omega_y \bar{K}' Y_b = \bar{K}X + \bar{K}X_g = \bar{K}X_g \tag{6-2}$$

in which the only unknowns are the branch across variables.

Equations (6-1) and (6-2) correspond to the circuit (or mesh) and vertex (or nodal) formulation techniques, respectively. In either case, if the resulting set of simultaneous equations can be solved by using appropriate mathematical tools, the remaining through and across variables are readily obtained from relationships given above. Note in particular that $\bar{K}'$ and $\bar{C}'$ can be written down by inspection from the graph, once a tree has been chosen.

When some system elements have more than two terminals, or when some elements serve as "couplers" relating different energy types in the same system, the matrix characterizing the constitutive equations has a more complex structure and the resulting simultaneous equations are correspondingly more difficult to solve. The role of the system graph is fundamentally the same, however. For extensive discussions of this topic, from various points of view and with varying degrees of emphasis on the system graph, the reader is referred to the following books listed at the end of Chap. 1: 14, 15, 17, 22, and 27.

6-16 COMMUNICATION NETWORKS

We first consider some qualitative aspects of communications between members of a group. The members of a social group can communicate by several means, e.g., orally, by writing, or by gesticulation. These means of expression when used by members of the group for communication give rise to a network for that group. The network in this case is a graph whose vertices represent the members of the group and whose edges (called communication channels) indicate whether two members of the group are in a position to communicate directly.

The directed graph of Fig. 6-38 indicates whether individuals or stations, represented by vertices, can communicate with each other, the arrows characterizing directions in which messages can be sent. The vertex matrix of the graph is given by

$$V = \begin{array}{c} \\ v_1 \\ v_2 \\ v_3 \\ v_4 \\ v_5 \end{array} \begin{array}{c} \begin{array}{ccccc} v_1 & v_2 & v_3 & v_4 & v_5 \end{array} \\ \begin{pmatrix} 0 & 1 & 0 & 0 & 0 \\ 0 & 0 & 1 & 1 & 0 \\ 0 & 0 & 0 & 0 & 0 \\ 0 & 0 & 1 & 0 & 1 \\ 0 & 0 & 0 & 0 & 0 \end{pmatrix} \end{array}$$

Exercise 6-22 Answer the following questions from the matrix.

(*a*) With how many stations does each station communicate directly?

(*b*) How many stations communicate directly with each station?

(*c*) In how many ways does each station communicate with any other through one intermediate station?

(*d*) Can every station communicate directly or indirectly with every other station?

In a communication network assume that a subset of communications (direct through arcs or indirect through paths) between the vertices is prescribed as a necessary condition for the performance of a task. It may then be desired to determine whether it is possible to carry out the task by means of the given network. Generally it may be possible to effect the communications in several ways. Each such way is known as an organization. An optimal organization may be chosen according

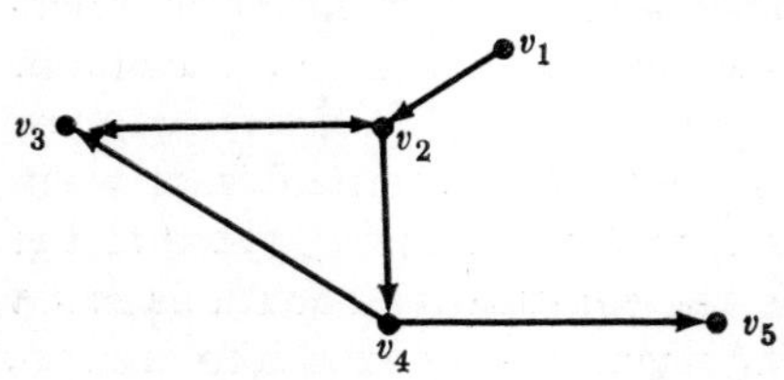

Figure 6-38

to some criterion, e.g., minimization of the global cost of operation using the organization.

Remark: An interesting measure defined on a strongly connected graph is the centrality index, which is a measure of the disparity between the vertices of the graph. If we define a deviation matrix whose coefficients m_{ij} give the minimum length of a path from vertex v_i to vertex v_j and if we form $\sum_{j=1}^{n} m_{ij}$, then the centrality index of vertex v_i is given by

$$\frac{\sum_{i,j=1}^{n} m_{ij}}{\sum_{j=1}^{n} m_{ij}}$$

The global index of the graph is obtained by summing over i.

Exercise 6-23 Compute the centrality indices and the global index of several graphs. Compare with the concept of radius.

Synthesis of a Communication Network

We now consider questions in which the quantity of communications is important. In a communication network (a connected simple graph) one may prescribe a numerical capacity on each edge to reflect the fact that the total communication load between any two points (cities, for example, in telephone communication) may be limited. In this manner we obtain a vertex-by-vertex matrix B, known as the *branch capacity matrix*, which is a symmetric matrix. Each element of B gives the capacity of the edge incident with the two vertices corresponding to the entry. Every main diagonal element is assigned the same constant value d, which is left arbitrary. From B one may determine a terminal capacity matrix T, another symmetric matrix giving the maximum possible communication flow between every pair of vertices. Its entries are obtained by taking all the cut-sets separating the two vertices regarded as terminals, computing the sum of the capacities of the edges in each cut-set, and taking the minimum of these capacities.

The present discussion anticipates certain results which are rigorously established in Chap. 7. In particular, it is shown there that the maximum feasible flow between two specified vertices, of a directed graph, is equal to the capacity of a minimal-capacity cut-set separating these vertices. Moreover, a systematic method for locating such a critical cut-set is presented. Thus B gives rise to a unique T. One question is to determine when a symmetric matrix is realizable as a terminal capacity matrix T. Another is to obtain in a systematic manner a B from a given T (there can be several B's) such that the sum of the branch capacities is as low as possible. This is known as the systematic synthesis of a communication network having given terminal characteristics. We now discuss this realizability problem.

Consider the cut-set corresponding to the minimum entry t_1 in T. This cut-set divides the graph into two components C_1, C_2 and enables the partitioning

$$T = \begin{pmatrix} T_{c_1} & T(t_1) \\ T'(t_1) & T_{c_2} \end{pmatrix}$$

where T_{c_1} and T_{c_2} are the terminal-capacity matrices for C_1 and C_2, and $T(t_1)$ and its transpose $T'(t_1)$ give the entry (everywhere t_1) indicating connection between pairs of vertices, one in each component (with a value t_1 the capacity of the cut-set).

We describe the following process given[12,63] as a necessary and sufficient condition for realizability. Rearrange the matrix as T above, where $t_{ij} \geq t_1$ and T_{c_1} and T_{c_2} are terminal-capacity matrices; i.e., two graphs whose cut-sets have the capacities indicated by these matrices can be constructed and then joined by the minimal-capacity cut-set. By appropriate rearrangement of rows and corresponding columns, T_{c_1} and T_{c_2} can each be brought into the same structural form as T. Hence each can be further partitioned into four submatrices (unless one is already a single element). The process of rearrangement and partitioning can be repeated until finally the diagonal submatrices of the partitioned T consist only of single elements and/or 2×2 symmetric matrices.

Chien gives the following simplified rule: At each partition stage the capacity of each edge which is connected to the subgraph to be partitioned is divided equally between the two partitioned subgraphs. The capacity of the new edge is the difference of the new terminal capacity and one-half the original branch capacity between all other subgraphs and the subgraph to be partitioned.

In the synthesis problem, the sum of the unknown capacities of edges in the branch capacity matrix which correspond to the minimal-capacity cut-set is equated to the capacity of the cut-set. An arbitrary 0 or 1 assignment is made to each of the entries consistent with the equation. Again by checking all possible cut-sets separating vertices in the partitions, the remaining edge capacities are determined. Chien gives a procedure for finding the total arc capacity without synthesis of B.

6-17 SIGNAL - FLOW GRAPHS

Here we introduce general concepts of signal flow in a directed graph or network that has sources $[\delta^-(v) = 0]$ and sinks $[\delta^+(v) = 0]$ and possibly cycles and loops. The presence of cycles and loops introduces the ideas of feedback in a network. In addition to quantities called gains associated with the arcs, we have the notion of a signal x_j transmitted from vertex v_j. The quantity x_j is called the weight of v_j. How the signals are combined with arc gains to derive expressions for the total flow from sources to sinks (often called the gain at the sink) is the object of the analysis of such networks. The relationship between the signals at the different vertices may be left in general functional form or specialized to linear relations. In the latter case, by introducing appropriate operations on the network, it is possible to establish a correspondence between these operations and, for example, the solution of a set of simultaneous linear equations.

The network itself may be the representation of a physical system. One may go directly from the system to the equations, or since it is often helpful in understanding the contribution of the various components of the system to the signal flow, one may represent the system by a network and, because of the correspondence with linear systems, attempt to determine the contribution of the network components to the flow. We begin by introducing network concepts which are pertinent to this type of analysis, although in this brief account none of them will receive great emphasis.

The arcs of a network may be divided into two classes: (1) feedback arcs, i.e., those which belong to cycles or which are loops, and (2) cascade or nonfeedback arcs.[62] In Fig. 6-39 v_2v_3, v_3v_2, v_4v_4 are the feedback arcs and v_1v_2, v_2v_4, v_4v_5 are the cascade arcs. The vertices may also be classified in the same manner depending on whether they belong to cycles or not. A cascade network is one in which every arc is a cascade arc. The vertices of such a network may be so labeled that the subscripts of vertices on every path are in increasing order. Thus one begins by labeling a source vertex (or vertices); then if such a vertex is removed with its connecting arcs, one has a new network with at least one source. This new source is labeled v_2 and removed with its arcs, etc., until isolated vertices which are the sinks of the original graph are reached and are labeled last. The labeling need not be unique. In a feedback network there is at least one feedback vertex. A feedback unit of the network is a subgraph consisting of only feedback arcs and vertices. The feedback units of Fig. 6-39 are given in Fig. 6-40.

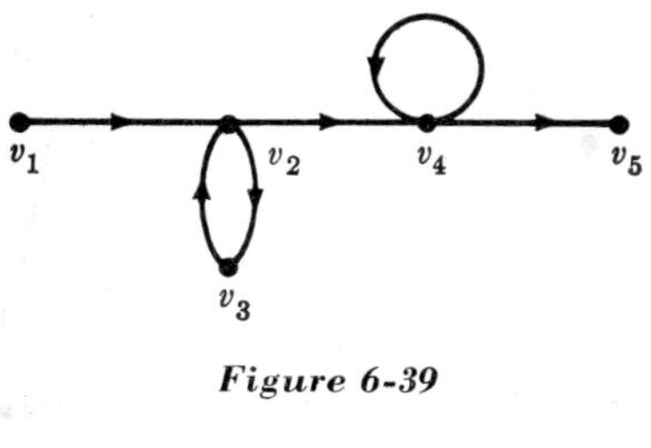

Figure 6-39

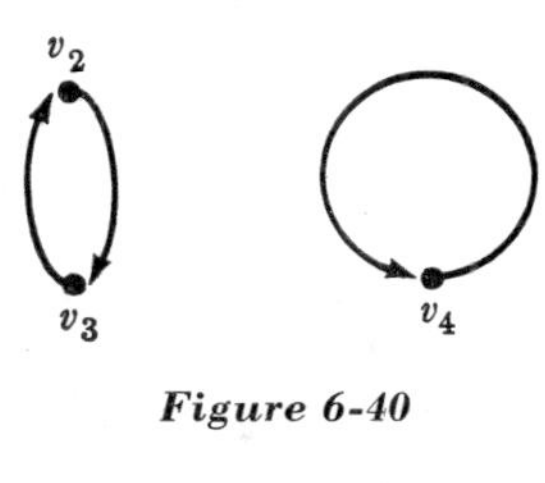

Figure 6-40

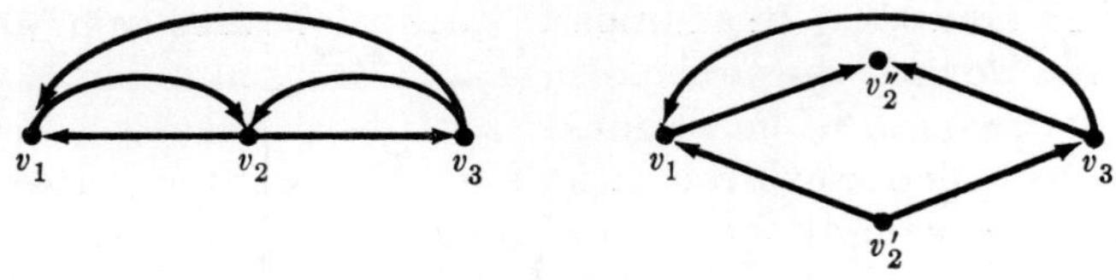

Figure 6-41

A vertex of a feedback unit may be split into two vertices, one of which is incident with only the outgoing arcs of the original vertex and the other with the incoming arcs. In this manner all feedback arcs incident with the original vertex become cascade arcs of the new network. In Fig. 6-41 we give a feedback unit on the left, splitting vertex v_2 on the right. The index of a feedback unit is the minimum number of vertices which must be split to transform all arcs to cascade arcs (i.e., intercept all feedback cycles). The splitting operation is essentially a simplification of the feedback network in order to compute the total flow without the effect of cycles. Once index vertices have been determined (there may be several collections with the same number of vertices that may produce the desired effect), a residual network is formed from these vertices and the sources and sinks. All other vertices are eliminated. Arcs between two vertices, and loops in the residual network, correspond to (1) arcs incident with these vertices in the original network, (2) paths connecting these vertices in the original network but passing through only non-residual vertices, and (3) loops which may be original loops or else cycles between a residual and a nonresidual vertex.

We again illustrate a network and its residual where we mark source and sink with $\times$ (Fig. 6-42). Here the index vertices are v_3 and v_4. A network may be condensed by shrinking each complete feedback unit to a vertex. It is then joined to the other vertices as indicated by the cascade arcs. If more than one such arc was present between the vertices of a unit and another vertex or another unit, they are all represented by a single arc.

As a final concept we mention path inversion in a network, whereby all the arcs of a path are reoriented in the opposite direction. However,

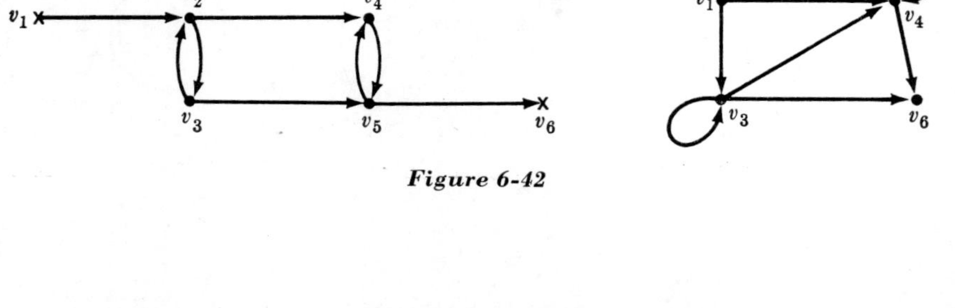

Figure 6-42

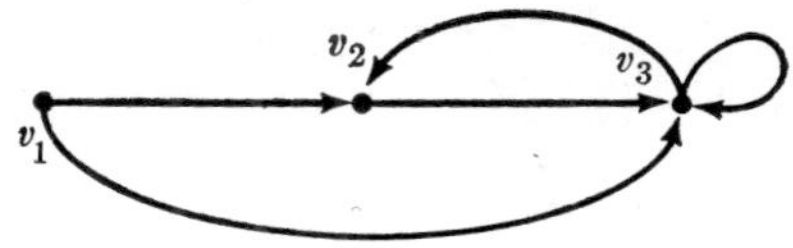

Figure 6-43

there is an additional requirement that all arcs originally incident with any arc of the path at its terminal vertex are disconnected from that vertex as their terminal vertex and are connected to the new terminal vertex (the original initial vertex) of the arc. This is done for all arcs of the path. The operation of arc inversion changes the incidence relationships of the graph.

We now introduce algebraic operations associated with the foregoing concepts. In a cascade network the weight at a vertex is a function of all the weights that are associated with initial vertices of arcs which terminate at the given vertex. These weights presumably undergo transformations through the arcs, determined by the arc gains. We may write, for example,

$$x_j = f_j(x_i,x_k)$$

indicating that the vertex v_j is the terminal vertex of two arcs starting at v_i and v_k. Now x_i itself has a similar functional relationship with other vertices, and so has x_k, etc. Substituting these relationships in f_j relates x_j to the weights at still earlier vertices, etc., back to the source vertices by use of substitution and the process of elimination of intermediate vertices. This type of simplification is not possible in a feedback network. For example, in Fig. 6-43 let x_1 be a given weight at v_1 and let

$$x_2 = f_2(x_1,x_3) \qquad x_3 = f_3(x_1,x_2,x_3)$$

Substitution gives

$$x_3 = f_3(x_1,f_2(x_1,x_3),x_3) \equiv F_3(x_1,x_3)$$

and cannot be explicitly simplified.

Now we turn to linear systems in which the f's are linear in the x's. Here we assume that the signal received at a vertex is the sum of the products of arc gains and the weights at their initial vertices of those arcs which terminate at the given vertex. Thus from Fig. 6-44a we have

$$x_3 = c_1x_1 + c_2x_2$$

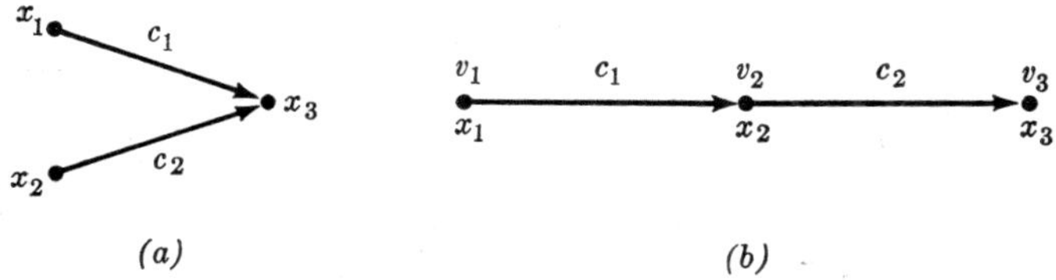

Figure 6-44

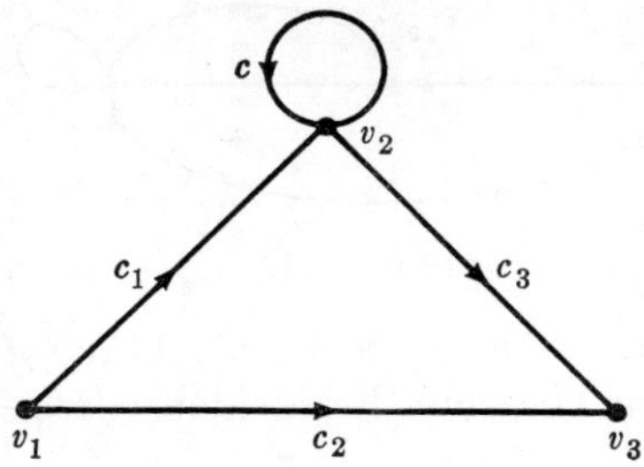

Figure 6-45

Also, from Fig. 6-44*b* we have

$$x_3 = c_2x_2 \qquad x_2 = c_1x_1$$

and hence $\quad x_3 = c_1c_2x_1$

that is, the arc gains multiply when the arcs are in series and may be replaced by a single arc between v_1 and v_3 with a gain of c_1c_2. The arc gains add when the arcs are in parallel between two vertices, and they may be replaced by a single arc with the sum of their gains between the two vertices.

These operations may be applied to a network to determine gains for the arcs of the residual network, whose index gives the minimum number of variables which cannot be eliminated by explicit operations. Each feedback unit corresponds to the solution of a set of simultaneous equations in the weights of its vertices, and each arc in the condensation graph corresponds to the elimination of a variable in terms of the variables of the initial vertices.

The effect of loops and cycles is studied by reducing the network to the residual network and assigning the arcs the appropriate gains according to the foregoing rules. Note that in the residual network, a loop may be obtained as the result of combining an outgoing and incoming arc and hence must have a gain corresponding to the product of their gains. Let v be a vertex with weight x and with a loop whose gain is c. Since the total signal entering v must yield x and since the loop introduces at v a signal cx, the signal from other arcs terminating at v must be $(1 - c)x$ so that the sum would be x. In that case we may replace the effect of the loop by the effect of an arc of capacity $1/(1 - c)$. Thus, for example, the weight at v_3 in the network of Fig. 6-45 takes the form

$$x_3 = c_2x_1 + \frac{c_1c_3}{1 - c}\,x_1$$

Similar reductions can be accounted for in the case of several loops.

Exercise 6-24 Show that the weight at v_4 in the network of Fig. 6-46 is given by

$$i + \frac{ab/(1 - g) + cd/(1 - h) + (afd + ceb)/(1 - g)(1 - h)}{1 - ef/(1 - g)(1 - h)}$$

Note that in order for these network operations to correspond to the solution of a system of simultaneous linear equations, one may formulate the equations of a signal-flow network as shown below and perform operations on the network corresponding to the operations used to solve the

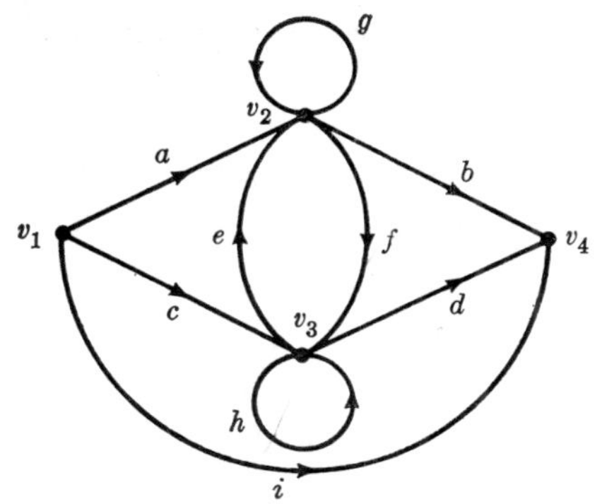

Figure 6-46

equations. Consider the network of Fig. 6-47 with its equations, where x_1 is given and

$$x_3 = 4x_1 - x_4$$
$$x_2 = x_1$$
$$x_4 = 6x_3 - 2x_2$$

If we wish to determine the gain at x_3, for example, we have by elimination:

$$x_3 = \tfrac{6}{7}x_1$$

On the other hand, we can also carry out the reduction by forming the residual graph shown in Fig. 6-48 by using v_3 as the index vertex. Since it is desired to have v_3 as the sink, we adjoin an artificial vertex v_3' with the transformation along the arc v_3v_3' equal to unity, i.e., that arc has unit gain, so that both vertices have weight x_3. The two parallel arcs v_1v_3 yield an arc of gain 6. When the loop whose gain is -6 is replaced by an arc of gain $1/[1 - (-6)]$ and the results are combined to obtain the net gain at v_3', we have $x_3 = \frac{6}{7}x_1$.

Actually, the elimination process and the graph reduction may be made to correspond in their steps. For example, if x_2 is replaced everywhere by x_1, the graph is reduced to another graph in which the gains of arcs passing through x_2 are multiplied by the unit gain of the arc v_1v_2, etc. In this manner we see the correspondence between the manipulation

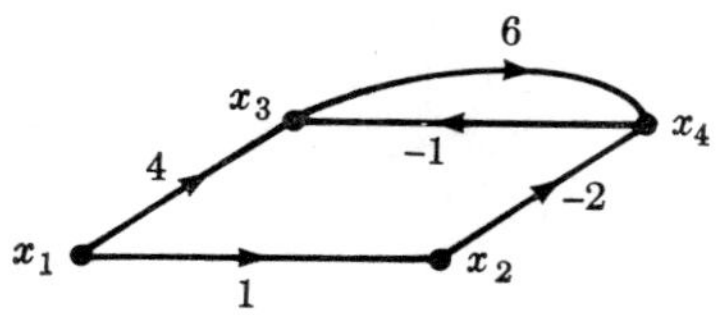

Figure 6-47

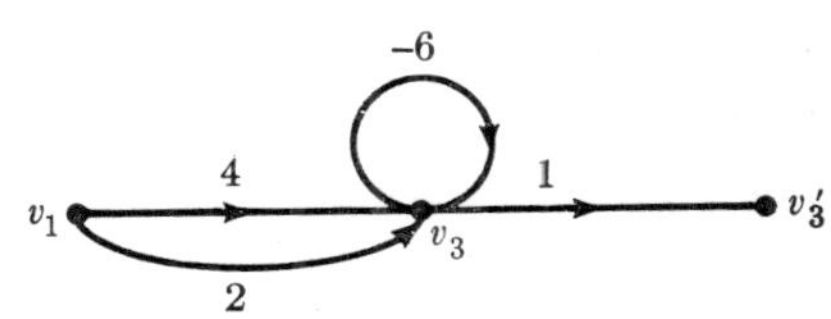

Figure 6-48

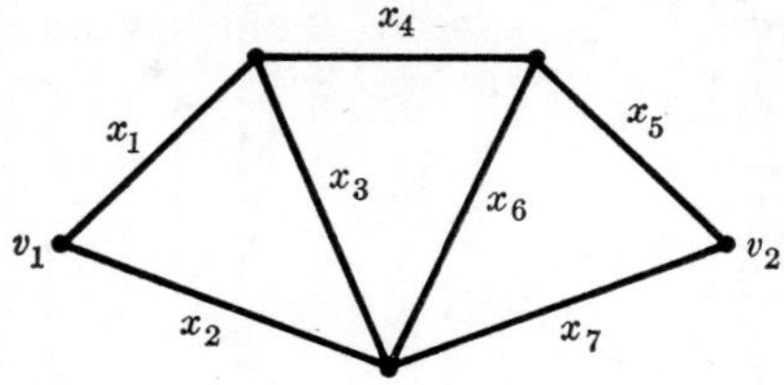

Figure 6-49

of signal-flow networks and the solution of linear equations.

6-18 SWITCHING NETWORKS

Consider a loop-free graph

$$G = (V,E)$$

each of whose edges e_i has an associated variable x_i which can assume only the values 0 and 1. Such a graph may be considered as a mathematical model of a set of interconnected physical devices, such as switches, each of which can be in either of two states: on ($x_i = 1$) or off ($x_i = 0$). Let v_1 and v_2 be two fixed, distinct vertices of G. The graph, together with the *switching variables* $\{x_i\}$, is called a *switching network*, and v_1 and v_2 are regarded as its *terminals*.

If there are n edges (i.e., switches) and if $X = (x_1,x_2 \ . \ . \ . \ ,x_n)$ is a particular assignment of values to the switching variables, the network is considered to be *closed relative to* X if and only if the set of edges for which $x_i = 1$ contains a simple chain joining v_1 and v_2. Otherwise, the network is said to be *open relative to* X. (Alternatively, the network is closed relative to X if and only if v_1 and v_2 lie in the same component of the subgraph determined by those edges for which $x_i = 1$.)

Consider, for example, the switching network of Fig. 6-49. The sets of switching variables corresponding to simple chains joining v_1 and v_2 are (in terms of their subscripts)

$$(1,4,5),\ (1,4,6,7),\ (1,3,6,5),\ (1,3,7),\ (2,7),\ (2,6,5),\ (2,3,4,5)$$

The vectors $X = (x_1,x_2, \ . \ . \ . \ ,x_7)$ for which the network is closed are precisely those vectors which have a 1 in each position corresponding to one of these simple chains and arbitrary values in the remaining positions.

Given a switching network N with n switches, the *switching function* $f(X)$ associated with N is the function defined on the 2^n possible values of X as follows:

$$f(X) = \begin{cases} 1 & \text{if } N \text{ is closed relative to } X \\ 0 & \text{otherwise} \end{cases}$$

Thus far we have implicitly assumed that x_i is independent of x_j whenever $i \neq j$, that is, that the n switches are controlled independently of one another. If this is not the case, not all 2^n values of X are permissible. For instance, suppose that

$$x_7 = x_1 \qquad \text{and} \qquad x_6 = \bar{x}_2 \equiv 1 - x_2$$

in the preceding example. Then the chain determined by subscripts

(2,6,5) is never "closed," since x_2 and x_6 cannot both be 1. Similarly, the chain (1,3,7) is closed whenever

$$x_1 = x_3 = 1$$

since then we necessarily also have

$$x_7 = 1$$

The following general problem arises: Given m independent switching variables $x_1, \ldots, x_m$ and a specified switching function

$$f(X) = f(x_1, \ldots, x_m)$$

under what conditions can a switching network which realizes $f(X)$ be found? Any switching function can be realized by a sufficiently large network each of whose switching variables is equal to one of the m independent variables or to its 1's complement. For example, if $m = 3$ and if $f(X) = 1$ for these values of X:

	x_1	x_2	x_3
X_1	1	0	1
X_2	1	1	0
X_3	0	1	1
X_4	0	0	1

the network of Fig. 6-50 clearly realizes this switching function. Unfortunately, the high degree of redundancy occurring in this type of construction is generally not acceptable. One normally wishes to use as few switches as possible: hopefully only m. In the above example, where $m = 3$, there are only three distinct configurations of interest, shown in Fig. 6-51. These do not provide sufficient variety of structure to realize all 2^8 possible switching functions which may be defined on three independent switching variables.

It is possible to use concepts from graph theory to realize specific switching functions which use a relatively small number of switches. In

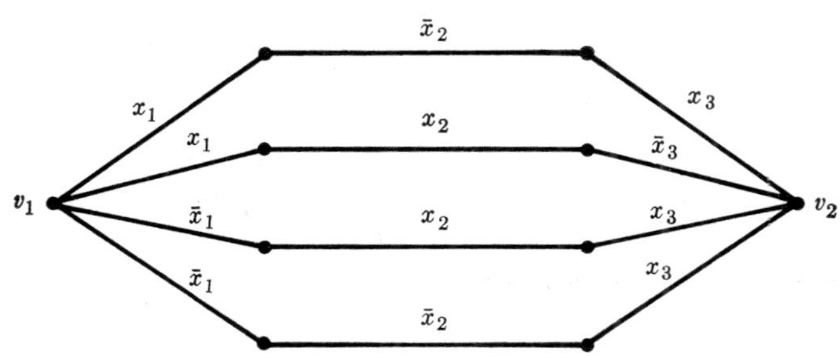

Figure 6-50

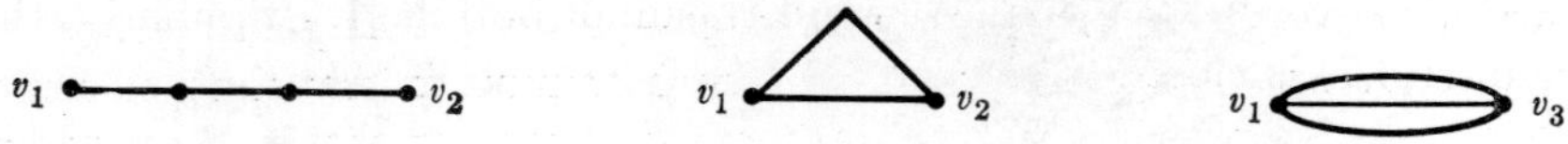

Figure 6-51

Ref. 64, for example, the properties of fundamental circuit and cut-set matrices, discussed in Chap. 5, are brought to bear on this problem. The question of the realizability of a switching function is shown to be related to the problem of whether a given matrix is the circuit matrix of an appropriate graph.

6-19 POWER STATION INTERCONNECTION

Suppose that a set of power stations v_i $(i = 1, \ldots, n)$ wish to be interconnected with each other so as to be capable of sharing each other's power at times of extra need. Depending on the size and potential need of the various stations, suppose that station v_i $(i = 1, \ldots, n)$ desires to be directly linked to at least d_i other stations. For distinct v_i and v_j the cost of directly linking these stations is a real number c_{ij}.

The "interconnection" problem is to design a system of interconnections having minimum total cost which satisfies the demands d_i of the various stations. This is, of course, equivalent to the following graph problem: Given a complete graph with vertices v_i $(i = 1, \ldots, n)$ and with weight c_{ij} attached to edge e_{ij} joining v_i and v_j, find a subgraph which has minimum weight-sum of edges and which has degree at least d_i at vertex v_i. Note that this in turn is equivalent to finding a subgraph having maximal weight-sum, each vertex having degree at most $n - d_i - 1$, and then taking the complementary graph. When viewed in this way, the problem is seen to involve both generalizations of the maximal matching problem discussed at the end of Sec. 6-14.

In the special case where the vertices partition into two classes so that every edge joining a pair of vertices in the same class has weight equal to zero, the interconnection problem is the same as the transportation problem. In general, its theory is much more complicated than in the transportation case. J. Edmonds has found a good algorithm for it (not yet published).

The interconnection problem, or general optimum degree-constrained subgraph problem, essentially includes most of the combinatorial extremum problems which are known to have a good algorithm in the sense that the work involved in applying the algorithm increases only algebraically with the size of the problem. In the case where all $d_i = 2$, interconnection comes rather close to the traveling salesman problem. In fact, a slight modification to the interconnection algorithm will yield a minimum weighted subgraph with degree exactly 2 at each vertex. If that subgraph is connected, it will be a minimum traveling salesman route for distances c_{ij}. Unfortunately, the problem obtained by adding to the interconnection problem the further constraint that the subgraph be connected is still unsolved.

6-20 PRINTED CIRCUITS

A printed circuit is an electrical network which is formed by printing the connecting wires on one or more plane surfaces of a nonconducting material. Most commonly a plastic plate is printed on both its surfaces, and appropriate connections are made between the two sides by means of connecting wires passing through holes in the plate. Since the printed wires are not insulated, no two of them must cross except where a junction is intended.

It is clear that such a network can be printed on a single plane surface if and only if it corresponds to a planar graph. The possibility of printing an arbitrary nonplanar network on two plane surfaces, with interconnections between them, depends on the nature of the interconnections which are permitted. Suppose that the interconnections may occur only at the original junction points (vertices) of the network. In other words, suppose that all vertices are printed on both sides, opposite each other, and each corresponding pair is joined, through the plate, by a conducting wire. We may print some lines on one surface, and some on the other, but we may not create additional holes in the plate.

With this restriction, not every network can be printed on the two sides of a plate. We saw in Chap. 4 that a simple graph G and its complement were not both planar if G had nine or more vertices. Thus we could not print the complete (simple) graph having nine vertices, since to do so would imply that some subgraph of G (the part of the network printed on one side) and its complement (the part printed on the other side) were planar.

If we are permitted to create additional holes and pass conductors through the plate at will, any network can be printed. We could, for example, start by drawing the network in a plane in such a way that at most two wires intersect at any point which is not an intended junction. We could then print the entire network in one plane, except that at each such intersection one of the wires would be brought to the other plane through two holes and a short segment of this wire would be printed on the other plane. The material in Sec. 6-7 establishes bounds on the number of wire intersections which must be treated in this way.

Assume now that it is considered desirable to print a given network in such a way that the fewest possible additional holes (other than those at the original vertices) are required. Any particular way of printing the circuit can be depicted by drawing the network in a plane and using two types of arcs, say, bold and light, to distinguish the edges printed on one side from those printed on the other. The problem can now be reformulated as that of inserting the fewest new vertices of degree 2 (corresponding to points at which a line is brought from one side of the plate to the other) and drawing the graph in such a way that at most two edges cross at any point other than a vertex and that, at such a point, the edges are of opposite types. Naturally, an edge is required to be entirely bold or entirely light.

PHYSICAL SCIENCES APPLICATIONS

6-21 CHEMICAL IDENTIFICATION

A fundamental unsolved problem in graph theory is to find a good algorithm for deciding whether two finite graphs are abstractly the same or, more generally, for deciding whether one graph is a subgraph of the other. The first task is called *graph identification,* and the second task is called *subgraph identification.* For the identification problem to be, strictly speaking, mathematical, "good" must be given some mathematical meaning. Let us say that, for an algorithm which deals with any pair of graphs, "good" means that the amount of work involved in applying the algorithm increases at most algebraically, rather than exponentially, with the number of edges in the pair of graphs to which the algorithm is applied.

J. Edmonds has constructed a good algorithm for tree identification (not subtree identification) which we shall describe. He conjectures that there is no good algorithm for general subgraph identification or even general graph identification. From the practical viewpoint, graph and subgraph identification are important in organic chemistry, where molecules are represented as graphs. Vertices represent atoms, and edges represent the bonds between atoms. The fact that there are several kinds of atoms adds no serious complication to the identification problem. Considerable scientific attention is being devoted to the design of "heuristic" algorithms for chemical identification—algorithms which are not good in our sense but which usually work well in practice. More than recognizing whether two chemical structures are the same or whether one is a piece of the other, chemists are interested in having some sensible system for cataloging chemicals so that a given chemical can be easily found in a catalog or perhaps added to it. (Chemical patent searching at present is not a very pleasant task.)

The algorithm for tree identification involves a cataloging system which determines a certain linear ordering of the set of all finite trees.

However, an identification algorithm may not present a cataloging procedure. For example, there is a good algorithm for deciding whether any two surface maps are combinatorially the same. Maps have a kind of combinatorial "rigidity," so that if there is an identification between map M_1 and M_2 where edge e_1 in M_1 identifies with edge e_2 in M_2, end point v_1 of e_1 identifies with end point v_2 of e_2, and face f_1 containing e_1 identifies with face f_2 containing e_2, then the identification of the rest of the parts of M_1 with parts of M_2 is uniquely and easily determined. Therefore, to see if a map M_1 is identical to a map M_2, one need only try to identify edge e_1 of M_1 with each edge of M_2 in four ways, one for each possible way of identifying vertex v_1 and face f_1 in M_2. This procedure does not immediately suggest a straightforward cataloging system for maps.

For trees T_1 and T_2, each drawn in some particular way in the plane, the same "rigidity" phenomenon can be used to determine easily whether T_1 and T_2 are identical and drawn identically. By "drawn identically" we mean that for each pair of mutually identified vertices, v_1 in T_1 and v_2 in T_2, the edges which meet v_1 are in the drawing of T_1 arranged around v_1 in the same order that the identical edges of T_2 are in the drawing of T_2 arranged around v_2. The difficulty here is that different drawings of an identical T_1 and T_2 provide not much help in verifying identity and the number of different ways to draw a tree generally increases exponentially with the number of edges in the tree.

One may (though we shall not explicitly carry through this line of thought) think of the tree identification algorithm as a procedure for assigning to every tree a canonical way to draw it.

We recall that a *rooted tree* is a tree with one of its vertices distinguished from the others by being called the *root.* The identification algorithm actually applies to rooted trees. Therefore, we first describe a good algorithm for assigning a canonical rooting to any unrooted tree T. Let T_0 be T. Until a T_i which consists of only a single vertex or single edge and its end points is reached, obtain subtree T_{i+1} of T_i by deleting all vertices of degree 1 from T_i and the edges which they meet. For any tree T this process stops with either a unique vertex or a unique edge of T which is called the *center* of T.

In the case of a vertex, let that vertex be the root of T. In the case of an edge, let one of the end points of that edge be the root of T—in particular, the one which results in the lower ranked rooted tree (either end point in case they result in identical rooted trees). We shall rank all rooted trees so that for any two which are not identical one will have lower rank. (Another system for assigning a canonical rooting to every tree T is to choose as root one of the vertices which results in the lowest ranked rooted tree. The vertex chosen is not uniquely defined, but the resulting rooted tree is. Whichever method for rooting is used, two trees are identical if and only if their corresponding rooted trees are identical. The first method seems easier.)

To every rooted tree will correspond a certain finite sequence of integers. No two different rooted trees will have the same sequence. The algorithm for deciding whether two rooted trees are identical is to compute the corresponding sequences and compare them. There is a good algorithm for comparing the sequences, since both the number of terms in a sequence and the greatest term in the sequence are equal to the number of nodes in the tree the sequence represents. It will be evident from the definition that there is a good algorithm for computing a sequence from its tree.

Rooted trees are ranked according to the lexicographical ordering of their sequences. That is, T_1 ranks less than T_2 when there is an integer k such that for $j < k$ the jth terms of the two sequences are equal and such that either there is a kth term in the T_2 sequence and no kth term in the T_1 sequence or the kth term in the T_1 sequence is less than the kth term in the T_2 sequence.

When its root r and the edges which meet r are deleted from a rooted tree T, what remains is several trees, one for each edge meeting r in T. These trees are called the *factors* of rooted tree T. Each factor of T contains exactly one vertex which in T is joined to r. This vertex is regarded as the root of the factor. Thus every rooted tree T which is not merely a single vertex decomposes uniquely into one or more factors which are smaller rooted trees.

A rooted tree T which is merely a single vertex has as its sequence a single term equal to 1. For every other rooted tree T, all but the first term of the sequence S representing T is obtained from the sequences S_i representing the factors T_i of T by arranging the S_i's one after the other in order of ascending rank. Of course, in the case of some identical S_i's it does not matter which of them comes first, but all are separately there in S. Thus all the terms of all the S_i's comprise all the terms of S except the first. The first term of S is 1 plus the sum of the first terms of the S_i's, that is, the first term of S is the number of vertices in T.

Notice that there is a unique 1-1 correspondence between the terms of S and the vertices of T such that the first term of S corresponds to the root of T and the other terms in S correspond to the same vertices as the terms in the S_i's from which they arose. Each vertex v in T is the root of exactly one rooted tree, call it T_v, which is involved in the inductive definition of S. In other words, T_v is a factor of a factor . . . of a factor of T. The magnitude of the term in S which corresponds to v equals the number of vertices in T_v.

The crucial feature of the sequence S for purposes of rooted tree identification is the uniqueness of the ordering of its terms even though tree T has no a priori ordering of its vertices. It is evident from the definition of S that in order for two trees to be identical it is necessary that the sequences representing them be identical.

It remains to be verified that in order for two trees to be identical it is sufficient that the sequences representing them be identical. We simply verify that the following uniquely determined construction for obtaining a tree T from the sequence S which represents it is valid.

If S consists of a single term, then the term equals 1 and T consists of a single vertex. Otherwise, discarding the first term of S, the rest of S partitions into disjoint subsequences S_i ($i = 1, \ldots, n$) of consecutive terms in S such that, where u_i denotes the first term of S_i, u_1 is the second term of S and u_i ($i > 1$) is the earliest term in S after u_{i-1} which is as large as u_{i-1}. There is no term after u_n which is as large as u_n. The S_i's thus defined are the sequences which represent the factors T_i of T. This follows from two facts: (1) The first term of a sequence representing a rooted tree is greater than all its other terms. (2) The sequences representing the factors of T are arranged in S in order of increasing rank.

After constructing the T_i's represented by the S_i's, rooted tree T is constructed from the T_i's by joining an additional vertex, the root of T, to the root of each T_i. Induction on the size of the sequence completes the description of how to construct a rooted tree from the sequence which represents it.

The above method for tree identification seems not to offer much help toward a good algorithm for deciding whether one given rooted tree contains a subgraph identical to another given rooted tree so that roots identify. It would be very interesting to have a good algorithm for this more general identification task.

Exercise 6-25 Show that the sequences obtained by deleting all terms equal to 1 from the sequences S, representing rooted trees, are just as adequate as the sequences S for purposes of tree identification.

6-22 A SIMPLE APPLICATION TO ORGANIC CHEMISTRY

Some organic molecules may be represented as planar graphs with the atoms as vertices and the bonds between the atoms as edges. The simplest such molecules are hydrocarbons of the paraffin series, C_kH_{2k+2}, where there are k carbon atoms considered to be vertices of degree 4 (including bonds to both hydrogen and other carbons) and $2k + 2$ hydrogen atoms considered to be vertices of degree 1. In the figures which follow, the hydrogen atoms will not be displayed as vertices, since they make no contribution to isomerism (i.e., to different types of bonds in a molecule with the same number of atoms).

The total number of vertices n will then be $3k + 2$, and the total

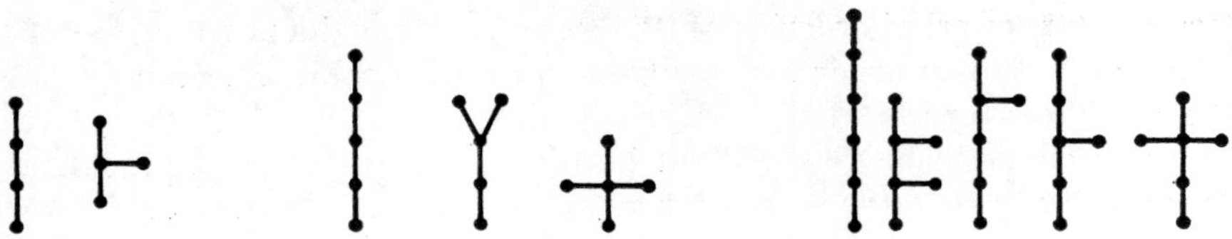

$k = 4$ (butane) $k = 5$ (pentane) $k = 6$ (hexane)

Figure 6-52

number of edges is $m = \frac{1}{2}(4k + 2k + 2) = 3k + 1$. Since the number of independent circuits $m - n + 1$ is equal to zero, such molecules may be regarded as trees. Hence, there can be no multiple bonds. The trees formed from k vertices include all possible carbon configurations in which the atoms are attached. When $k \geq 4$, more than one distinct configuration of the carbon atoms is possible. In each of the parts of Fig. 6-52 it will be seen that the requisite number of hydrogens is present if we merely add hydrogen bonds until each carbon atom vertex is of degree 4. Each distinct configuration is called an isomer.

The first diagram in each group of Fig. 6-52 represents what is called a *straight-chain hydrocarbon*, for obvious reasons, and the remaining diagrams represent *branched-chain hydrocarbons*. The names of the latter are often prefixed with iso-, for isomeric. These substances appear naturally as mixtures of all the possible stable isomers. With increasing k, the properties of the various isomers become quite different, and hence, in order to distinguish among them, it becomes necessary to know how many isomers can be present. Cayley[11] published, in 1875, the first paper applying graph theory to chemistry to solve this problem "without mistake or repetition." He represented a molecule by a rooted tree, considered all possible configurations, and then decided which forms were chemically identical. For example, for $k = 5$, there are nine rooted trees, as in Fig. 6-53; but six of them are seen to be chemically identical with others, so that the three figures shown for $k = 5$ in Fig. 6-52 are really the only possible isomers.

The problem of repetition can be solved by representing all the figures as "centered" or "bicentered" trees, as in Fig. 6-54. A centered

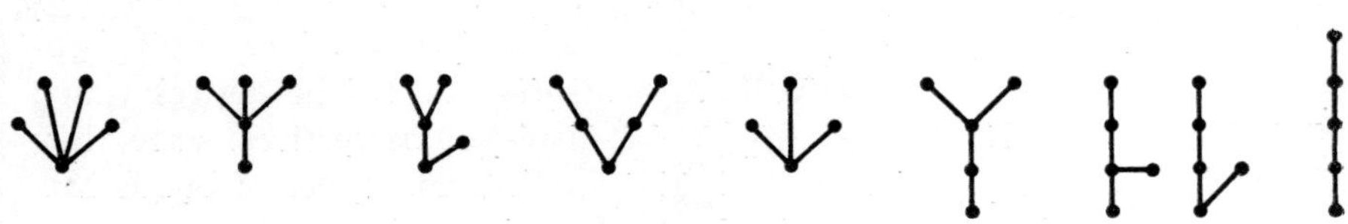

Figure 6-53

tree is understood to have two or more main branches, all of equal height, incident with its root; while a bicentered tree has two roots with one or more main branches incident with each root. It is essential that the branches here also be of equal height. When this is so, we find that the duplication is removed, and three chemically distinct isomers are again present.

Centered trees Bicentered tree

Figure 6-54

By drawing all possible isomers according to this simple rule, Cayley determined the number of structural isomers for the paraffin series up to $k = 13$. His results are shown in Table 6-1. A more sophisticated result is that of Schiff,[81] who published the second paper in this area, also in 1875. His method requires that these single-bonded, or saturated hydrocarbons, with $k \geq 5$ be written according to a symmetric scheme of main chains and shorter side chains as illustrated in Fig. 6-55. When trees are arranged in this manner, it is found without difficulty that the total number of vertices k may be expressed as a function of the number of vertices K on the main vertical stem (using four right triangles taken in pairs) as follows:

$$k = 2\left(\frac{K}{2}\right)^2 = \frac{K^2}{2} \qquad K \text{ even}$$

$$k = \left(\frac{K+1}{2}\right)^2 + \left(\frac{K-1}{2}\right)^2 = \frac{K^2+1}{2} \qquad K \text{ odd}$$

Table 6-1

$k =$	1	2	3	4	5	6	7	8	9	10	11	12	13
Centered.......	1	0	1	1	2	2	6	9	20	37	86	183	419
Bicentered......	0	1	0	1	1	3	3	9	15	38	73	174	380
Total........	1	1	1	2	3	5	9	18	35	75	159	357	799

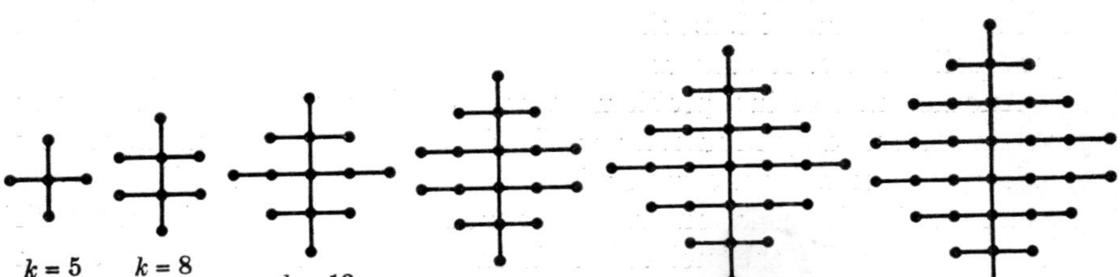

$k = 5$ $k = 8$ $k = 13$ $k = 18$ $k = 25$ $k = 32$

Figure 6-55

A rather complex method can be evolved for calculating the number of isomers for k up to 10; then additional refinements are necessary. The method may be extended to double-bonded molecules with C_kH_{2k} and triple-bonded molecules with C_kH_{2k-2}. Other research has determined the number of structural isomers for much more complicated substituted molecules.

6-23 GRAPH THEORY AND STATISTICAL MECHANICS: TWO EXAMPLES

We have already alluded to the fact that a number of combinatorial problems of graph theory have a statistical mechanics origin. Since any specific physical example requires substantial background, we shall limit our illustration of the transfer of ideas from physics to graph theory to the well-known (1) Ising and (2) dimer problems. A ferromagnetic substance has been observed to lose its magnetic intensity as a function of temperature according to a decreasing curve which exhibits a sharp drop at a certain temperature level known as the Curie point. The problem is to develop a model which explains this phenomenon of sharp drop. The problem can be defined on a lattice, and the hope is to write down the partition function which for a finite lattice is an analytic function of the temperature and does not account for the sharp drop. However, in the limit this function can reveal this property.

The above problem is defined on a three-dimensional lattice and is based on the assumption that ferromagnetism is caused by interactions between the spins of certain electrons in the atoms making up a crystal. These interactions give rise to the following generalized graph-theoretic version of the Ising problem. It is required to find the generating function of the number of admissible labeled subgraphs with k edges of a labeled graph which is an n-dimensional lattice. A subgraph is admissible if all its vertices are of even degree. So far this problem has been solved only for $n = 1$ (Ising) and $n = 2$ (Onsager). Figure 6-56 is an example of an Ising graph. Reference 65 is an excellent expository paper on Ising's problem. Montroll has other papers on lattice statistics with a good collection of references.

Another interesting problem in statistical mechanics is the dimer problem (attacked by Montroll with Pfaffians), which arises in the theory of adsorption of diatomic molecules (dimers) on surfaces. Here every diatomic molecule sticks on the two-dimensional lattice such that each of the two atoms of the molecule falls on a lattice point. The problem is to find the number of ways nearest neighbors can be connected on a doubly periodic lattice (a rectangular lattice each opposite pair of whose sides are identified, yielding a lattice on a torus) so that no point is left

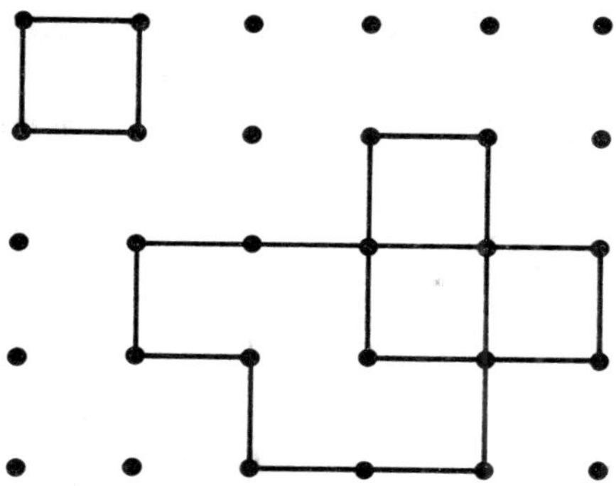

Figure 6-56

free of some atom. This problem is equivalent to that requiring the number of ways to cover a rectangular board divided into small squares with dominos each of which consists of two squares (of the same size as those on the board) glued together. Clearly, the board must have an even number of squares and must also have an even number at least along one side. A more elaborate mathematical version of the lattice problem is the paving problem, which requires the number of ways to cover a two-dimensional lattice of n squares with n_1 single squares and n_2 double squares, where $n_1 + 2n_2 = n$. (See also Sec. 6-13.)

6-24 A GENETICS PROBLEM

The following problem due to Benzer[2] is related to the possible linkage of allowable configurations of the chemical constituents of genes. If these constituents are molecules in three dimensions, they may be linked together according to some pattern. This linkage may be represented in three dimensions by assigning a vertex to a molecule and joining the vertices by straight lines if the corresponding molecules are linked together. The question now is whether it may be possible to arrange the molecules on the same straight line so that they still have the same linkages. On the line, each molecule is represented by an interval, and the fact that it is linked to another is shown by intersecting the intervals. Thus the problem is to find conditions for representing the linkage graph on a line in terms of appropriately overlapping intervals, i.e., to associate intervals with the vertices in such a way that two intervals intersect if and only if they correspond to adjacent vertices. A graph which can be represented in this manner will be called an *interval graph*. A triangle may be represented on the real line by the overlapping intervals

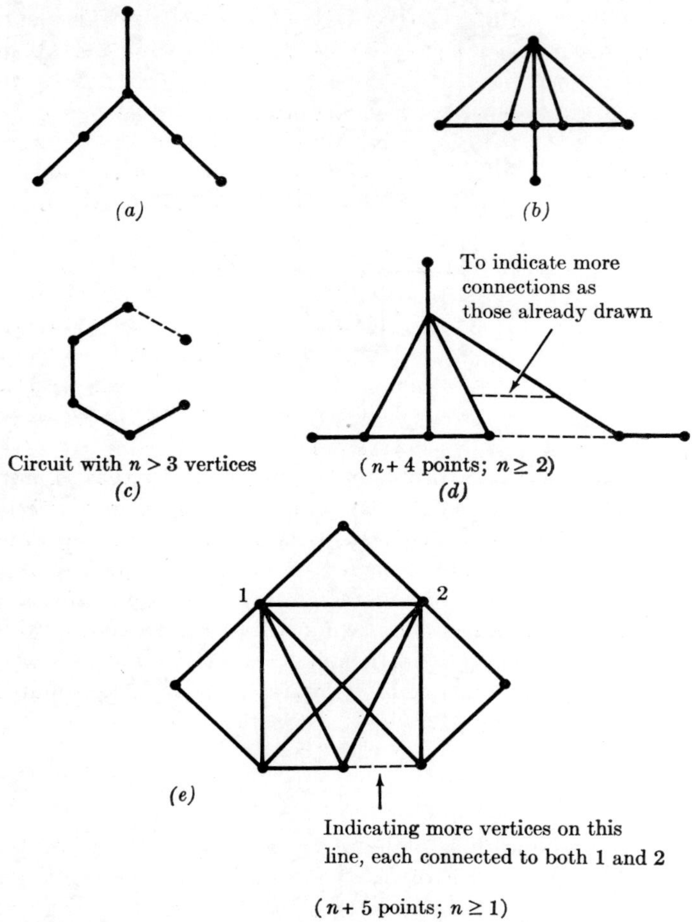

Figure 6-57

(0,3), (1,4), and (2,5). On the other hand, a circuit of more than three vertices cannot be represented in this manner.

It has been shown[59] that if a graph does not contain one of the five types of subgraphs embedded exactly as shown in Fig. 6-57 (i.e., the vertices of the subgraph have only the connections indicated here) then it can be represented by intervals on a line. The biological problem in its generality is concerned with mutant genes where one is given information as to which pairs intersect (i.e., have linkages) and it is desired to test the compatibility of this information with a linear model of the gene.

Fulkerson and Gross[25] have shown that the problem of characterizing an interval graph is a special case of the following problem which they have solved: Given a (0,1) matrix A (that is, a matrix whose entries are 0's and 1's), when can its rows be permuted so that the 1's in every column of the resulting matrix appear consecutively? Note that one can associate a graph with an arbitrary matrix A in various ways. One way, for example, is to form the intersect graph of the columns of A. (The intersect graph of a family of n sets is obtained by associating a vertex with each set and joining a pair of vertices by an edge if their corresponding sets have a nonempty intersection.)

In solving the problem, the concept of an "overlap" (to be distinguished from "intersect") graph and that of a component graph are introduced. When applied to interval graphs, A must be taken to be the dominant clique-vertex incidence matrix of the graph. (A set of vertices every pair of which is joined by an edge is known as a clique of the graph. If the family of all such sets of vertices are partially ordered by set inclusion, the maximal elements are called dominant cliques of the graph. Since two vertices are joined by an edge if and only if they belong to some dominant clique, the dominant clique vs. vertex matrix characterizes the graph.) The result is that a graph is an interval graph if and only if the dominant clique-vertex incidence matrix of the graph has the consecutive 1's property. Gilmore and Hoffman[31] have proved that a graph G is an interval graph if and only if every quadrilateral in G has a diagonal and every circuit of odd length in the complementary graph has a triangular chord. A triangular chord of a circuit through vertices $v_1, \ldots, v_k$ is any of the edges (v_i,v_{i+2}), $1 \leq i \leq k-2$, (v_{k-1},v_1), or (v_k,v_2).

HUMAN SCIENCES APPLICATIONS

6-25 GRAPHS AND CYBERNETICS[1]

In reading the present section, the reader should compare and contrast the concepts presented here with those discussed in Sec. 6-30. He will find an overlapping of concepts, but the common ideas appear in two quite different settings in these two sections. Also, the transformations discussed here are more general than the special transformation in which each member of a group is mapped into a new member by adding (assuming an additive group) a fixed member of the group.

The concepts of change and of selection are of paramount importance in the study of biology and evolution. Let us consider first the concept of change, e.g., the change in color of human skin in the presence of sunlight. We call the pale skin the *operand*, the dark skin the *transform*, and the sunlight the *operator*. Let the change itself be called the *transition*. The transition may be represented as

$$\text{Pale skin} \rightarrow \text{dark skin}$$

A set of transitions, by one operator, over a set of operands will be called a *transformation*. It may be represented schematically as

$$\downarrow\begin{matrix} a & b & c \\ a' & b' & c' \end{matrix}$$

The transformation is *closed* if the set of transforms contains no elements not already present in the set of operands; thus

$$\downarrow\begin{matrix} a & b & c \\ c & a & a \end{matrix}$$

is closed. A transformation is *single-valued* if it converts one operand to precisely one transform. (The set of transforms may contain duplicate elements). Thus

$$\downarrow\begin{matrix} a & b & c \\ c & a & a \end{matrix}$$

is single-valued. On the other hand,

$$\downarrow\begin{matrix} a & b & c \\ c \text{ or } d & e \text{ or } f & g \text{ or } h \end{matrix}$$

is not. Furthermore, the transformation is one to one if the set of transforms contains no duplicate elements. For example,

$$\downarrow\begin{matrix} a & b & c \\ c & a & b \end{matrix}$$

is one-to-one but

$$\downarrow\begin{matrix} a & b & c \\ c & a & a \end{matrix}$$

is not.

Let us now consider an example of successive single-valued transformations on a set of operands. This example will demonstrate "selection" on a set of operands:

A	B	C	D	E	F	G	H	I	J	K	L	M	N	O	P	Q	R	S	T	U	V	W	X	Y	Z
D	I	F	F	E	R	E	N	C	E	B	E	T	W	E	E	N	A	M	E	R	M	A	I	D	A
F	C	R	R	E	A	E	W	F	E	I	E	E	A	E	E	W	D	T	E	A	T	D	C	F	D
R	F	A	A	E	D	E	A	R	E	C	E	E	D	E	E	A	F	E	E	D	E	F	F	R	F
A	R	D	D	E	F	E	D	A	E	F	E	E	F	E	E	D	R	E	E	F	E	R	R	A	R

(The second line is a fragment of a sentence selected from Mark Twain's "Letters from Earth"). Note that after only four transformations the original set of 26 elements has been reduced to 5. The foregoing successive transformations may be represented by a directed graph which Ashby calls the kinematic graph (Fig. 6-58). Note that the vertices

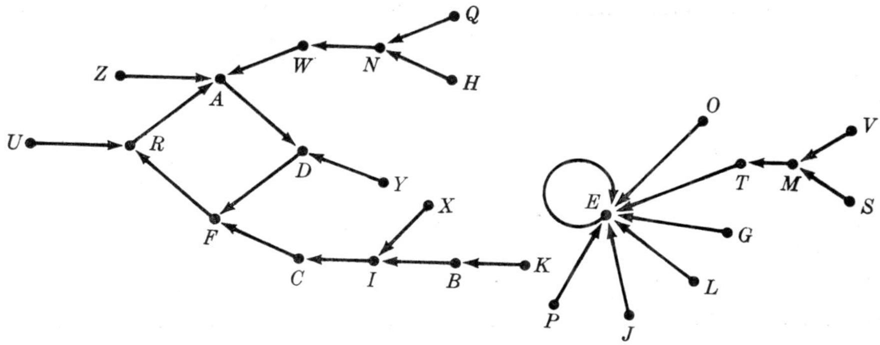

Figure 6-58

incident with the cycles (or loops) will not be eliminated under successive transformations. The cycles of a kinematic graph are called "equilibrium basins," since they are invariant or periodic under successive transformations. In evolutionary terminology, the elements incident with cycles are the "survivors" under this transformation. (The elements in the original set of operands might be the elements in a gene pool under radiation.) Note that no element will be lost under transformation only in the special case of a 1-1 transformation. In graph-theoretic terms we have the following theorem:

Theorem 6-11 The kinematic graph consists of a number of vertex-disjoint simple cycles if and only if the transformation is one to one.

Proof: If the transformation is one to one, then the negative and positive degree of each vertex is unity. Thus the arcs can be partitioned into simple cycles, which are necessarily vertex-disjoint. If the arcs form a set of vertex-disjoint cycles, the positive and negative degree of each vertex is unity, and so the transformation is one to one.

The identity transformation $\begin{array}{c|ccc} & a & b & c \\ \downarrow & a & b & c \end{array}$ will have a kinematic graph consisting of $k = 3$ loops (Fig. 6-59). When the transformation is not one to one, the number of times it must be applied to select the minimal number of elements equals the number of arcs in the longest of the paths from a vertex not in a cycle to the first vertex in a cycle.

The important point to Ashby as a biologist is that under any other than the 1-1 transformation, the variety in the set of elements will decrease and that it can never increase. Some subset of the original set will be "selected" by the transformation.

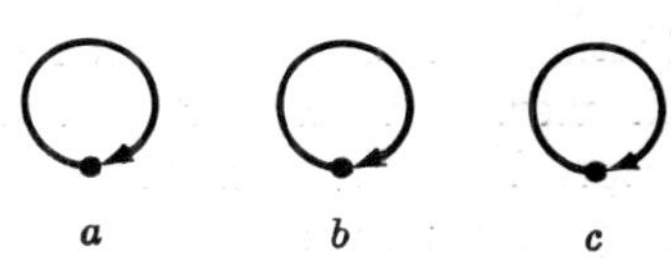

Figure 6-59

Transformations can be represented

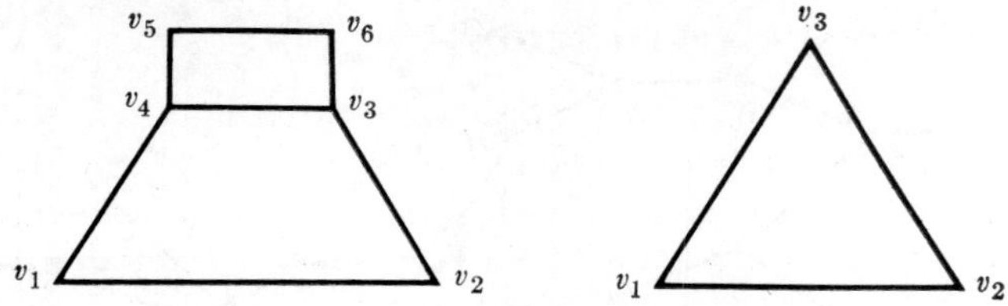

Figure 6-60

by means of the vertex matrix, for example,

$$\downarrow\begin{array}{ccc} a & b & c \\ c & a & a \end{array} \quad \text{becomes} \quad \begin{array}{c|ccc} & a & b & c \\ \hline a & 0 & 0 & 1 \\ b & 1 & 0 & 0 \\ c & 1 & 0 & 0 \end{array}$$

The problem of which elements will be selected by the mth repetition of a transformation is equivalent to asking which columns of the matrix will contain positive elements after raising the matrix to the mth power.

The matrix of a 1-1 transformation is the only matrix of a single-valued transformation having unit entry in each row and column; therefore it is the only matrix capable of being primitive. Thus

$$\downarrow\begin{array}{ccc} a & b & c \\ c & a & b \end{array} \quad \text{yields} \quad \begin{array}{c|ccc} & a & b & c \\ \hline a & 0 & 0 & 1 \\ b & 1 & 0 & 0 \\ c & 0 & 1 & 0 \end{array}$$

The third power of this matrix is a matrix all of whose elements are equal to unity. The matrix will be primitive only if the kinematic graph is connected; otherwise, there will be k primitive subgraphs. This treatment can be extended to multivalued transformations equivalent to Markov chains.

In addition to the notion of an isomorphism for transformations which are consistent with our definition for graphs, Ashby also defines a homeomorphism for transformations. The requirements for homeomorphism are less rigid than those for isomorphism. Two sets of operands are homeomorphic if a many-to-one transformation applied to the more complex can reduce it to a set which is isomorphic with the simpler. In graph terms, two graphs would be homeomorphic if an appropriate collapsing of subgraphs of the graph with the greater number of vertices to single vertices yields a graph which is isomorphic with the simpler graph. The two graphs of Fig. 6-60 are homeomorphic.

6-26 SOCIAL SCIENCE APPLICATIONS

A directed graph can be used to represent a social hierarchy or kinship. Occasionally, studies in anthropology reveal complicated kinships for which the following formulation is applicable. We give a simple illustration of the ideas. For example,[91] we consider a family group consisting of David (D), his son John (J) and daughter Grace (G), John's wife Sylvia (S) and two sons Michael (M) and Richard (R) and daughter Emily (E), and Grace's son Ben (B). The relationships P (being a son of) and Q (being a child of) are illustrated by the directed graphs of Fig. 6-61. The associated matrices of these two graphs are given by

$$P = \begin{array}{c} \\ \text{D} \\ \text{G} \\ \text{B} \\ \text{J} \\ \text{S} \\ \text{E} \\ \text{R} \\ \text{M} \end{array} \begin{array}{c} \begin{array}{cccccccc} \text{D} & \text{G} & \text{B} & \text{J} & \text{S} & \text{E} & \text{R} & \text{M} \end{array} \\ \begin{pmatrix} 0 & 0 & 0 & 0 & 0 & 0 & 0 & 0 \\ 0 & 0 & 0 & 0 & 0 & 0 & 0 & 0 \\ 0 & 1 & 0 & 0 & 0 & 0 & 0 & 0 \\ 1 & 0 & 0 & 0 & 0 & 0 & 0 & 0 \\ 0 & 0 & 0 & 0 & 0 & 0 & 0 & 0 \\ 0 & 0 & 0 & 0 & 0 & 0 & 0 & 0 \\ 0 & 0 & 0 & 1 & 1 & 0 & 0 & 0 \\ 0 & 0 & 0 & 1 & 1 & 0 & 0 & 0 \end{pmatrix} \end{array}$$

$$Q = \begin{array}{c} \\ \text{D} \\ \text{G} \\ \text{B} \\ \text{J} \\ \text{S} \\ \text{E} \\ \text{R} \\ \text{M} \end{array} \begin{array}{c} \begin{array}{cccccccc} \text{D} & \text{G} & \text{B} & \text{J} & \text{S} & \text{E} & \text{R} & \text{M} \end{array} \\ \begin{pmatrix} 0 & 0 & 0 & 0 & 0 & 0 & 0 & 0 \\ 1 & 0 & 0 & 0 & 0 & 0 & 0 & 0 \\ 0 & 1 & 0 & 0 & 0 & 0 & 0 & 0 \\ 1 & 0 & 0 & 0 & 0 & 0 & 0 & 0 \\ 0 & 0 & 0 & 0 & 0 & 0 & 0 & 0 \\ 0 & 0 & 0 & 1 & 1 & 0 & 0 & 0 \\ 0 & 0 & 0 & 1 & 1 & 0 & 0 & 0 \\ 0 & 0 & 0 & 1 & 1 & 0 & 0 & 0 \end{pmatrix} \end{array}$$

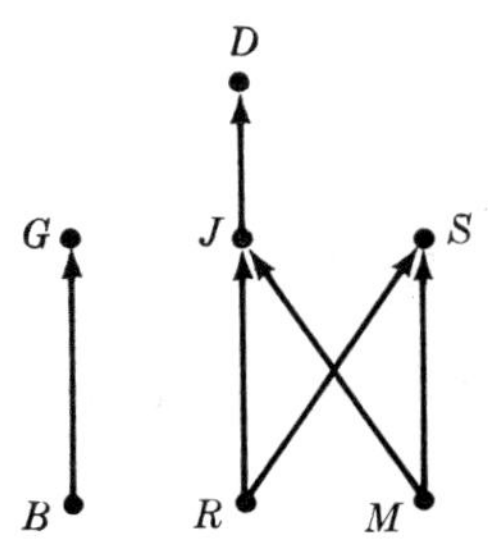

Graph of relation P (son of)

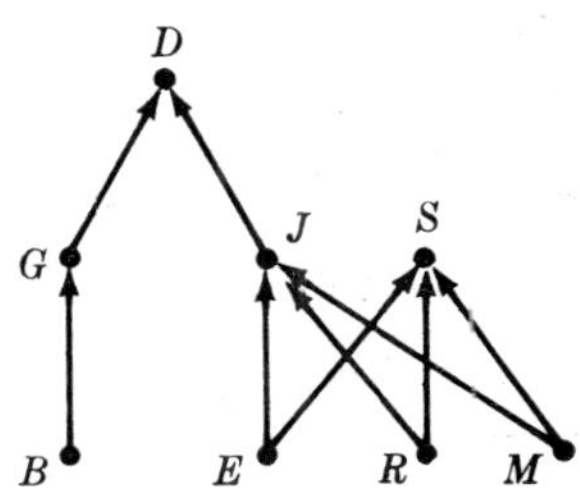

Graph of relation Q (child of)

Figure 6-61

We can then compute the product of these two matrices to obtain the matrix for the relation PQ (being a son of a child of, i.e., a grandson of). We have

$$PQ = \begin{array}{c} \\ D \\ G \\ B \\ J \\ S \\ E \\ R \\ M \end{array} \begin{array}{l} \quad D \quad G \quad B \quad J \quad S \quad E \quad R \quad M \\ \left(\begin{array}{ll} 0 & \\ 0 & \\ 1 & \\ 0 & \qquad\qquad \text{0 elsewhere} \\ 0 & \\ 0 & \\ 1 & \\ 1 & \end{array}\right) \end{array}$$

Here is another illustration of the applicability of graph theory to the social sciences. In an experiment in the study of the effectiveness of communication patterns, each of five subjects was given five symbols out of a set of six symbols[57] shown here: △, +, ∗, □, ○, ◇. Only one symbol was held in common by the five subjects during each trial. Records were kept on the number of messages and the amount of time necessary for each subject to determine, by communicating with others, which was the common symbol.

The four communication patterns of Fig. 6-62 were used. (Here communication is assumed possible in both directions along each edge.) These patterns provided increasing degrees of centrality for subject C and increasing degrees of isolation for the other four subjects as a group. The structure of the graphs permits communication flexibility; i.e., within any given pattern there are several ways to transmit information and a leader could have developed at any vertex (but C was most often the leader). These patterns allowed personal evaluation by the subjects on their satisfaction with different vertex positions. It was found that the wheel is the best organized for communicating, i.e., required the smallest number of information messages, and the circle the least organized, i.e., required the largest number. The time for solution on the

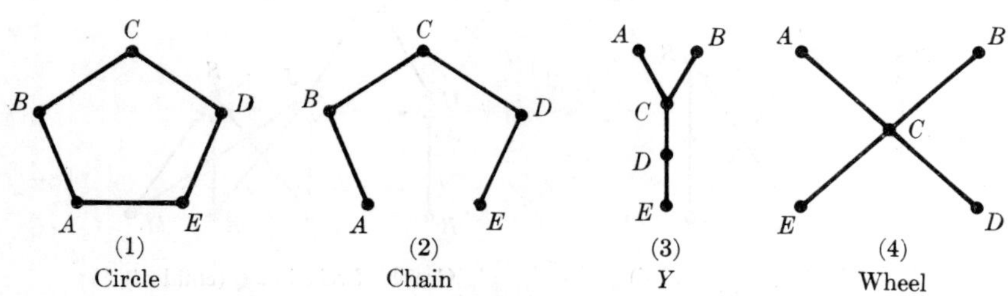

Figure 6-62

wheel was less than for the other patterns, all three of which required nearly the same time. The greatest satisfaction of the whole group was with the circle and the least with the wheel. The position occupied by centerman *C* was most favored for all patterns where position made a difference.

In another experiment three subjects *A*, *B*, *C* were assigned the task of reconstructing a total list of 25 words where each of them had a partial list.[45] Their lists contained sequential word couples such that the second of a couple on, for example, *A*'s list was the first word of a couple on *B*'s or *C*'s list. This idea is illustrated in the accompanying table.

Total	*A*	*B*	*C*
1. South	south	south	
2. Through		through	through
3. Fine	fine		fine
4. Gate	gate	gate	
5. Do		do	do
6. West	west		west
7. Have	have	have	
8. Rain		rain	rain
. . . etc.			etc.

Another task the subjects were assigned was to reconstruct a sentence 25 words long (where no word has more than three syllables) when each subject had a partial list of the words as they appear in sequence in the sentence. One such sentence was: "The picture we saw was painted by an old woman who had been taught how to mix the colors by one of the native artists." One partial list might have been:

picture was an who been to the colors of native

The third task assigned was an anagram of three-letter words made from words of nine letters each with four vowels, e.g., abolished, courtesan. Subjects had to maximize the group score, and the same word on each paper counted three times. The five communication patterns of Fig. 6-63 were used. Note that in this case communication was limited to specified directions. For the different patterns the length of time, the number of messages, and the difficulty in correcting errors in transmission were measured. The subjects were in different rooms, connected only by earphones and microphones. The wires were supplied with white noise, i.e., all frequencies at the same intensity, to dampen sounds that went through the walls. In addition, each task was tested under three different signal-to-noise ratios, +6, −2, and −10 decibels, which gave

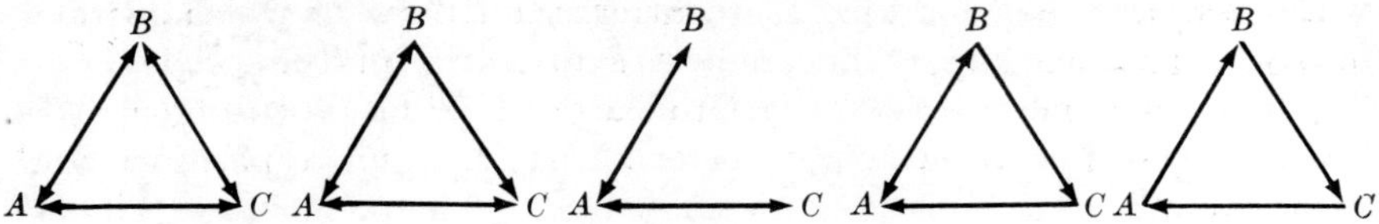

Figure 6-63

correct reception 85, 66, and 24 per cent of the time, respectively. It was found that the first pattern was best for the first task, the third pattern was best for the second task, and all nets were the same for the third task. The fifth pattern was the worst for the first and the second tasks, and the different noise levels only accentuated the disadvantage of this pattern.

6-27 A MATHEMATICAL MODEL OF DISARMAMENT[80]

One may define the set Σ consisting of a finite number of states, where each state represents a level of arms available to two competitive opponents X and Y under conditions of stability. Stability, balance, or equilibrium is an essential criterion in this formulation. It requires that neither opponent judges his position, i.e., state (defined below), to be weaker than his opponent's.

The elements E_j $(j = 1, \ldots, p)$ of Σ are vectors

$$E_j = (a_{1j}, \ldots, a_{nj}; b_{1j}, \ldots, b_{nj})$$

where a_{kj} denotes the number of units in a weapons system (e.g., a gun, an amount of information, an economic factor, etc.) of type k available to X at step j of the disarmament process and b_{kj} is a similar quantity available to Y. Each of X and Y will select a set of rules to be applied to an initial state of arms to produce a new state. The same rules or different rules may be applied to the new state to obtain a third state, etc. X's total scheme of arms reduction will produce a set of states any of which need not be acceptable to Y. The object is to find those states on which there can be agreement and then establish rules of reducing arms on these states. It is assumed that the initial state to which the rules are applied will be considered as an equilibrium state by both sides not necessarily for military reasons only but also for political, economic, or other reasons. Our subsequent discussion will show the dependence of the entire process on compensating factors used by both sides. Our purpose now is to show how one obtains Σ.

An equilibrium or stable state is a state that is an admissible state of both sides. A natural criterion for selecting admissible states is for

X to take $a_{kj} = \alpha_{kj}b_{kj}$, where α_{kj} is known as the compensating factor. In fact, comparisons must be made with all of Y's weapons.

It is clear that a common denominator for units of arms reduction is essential to this formulation. Thus if there is a numerical superiority of one weapon type, this may be compensated for by superiority in the reverse order by another weapon. The deficiency (or advantage) α_{kj} must be measured in the basic unit of both weapons. Indeed, the compensation may be on the basis of several other weapon types, rather than one other weapon type, and hence a unit of common measurement is required.

A single judgment factor may not be adequate to determine whether a state is admissible; hence, we define a state E_j to be in the set Σ_x of states admissible to X if $\|\alpha_j\|$—called the norm of

$$\alpha_j = (\alpha_{1j}, \ . \ . \ . \ ,\alpha_{nj})$$

the vector of compensating factors—is not less than a specified value α selected by X. The norm $\|\alpha_j\|$ is a measure of all α_{kj} $(k = 1, \ . \ . \ . \ , n)$. Since the importance of weapons will differ from one type to another, a useful norm to take for α_j is

$$\|\alpha_j\| = \sum_{k=1}^{n} w_k\alpha_{kj}$$

where w_k is a weight of the importance of weapon type k averaged over different conflict situations. Similarly, one may introduce β_{kj} and $\|\beta_j\|$ for the definition of Σ_y, the set of Y's admissible states. Note, for example, that the state $(0, \ . \ . \ . \ ,0;1, \ . \ . \ . \ ,1)$ is admissible to Y but not to X, and hence it is in Σ_y but not in Σ_x. Similarly, $(1, \ . \ . \ . \ ,1;0, \ . \ . \ . \ ,0)$ is in Σ_x but not in Σ_y. It is easy to assume that such states are admissible, since one side will have zero arms. Finally, the set of admissible equilibrium states (of both X and Y) is $\Sigma = \Sigma_x \cap \Sigma_y$; that is, it is the common part of both.

One problem of arms control is to prescribe rules for reducing arms. (Note that we are using reduction in the broad sense, since certain individual weapon types may actually require increase.) Whatever the rule, its purpose is to effect a transition from one state in Σ to another state in Σ. The rules developed by each competitor need not coincide with those of the other, because, for example, Σ_x will naturally contain states not in Σ_y, and conversely. Thus the problem is to find those rules which provide a sequence of transitions between states of Σ without ever going into a state not in Σ. In practice, it is usually the rules that define the states.

Assume now that we have a listing of all the elements of the set Σ.

(Clearly, this is difficult in practice, because neither side would be willing to acknowledge his compensating factors. However, one of the advantages of this approach is to show how, by offering the other side various rules which he accepts or rejects, it is possible to guess at approximate magnitudes of his compensating factors.) These elements are obviously finite, although the escalation of weapons increases the size of Σ in time. For simplicity, let us assume that the states of Σ are $E_1, \ldots, E_r$.

If the problem of finding transition rules is solved, the next problem is to find a method of applying these rules to obtain all those states which fall on a path of arms reduction from a given initial state, for example E_1, to any intermediate state E_q, $q \leq r$. If there is no such path, the rules are inadequate and must be altered to make the steps possible. It is clear that the transition from the initial to a designated intermediate state can be effected in a single step, but because of the many possibly adverse effects of large reduction steps in arms it is essential to carry out the reduction in tolerably small steps. In addition, a single step need not be acceptable to both sides or feasible as far as safety is concerned, because it takes time to effect arms reduction and monitor its execution. A useful outcome of this approach is the fact that it can be used to determine whether it is impossible to reach a prescribed state from an initial state according to a given set of rules. In other words, not every method which one may advance, even a seemingly good one, is guaranteed to lead to a stable predetermined state, assuming that this method is to be used over and over again.

The first problem of choosing the rules of transition is nonmathematical. It depends on several political, military, and economic factors. Its solution requires detailed information and judgment from these areas. However, the problem of applying the rules to determine the possible intermediate steps of disarmament can be studied mathematically even if the rules are changed between the steps; for then all that need be done is again to apply the method using the state at which the rules are changed as an initial state. One can now associate vertices with the states and apply the approaches to transition problems discussed earlier.

6-28 GRAPHS AND LINGUISTICS†

A language is composed of a finite set of recognizable characters comprising an alphabet, together with a finite set of rules for combining the characters. The set of rules is called a grammar, Γ, and the sequences of characters which can be formed according to Γ are called the strings, Σ, of the language. In particular, the characters of the alphabet may be

† For references, see end of the section.

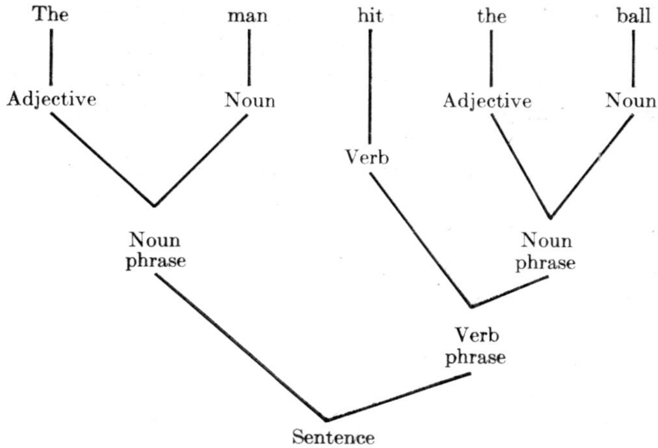

Figure 6-64

considered as special one-character strings. A language Λ is then completely characterized as $\Lambda = (\Sigma, \Gamma)$.

A typical problem of computational linguistics is the recognition of a fixed string Σ_0 as belonging to a language. The recognition process involves diagramming the string, i.e., finding grammatical types such as nouns, verbs, and phrases from the string (Fig. 6-64). In general, natural languages cannot be completely characterized by grammars alone (Chomsky, 1962). One of the foremost problems is finding:

1. Subsets which can be characterized by grammar.
2. An adequate (nongrammatical) model for natural language.

Two problems arise in analyzing natural-language strings:

1. Admission of nonsense strings such as: The green ideas sleep furiously.
2. Admission of grammatical ambiguity: Pretty little girls' camp.

This second example can be diagrammed as several distinct noun phrases, while noun phrases can always be used equivalently.

Definition of a Grammar[1]

A *category* is a set of strings which are denoted by the same name. There are three kinds of categories:

1. Arbitrary sets of one character strings
2. Products of categories; $c = AB \Leftrightarrow C$ contains all strings $c = ab$, where $a \epsilon A$, $b \epsilon B$
3. Unions of categories; $c = A \cup B \Leftrightarrow c \epsilon A$ or $c \epsilon B$

A grammar is a list of categories and their derivation. No ordering is implied, except that category derivations which are unions must appear lower in the list than the categories which form the union. This restriction avoids circularity of the form

$$A = B \cup C \qquad B = C \cup A \qquad C = A \cup B$$

Example of grammar

1. V
2. U
3. B
4. $P = BF$
5. $Q = P \cup U$
6. $R = QF$
7. $F = V \cup R$

Let α, β, and γ be types of vertices corresponding to the three kinds of categories. The grammar above can be represented by a graph[4] (Fig. 6-65). Note that the graph has names for only α-type vertices and the vertex which is used recursively. This suggests the following distribution and substitution:

$$\begin{aligned} F &= V \cup R \\ &= V \cup QF \\ &= V \cup (P \cup U)F \\ &= V \cup (BF \cup U)F \\ F &= V \cup BFF \cup UF \end{aligned}$$

This may be interpreted as a grammar for a simple prefix language where F means formula, V means variable, B means binary operator, and

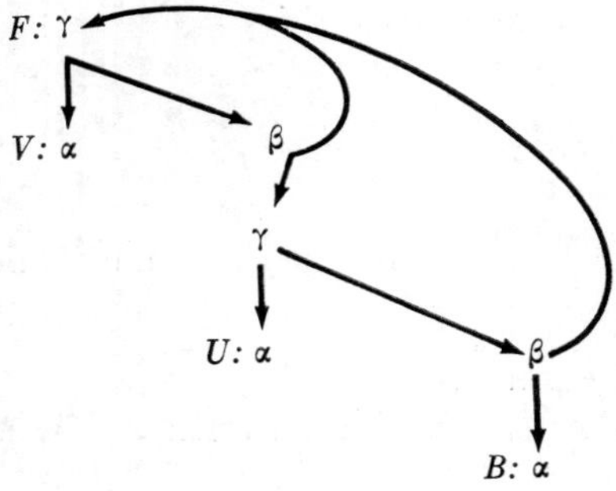

Figure 6-65

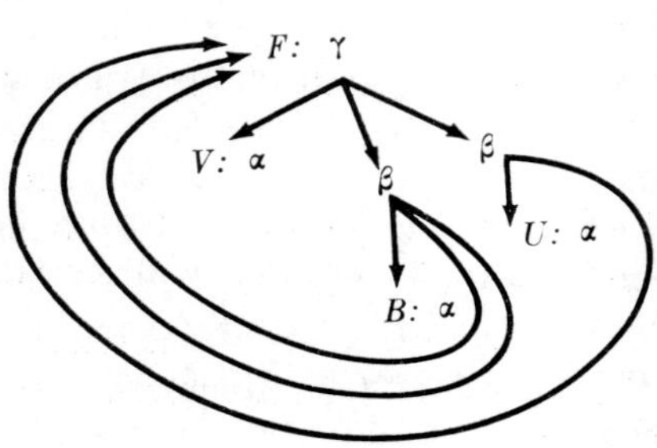

Figure 6-66

U means unary operator. The recursive definition for formula is:

1. A variable
2. A binary operator followed by two formulas
3. A unary operator followed by one formula

The corresponding graph is given in Fig. 6-66. The graph of a grammar is called the Γ graph of a language and is used to (1) generate all strings or (2) recognize arbitrary given strings. The strings of the prefix language are:

1. V
2. BVV
3. UV
4. $BBVVV$
5. $BVBVV$
6. $BBVVBVV$
7. $UBVV$
8. $BUVV$

. . . .

The following points may be made:

1. The Γ graph of a language is finite.
2. The set of all strings Σ is infinite.
3. The set of all strings Σ is enumerable.
4. An arbitrary string Σ_k can be represented by a tree. To see this, consider Σ_6, the sixth string above, Fig. 6-67.

Note that Σ graphs are directed, rooted trees, each vertex of which is named by the corresponding category name. It is usually required that the Σ graphs of formal languages be connected and single-headed, i.e., trees and not forests in the terminology of graph theory. It is also usual to expect that the strings of a language are unambiguous; that is, each string has a unique Σ graph. While it is easy to tell if a given string is ambiguous with respect to a given grammar, it is difficult to determine whether or not a grammar is ambiguous or a language (which may have many different grammars) is inherently ambiguous. In fact it has been shown[2] that whether or not a formal grammar (of the type described here) is ambiguous is unsolvable. On the other hand, if the

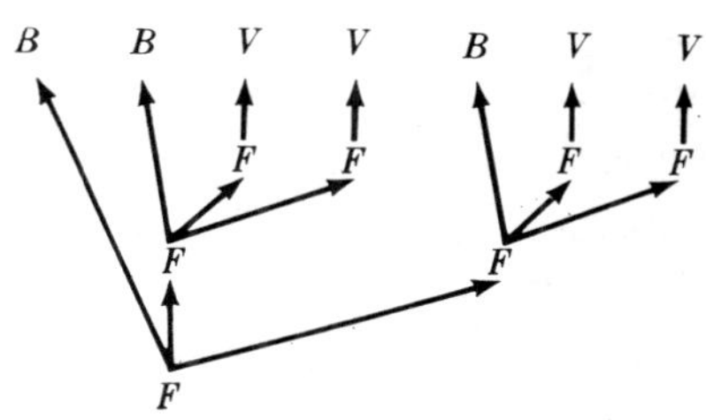

Figure 6-67

strings of a language are enumerable, they may be checked for ambiguity for sufficiently large numbers.[4]

The efficient recognition of strings by machine has been a central problem of computational linguistics. Many different devices have been used, including no-back-up indicators and precedence functions for each category. (For example, do not back up past a sentence structure, do not look for a verb phrase unless a noun phrase has been found.) Some investigation has been made into special restrictive grammars.[3]

It might be more fruitful to find special restrictions or extensions on the Γ graph of a language. One extension which appears interesting is the assignment of length functions to the arcs of the graph. Initially, all lengths are equal, but as each arc is successfully used, its length function is increased. Alternative paths are always chosen on the basis of length, so that the machine, based on "experience," improves the efficiency of its operation. If such a weighting scheme were built into a FORTRAN compiler, the compiler would (in some sense) learn the style and programming idiosyncracies of the individual programmer.

It should be clear that computational linguistics is not concerned with languages as much as it is concerned with structure and pattern recognition in linear strings. Thus far, the focus has been on artificial languages and special subsets of natural language, and the objective has been mechanical analysis.

Recently, a project which will attempt to characterize large stores of tabular data structurally has been initiated. In this instance, the alphabet consists of table entries—sets of related properties—which can be recognized. Categories of information are defined structurally from entries. It is hoped that this will prove to be an effective technique for dealing with nonhomogeneous data which are stored sequentially (as on a magnetic tape).

REFERENCES FOR SEC. 6-28

1 Floyd, R. W.: A Note on Mathematical Induction on Phrase Structured Grammar, *Inform. Control,* **4:** 353–358.

2 Floyd, R. W.: On the Non-existence of a Phrase Structured Grammar for ALGOL-60, *Commun. Assoc. Computing Machinery,* **5:** 483 (1962).

3 Floyd, R. W.: Syntactic Analysis and Operator Precedence, *J. Assoc. Computing Machinery,* **10:** 3 (1963).

4 Gorn, S.: Detection of Generative Ambiguities in Context-free Mechanical Languages, *J. Assoc. Computing Machinery,* **10:** 196–208 (1963).

GENERAL REFERENCES FOR SEC. 6-28

1 Chomsky, N.: Syntactic Structures, Moutan, 1962.

2 Chomsky, N.: On Certain Formal Properties of Grammars, *Inform. Control,* **2:** 137–167.

3 Luce, Bush, and Galanter (eds.): "Handbook of Mathematical Psychology," John Wiley & Sons, Inc., New York, 1964.
4 Proceedings of a Working Conference on Mechanical Language Structures, August, 1963, *Commun. Assoc. Computing Machinery,* **7:** 2 (1964).
5 Nour, P. (ed.): Revised Report on the Algorithmic Language ALGOL-60, *Commun. Assoc. Computing Machinery,* **6**(1): 1–17 (1963).
6 Meeting on IR-Oriented Languages, October, 1961, *Commun. Assoc. Computing Machinery,* **5:** 1 (1962).

ADDITIONAL APPLICATION

6-29 MATHEMATICAL MACHINES AND MARKOV CHAINS

Many physical systems can be characterized by itemizing the various "states" which they may assume and describing the way in which they will respond when in any given state and when subject to each of a number of possible "input stimuli." In general, the response consists of a change of state together with the production of appropriate "output." The concept of a mathematical machine results when one formalizes this idea.

Recent work in the theory of mathematical machines (sometimes referred to as the theory of automata) seems to bear little resemblance to the deep researches of Gödel, Turing, and other mathematical logicians. The new theories apparently consist largely of elementary theorems, but suggest, however, difficult finite combinatorial problems, possibly amenable to graph-theoretic techniques.

Definition A *machine* is a mathematical system consisting of:

1. A finite set $S = \{s_1, \ldots, s_m\}$ of elements called *states*
2. A finite set $X = \{x_1, \ldots, x_n\}$ of elements called *inputs*
3. A finite set $Y = \{y_1, \ldots, y_k\}$ of elements called *outputs*
4. A *transition function* T which maps $S \times X$ into S
5. An *output function* Ω which maps $S \times X$ into Y

If $s \in S$ and $x \in X$, then $s' = T(s,x)$ is interpreted as the next state which the machine will be in if its present state is s and if it is presented with input x. Similarly, $y = \Omega(s,x)$ is the output produced when the machine is in state s and is presented with input x. The sets X and Y are referred to as the input and output *alphabets* (although the nature of their elements vary greatly from one application to the next).

The above definition of "machine" should be qualified in several respects. Such machines are *deterministic,* since the output and next state are completely determined by the input and current state. They are *sequential,* since we are assuming that inputs are presented, and reacted to, at discrete points in time $t_1, t_2, t_3, \ldots$ rather than continuously. They are *complete,* in the sense that every combination of state and input is considered meaningful and produces a known output and next state. They are *memoryless* in the sense that the current output and next state do not depend on past inputs, states, or outputs. Finally, they are *stationary* in the sense that the transition function T and output function Ω do not depend on the point in time t_i under consideration. One can formulate more general machines by altering some or all of these assumptions.

For some purposes, it is convenient to represent a machine by a directed graph whose vertices are the states and whose arcs—with appropriate adornments—characterize X, Y, T, and Ω. This is most easily explained in terms of an example. Consider a machine for which

$$S = \{s_1, s_2, s_3, s_4\} \qquad X = \{0,1\} \qquad \text{and} \qquad Y = \{a,b,c\}$$

The graph in Fig. 6-68 describes one possible machine associated with state set S and alphabets X and Y. Each vertex corresponds to one state and is the initial vertex of precisely two arcs (in general, k arcs, where k is the number of distinct inputs). If an arc having initial vertex s has terminal vertex s' and associated ordered pair (x,y), the interpretation is that

$$T(s,x) = s' \qquad \Omega(s,x) = y$$

For example, if the system is currently in state s_3, then input 0 will yield output b and transform the system to state s_4. On the other hand, input 1 will produce output a and the system will remain in state s_3. (In practice, one may reduce the number of arcs by labeling some arcs with several ordered pairs if several inputs will yield the same next state.)

One can classify individual states, and sets of states, in terms of structural features of the associated graph. For example, a *strongly*

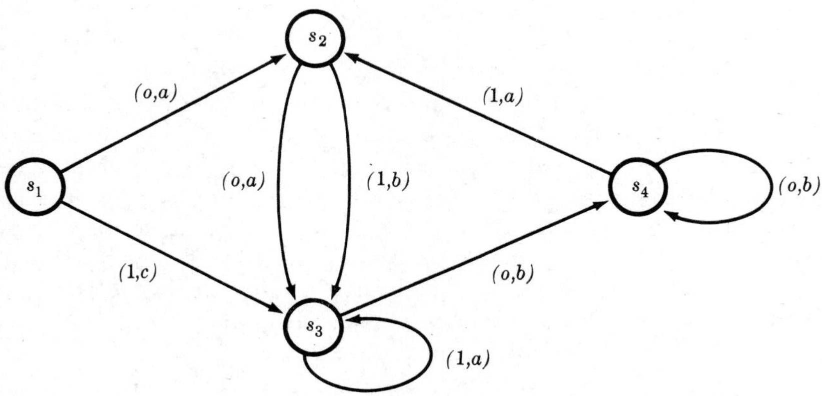

Figure 6-68

connected machine is one whose graph is strongly connected. For such a machine, one can start with any state s and transform the machine into any other state s' (but not necessarily in one "step") through an appropriate sequence of inputs. A *transient* state s is one which is the initial vertex of at least one arc (s,t), where $t \neq s$, but is not the terminal vertex of any arc (u,s), where $u \neq s$. For example, s_1 is a transient state. A *persistent* state s is one which is the terminal vertex of at least one arc (t,s), where $t \neq s$, but is not the initial vertex of any arc (s,u), where $u \neq s$. (The machine under consideration has no persistent individual states. However, states s_2, s_3, and s_4 jointly constitute a persistent *set* of states in an appropriate sense.)

Given any initial state s and any *tape,* or finite sequence $x_1, \ldots, x_t$ of inputs, one can use the graph to determine the resulting final state s' (after t transitions) and the corresponding output sequence $y_1, \ldots, y_t$ readily. For example, if the machine is currently in state s_1 and if the next five inputs are 1, 1, 0, 1, 0, then the machine will pass successively into states s_3, s_3, s_4, s_2, s_3 and yield outputs c, a, b, a, a.

Among the classes of problems which arise in machine theory are these:

1. Analysis of the responses (transitions and outputs) of a given machine.
2. Synthesis or construction of machines having specified response characteristics.
3. Reduction of machines to "simpler" machines which are in some sense equivalent.

The reader is referred to Gill[30] and Ginsburg[32] for extensive discussions of these and other topics in the theory of mathematical machines. It has been our objective here merely to indicate that one convenient way to present a machine is by means of a directed graph, with associated symbols, and that this point of view is useful in connection with the classification of machines and with some kinds of analyses of machines.

The notion of a Markov chain is, in some respects, a probabilistic counterpart of the (deterministic) mathematical machines. Here again we have a system which may be in one of a finite number of states and which is subject to state transitions at discrete points in time. Now, however, the transitions do not depend on any (controllable) inputs, but are governed by probability distributions. No outputs are associated with the process. One is frequently interested in the probability distribution of states, as a function of time, given a particular initial state.

Expressed formally, a *Markov chain* is a mathematical system consisting of

1. A finite set $S = \{s_1, \ldots, s_n\}$ of elements called *states*.

2. An $n \times n$ transition matrix $P = \{p_{ij}\}$, where p_{ij} is the probability that at the next observed point in time the system will be in state s_j, given that it is currently in state s_i. It is required, of course, that

$$\sum_j p_{ij} = 1 \qquad \text{for } i = 1, 2, \ldots, n$$

A Markov chain, as defined here, is sometimes called a *stationary* (or time-invariant) Markov chain to suggest the fact that there is a more general notion in which the transition probabilities may be a function of time.

The graph of a Markov chain is the directed graph having the states as vertices and having an arc from s_i to s_j, labeled p_{ij}, whenever $p_{ij} > 0$ (that is, whenever a "one-step" transition from s_i to s_j is possible). Figure 6-69, for example, depicts a five-state Markov chain. Given that the system is currently in state s_2, the probabilities are 0.2, 0.3, and 0.5, respectively, that the system will pass to state s_3, pass to state s_5, or remain in state s_2. The other arcs have similar interpretations.

A qualitative classification of states, and sets of states, can be made, based on structural characteristics of the graph and ignoring the specific values of the probabilities (except that nonzero probabilities have been distinguished from zero probabilities). For example, a set of states $T \subset S$ is a *trapping* (or absorbing) *set* if the directed cut-set $\{T, S - T\}$ is empty (i.e., there are no possible transitions from a state in T to a state not in T). [In particular, an individual state s_i constitutes a trapping (or absorbing) state if and only if $p_{ii} = 1$.] A Markov chain is said to be *ergodic* if the associated graph is strongly connected. Thus an ergodic chain is one such that, given any current state s_i, there is a nonzero prob-

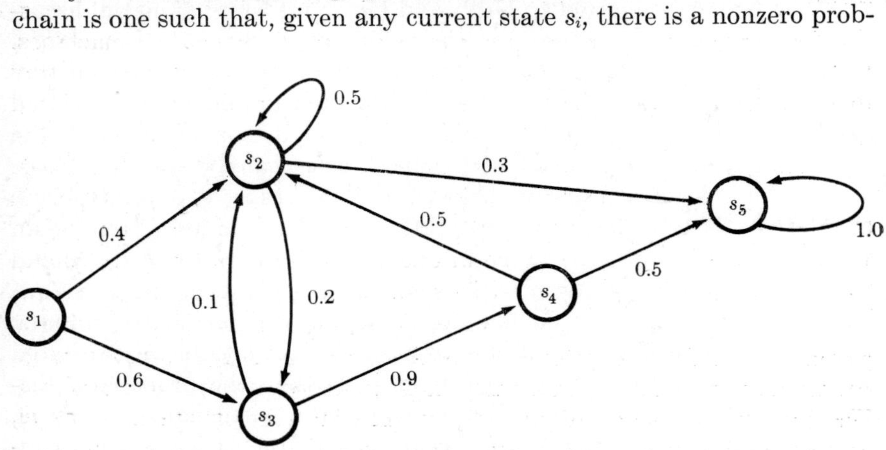

Figure 6-69

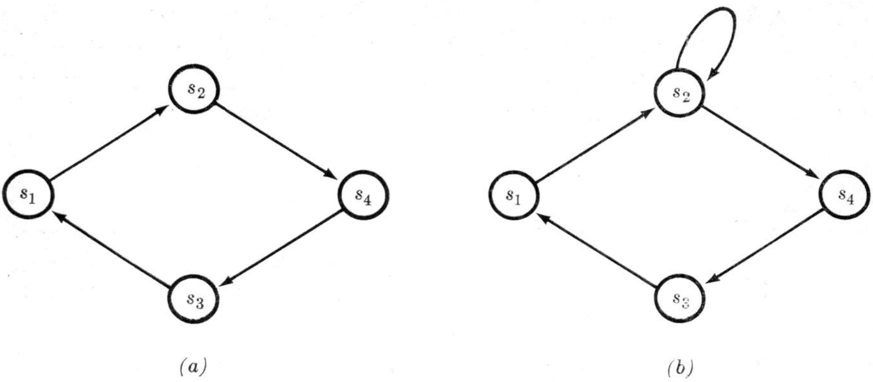

Figure 6-70

ability of reaching any other state s_j after an appropriate number of transitions. An ergodic chain is further said to be *regular* if there is a positive integer t_0 such that for any states s_i and s_j (with possibly $i = j$) there is an arc progression from v_i to v_j having precisely t arcs for every $t \geq t_0$. In Fig. 6-70, for example, the chain shown in (a) is ergodic but not regular. (Starting with state s_1, for example, one can be in state s_3 only after an even number of stages.) In contrast, the chain shown in (b) is regular.

Exercise 6-26 Referring to the definition of a regular Markov chain, what is the smallest value of t_0 which will suffice for the chain shown in Fig. 6-70*b*?

Exercise 6-27 Give a graph-theoretic proof of the fact that if the graph of an ergodic chain has at least one loop, then the chain is necessarily regular.

The reader who has an interest in Markov processes is encouraged to formulate, in graph-theoretic terms, additional concepts associated with the classification of states (as discussed, for example, in Refs. 21, 55, and 73).

6-30 GROUPS AND SIMPLE GRAPHS

Every simple graph has at least one self-isomorphism, namely, the trivial isomorphism in which each vertex and edge corresponds to itself. However, there may be additional ways to set up an isomorphism of the graph with itself so that one has other than the identity isomorphism. An isomorphism of a graph with itself is known as an *automorphism*. The totality of automorphisms of a graph form a group which is known as the *group of the graph*. This group can always be considered as a group of permutations of the vertices of the graph. The automorphisms

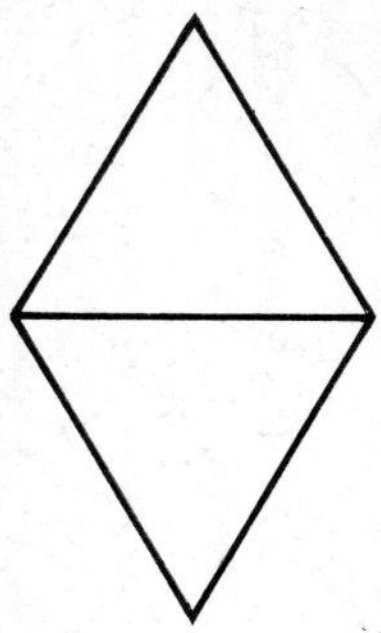

Figure 6-71

of a polygon with $2n$ sides (an n-gon) give rise to a group known as the dihedral group of order n, whereas the group of automorphisms of the complete graph on n vertices is the symmetric group of order n. The order of the group is called the *symmetry number* of the graph. It is as yet not known whether in general every permutation group is a graph group. This problem is contained in that of finding the number of nonisomorphic graphs with a given permutation group.

Exercise 6-28 Find the automorphisms of the graph in Fig. 6-71.

Consider the operation of addition of integers modulo a given integer, and let us indicate by means of a graph the relations among the members of the residues when a certain member is added to all of them. For example, take the integers mod 6. This leaves the residues 0,1,2,3,4, and 5. Let us add 4 to all of them and indicate the subsequent relations. We start with zero. This produces two cycles, Fig. 6-72.

On the other hand, multiplication by 4 yields the graph of Fig. 6-73. Because they map into themselves, 0,2, and 4 are called fixed points. One can similarly obtain a graph for a function $f(x) = ax + b \bmod 6$, for example, where a and b are members of the residue class and x takes on values in that class.

Exercise 6-29 Draw the graphs corresponding to the squaring and the cubing of each of the residues mod 6.

Remark: A fraction $\frac{4}{3}$ mod 7 is resolved as follows: We first determine the member x of the residue class which when multiplied by 3 yields 1 mod 7; that is, $\frac{1}{3} \equiv x \bmod 7$, and we find that $x = 5$. We multiply 4×5 and obtain 6 mod 7; hence $\frac{4}{3} \equiv 6 \bmod 7$.

Exercise 6-30 Obtain the graph of the residue class mod 7 mapped according to $f(x) = (2x + 3)/(x + 2)$.

The foregoing ideas give rise to graphs that are known as Cayley color graphs or Vehn diagrams. We start with a finite group and select a subset of the group (e.g., a set of elements which generates the group). We associate a vertex with each element of the group and an arc terminating at the ver-

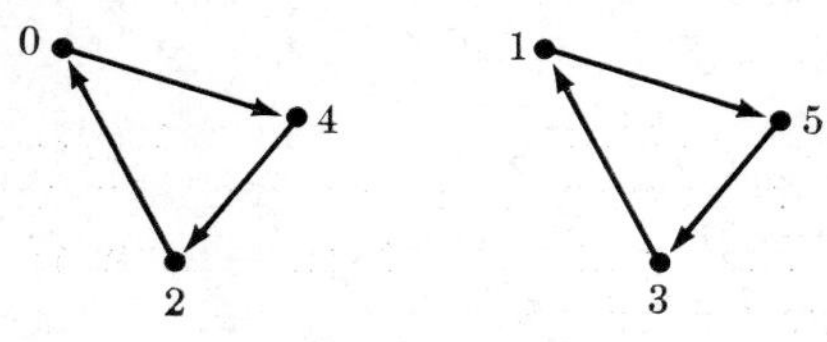

Figure 6-72

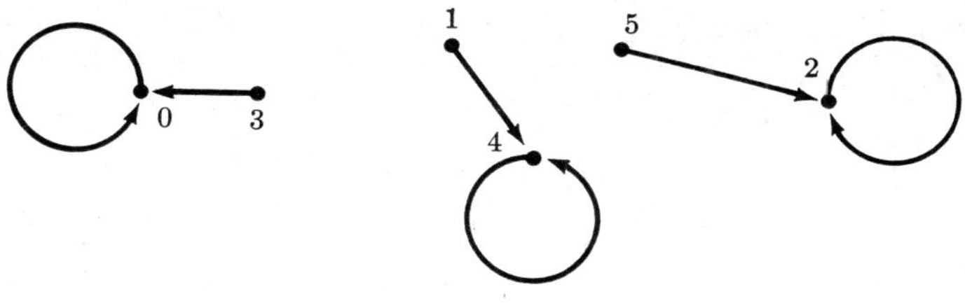

Figure 6-73

tex to which the element is transformed on being transformed (e.g., by multiplication or addition) by a member of the subset.

Thus each vertex is the initial vertex of as many arcs as there are members of the operating subset. Each arc is colored with a specific color associated with each element of the subset. Note that a loop is associated with a vertex when the corresponding element is operated on by the identity. This yields a Cayley color graph. It is known that such a graph is connected if and only if this generates the group, i.e., there is a path between any two vertices, since, by successive multiplication by the subset elements, one can operate on any element to get any other element. Thus 4 is not a generator of the group defined by the residues mod 6 under addition, since the graph is not connected.

Exercise 6-31 Show that 5 is a generator of this group by showing that the corresponding graph is connected.

6-31 CONSTRUCTION OF TREES OF MINIMUM TOTAL LENGTH[56]

It is easy to visualize problems in which it is desired to build roads between several centers where there is one and only one path connecting any two centers. Of all such possible road systems between the centers, we seek one with minimum total length. This is the problem of finding a spanning tree of a graph with minimum total length. Note that a necessary condition for a tree to have minimum total length is that the length of every chord be greater than or equal to the maximum of the lengths of branches in the fundamental circuit which it determines. Otherwise, a single replacement using this chord could be made. It turns out that this is also a sufficient condition, but the proof is not immediate.

To choose a tree of minimum total length, we first index the edges according to increasing lengths so that the length of e_i is less than or equal to the length of e_j whenever $i < j$. We then start by selecting e_1 and add e_2 if e_2 does not form a circuit with e_1. In general, we continue to consider edges of successively higher indices, selecting an edge whenever it does not form a circuit with the set previously selected and rejecting it otherwise. This process always yields a tree of minimum total length. (See Ref. 56 for proof.)

6-32 GRAPHS AND EIGENVALUES OF NONNEGATIVE MATRICES

Directed graphs have been used by Dulmage and Mendelsohn[17] to study various properties of the characteristic equation and eigenvalues of matrices which occur in stochastic processes, economics, and numerical analysis. A directed graph is *cyclically k-partite* ($k \geq 2$) if and only if the set of vertices V can be partitioned into k sets $V_1, \ldots, V_k$ such that all arcs satisfy the following property: If an arc has its initial vertex in V_p, then its terminal vertex is in V_{p+1} if $p < k$ and V_1 if $p = k$. We can associate with any $n \times n$ matrix $A = \{a_{ij}\}$ a directed graph, the coefficients of whose adjacency matrix are unity if the corresponding $a_{ij} \neq 0$. If this graph is cyclically k-partite, then corresponding to a partitioning $V_1, \ldots, V_k$ there is a permutation matrix P such that

$$P^{-1}AP = \begin{pmatrix} 0_1 & B_1 & & & & \\ & 0_2 & B_2 & & & \\ & & 0_3 & B_3 & & \\ & & & \cdot & & \\ & & & & \cdot & \\ & & & & \cdot & B_{k-1} \\ B_k & & & & & 0_k \end{pmatrix}$$

where $0_1, \ldots, 0_k$ are zero matrices and every element of $P^{-1}AP$ not in $B_1, \ldots, B_k$ is zero. The matrices $B_1, \ldots, B_k$ are known as the cyclic components of A.

Thus, a cyclically k-partite graph has the structure exemplified in Fig. 6-74. Note that every arc progression (and in particular every path and cycle) having n arcs has the property that if its initial and terminal vertex are in V_i and V_j, respectively, then $j \equiv i + n \pmod{k}$. In particular, the length of every cycle is a multiple of k.

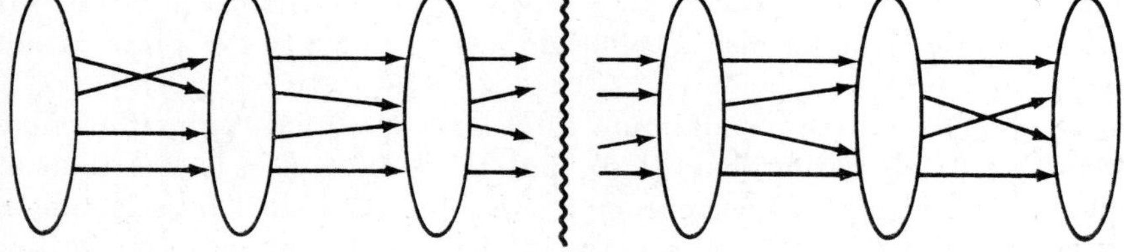

Figure 6-74

The $n \times n$ matrix A has a set of k diagonal components $A_1, A_2, \ldots,$ A_k if there exists a permutation matrix P such that $P^{-1}AP =$ diagonal $(A_1, \ldots, A_k)$. Such a P exists, for example, if the matrix A is irreducible (this is identical with the requirement of strong connectedness of the graph of A).

A monic polynomial is one whose highest-power term has a coefficient of unity.

We have the following theorem:[17]

Theorem 6-12 If A is a matrix such that its directed graph is cyclically k-partite and if $A_1, A_2, \ldots, A_k$ constitute a set of diagonal components of A^k, then there exist a monic polynomial $f(\lambda)$ and a nonnegative integer p such that $f(0) \neq 0$, the characteristic polynomial of A is $f(\lambda^k)\lambda^p$, the characteristic polynomial of A^k is $[f(\lambda)]^k\lambda^p$, there exist nonnegative integers $p_1, \ldots, p_k$ with sum p such that the characteristic polynomial of A_i is $f(\lambda)\lambda^{p_i}$, $i = 1, \ldots, k$, and for any nonzero root of $f(\lambda)$ the elementary divisors are the same for each A_i, $i = 1, \ldots, k$.

Exercise 6-32 By using the illustration of Dulmage and Mendelsohn with

$$A = \begin{pmatrix} 0 & 1 & 0 & 0 \\ 2 & 0 & 0 & 0 \\ 0 & 0 & 0 & 3 \\ 0 & 0 & 4 & 0 \end{pmatrix}$$

show that the associated directed graph is cyclically 2-partite with partitioning

$$V_1 = \{v_1, v_3\} \qquad V_2 = \{v_2, v_4\}$$

and with corresponding cyclic components

$$B_1 = \begin{pmatrix} 1 & 0 \\ 0 & 3 \end{pmatrix} \qquad B_2 = \begin{pmatrix} 2 & 0 \\ 0 & 4 \end{pmatrix}$$

Show that the graph is also cyclically 2-partite with the partitioning

$$V_1 = \{v_1, v_4\} \qquad V_2 = \{v_2, v_3\}$$

and that in this case

$$B_1 = \begin{pmatrix} 1 & 0 \\ 0 & 4 \end{pmatrix} \qquad B_2 = \begin{pmatrix} 2 & 0 \\ 0 & 3 \end{pmatrix}$$

Verify that the diagonal components of A relative to both partitionings are the same with

$$A_1 = \begin{pmatrix} 2 & 0 \\ 0 & 12 \end{pmatrix} \qquad A_2 = \begin{pmatrix} 2 & 0 \\ 0 & 12 \end{pmatrix}$$

Finally, show that the desired polynomial is

$$f(\lambda) = \lambda^2 - 14\lambda + 24$$

and that all the p's for both cases are zero.

6-33 A RANKING PROBLEM

Consider the problem of ranking the elements of a set

$$S = \{s_1, s_2, \ . \ . \ . \ , s_n\}$$

according to a linear ordering which is known to exist, but whose detailed structure must be learned by appropriate comparisons of pairs of elements of S. Suppose we wish to rank a set of physical objects in order of increasing weight under the assumption that no two objects have identical weights. Or suppose that S is a set of individuals who are to be ranked with respect to their capability to defeat one another in some given form of competition. In this context it is necessary to assume that (1) any given individual invariably defeats, or is invariably defeated by, any given second individual and that (2) the relation "can defeat" is transitive.

In the first example the means for discovering the linear order of the elements is a succession of weighings with a balance scale. In the second example the means is a sequence of individual competitions between appropriate pairs of individuals. If t consecutive competitions involve $2t$ distinct individuals, these competitions may be regarded as being concurrent. However, we are interested here in the total number of individual competitions, not in the number of levels in a hierarchy of partially concurrent competitions.

Returning to the general problem, we wish to find a procedure which, in the worst case, requires the fewest individual comparisons in order to completely rank the elements. Let $S_P(n)$ be the maximum number of comparisons required to rank n elements, using procedure P. We seek a procedure which minimizes $S_P(n)$. While we shall not formally define "procedure," we roughly mean a rule for making the next comparison (or for terminating the process) which is completely determined by the outcomes of prior comparisons and which is implementable for any possible set of prior outcomes.

The following procedure was given by Steinhaus.[84] Initially, pick and compare any two elements. In general, having completely ranked some subset of k elements, pick any $(k+1)$st element s_{k+1} and compare it with the middle element of the k already ranked (either of the middle two if k is even). Depending on the outcome, next compare s_{k+1} with the middle element in the set of elements which rank above or below the element on which the first comparison was based. By successive "halvings" of this type, determine the precise ranking of s_{k+1} in the set of $k + 1$ elements. With this ranking of $k + 1$ elements as a new departure point, insert some next element s_{k+2} in the ranking, etc.

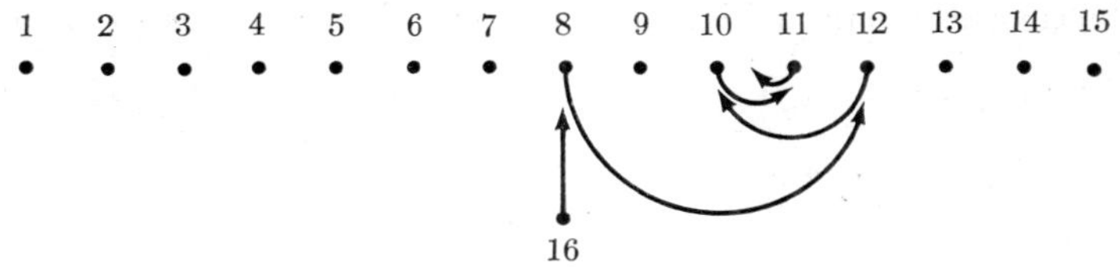

Figure 6-75

As an example, suppose that we have already ranked s_1 to s_{15} in the order

$$s_1 > s_2 > s_3 > \cdots > s_{14} > s_{15}$$

and that the true rank of s_{16} is between s_{10} and s_{11}. The sequence of comparisons is indicated by the arcs in Fig. 6-75, in which the elements are represented by their subscripts. Thus s_{16} is compared successively with s_8, s_{12}, s_{10}, and s_{11}.

In general, one can show that at most

$$S(k) = 1 + [\log_2 k]$$

comparisons are required in order to add a $(k + 1)$st element to a completely ranked set of k elements, where $[\log_2 k]$ denotes the integral part of $\log_2 k$. If the above procedure is applied to rank n elements, by inserting them one at a time, it can be shown that at most

$$M(n) = 1 + nS(n) - 2^{S(n)}$$

and at least

$$L(n) = 1 + [\log_2 (n!)]$$

comparisons are required.

Ford and Johnson[23] have proposed the following more efficient ranking method, which employs the Steinhaus method. Assuming that there are $2r$ or $2r + 1$ elements to be ranked, first form r disjoint pairs and compare them. Next, use the Steinhaus method to rank the r winners of these comparisons (i.e., the higher-ranked members of the r pairs). Figure 6-76, taken from Ref. 23, depicts the results of these rankings in the case of a 19-element set. The nine winners of the first-round comparisons (in decreasing order of rank) are $J, I, H, \ldots, C, B$. The nine vertices below these are the losers in the first-round comparisons, each being shown below the element with which it was compared. (The leftmost element was not considered in the first round.) Since the 10 elements $A, B, C, \ldots, J$ have already been ranked, it remains to insert the remaining 9. The Steinhaus procedure is used for this purpose. The order in which the elements are inserted is indicated by the non-parenthetical numbers in Fig. 6-76. This order is chosen in such a way that the maximum number of comparisons required is as small as possible. The numbers in parentheses indicate the maximum number of comparisons required to insert the corresponding elements in their turn.

It is shown in Ref. 23 that this method requires (for n elements) at most $U(n)$ comparisons, where

$$U(1) = 0 \qquad U(2) = 1$$

$$U(2k) = k + U(k) + \sum_{2}^{k} T(i)$$

$$U(2k+1) = k + U(k) + \sum_{2}^{k+1} T(i)$$

and where

$$T(i) = 2 \qquad \text{for } 1 < i \leq 3$$
$$T(i) = 3 \qquad \text{for } 3 < i \leq 5$$
$$T(i) = j \qquad \text{for } \frac{2^j + (-1)^{j-1}}{3} < i \leq \frac{2^{j+1} + (-1)^j}{3}$$

For $1 < n \leq 11$ and for $n = 20$ or 21, it is known that $U(n) = L(n)$. It is not known whether or not this procedure is in general optimal.

Exercise 6-33 Assume that the true rankings of A to O are

A	B	C	D	E	F	G	H	I	J	K	L	M	N	O
6	7	2	9	13	15	8	14	1	11	4	3	10	12	5

Starting with a comparison of A and B, apply the Steinhaus method, adding C, D, . . . , N, O in that order. (When there are two middle elements use the leftmost one.) Count the total number of comparisons required, and compare with $M(15)$ and $L(15)$.

The ranking problem can be viewed from the standpoint of graph theory as follows. Let the elements of the set to be ranked be regarded as the vertices of a directed graph. Initially, the graph has no arcs. As each comparison is made, an arc is added from the higher- to the lower-ranked member of the pair compared. (Owing to the transitivity of the order relation, additional arcs are implied by those directly resulting from comparisons, but these need not be drawn.) The objective is to add enough arcs so that one discovers a path which includes all of the

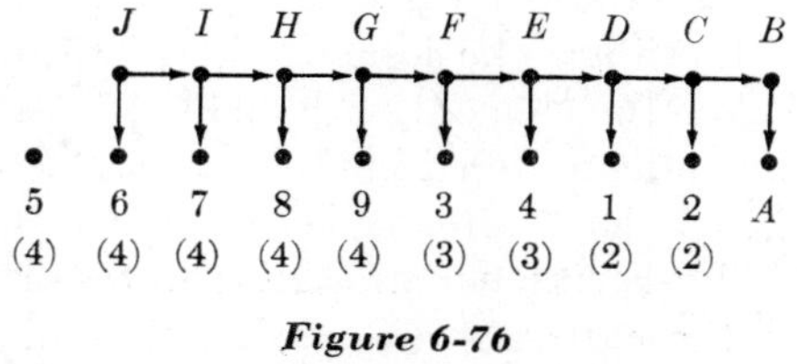

Figure 6-76

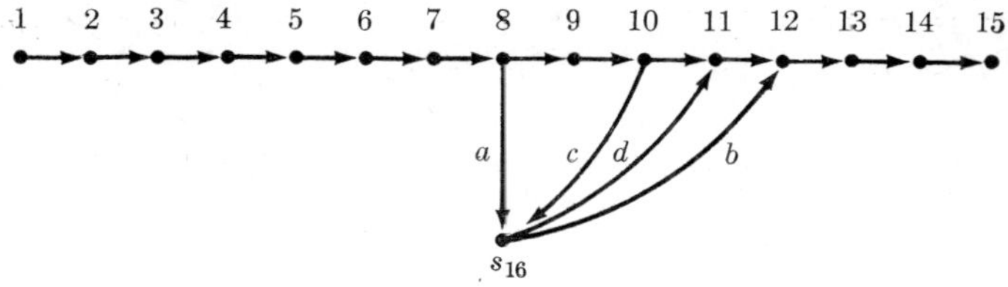

Figure 6-77

vertices (i.e., a hamiltonian path). The order in which the vertices are encountered along this path then determines their proper rank. Returning to the example of Fig. 6-75, after s_1 to s_{15} have been ranked, as indicated by the horizontal arcs in Fig. 6-77, the rank of s_{16} is determined by adding arcs *a*, *b*, *c*, and *d* (in that order). Note that the vertex sequence

$$1, 2, 3, 4, 5, 6, 7, 8, 9, 10, 16, 11, 12, 13, 14, 15$$

determines a hamiltonian path. A good ranking procedure is one which tends to determine such a path after adding the fewest possible arcs.

REFERENCES

1 Ashby, W. R.: "An Introduction to Cybernetics," John Wiley & Sons, Inc., New York, 1956.

2 Benzer, S.: On the Topology of the Genetic Fine Structure, *Proc. Natl. Acad. Sci. U.S.*, **45:** 1607–1620 (1959).

3 Berge, C.: Two Theorems in Graph Theory, *Proc. Natl. Acad. Sci. U.S.*, **43:** 842–844 (1957).

4 Bloch, A.: On Methods for the Construction of Networks Dual to Nonplanar Networks, *Proc. Phys. Soc.* (*London*), **58:** 677–694 (1946).

5 Bose, R. C.: Paired Comparison Designs for Testing Concordance between Judges, *Biometrika*, **43:** 113–121 (1956).

6 Bose, R. C.: Strongly Regular Graphs, Partial Geometries and Partially Balanced Designs, *Pacific J. Math.*, **13:** 389–419 (1963).

7 Bose, R. C., and T. Shimamoto: Classification and Analysis of Partially Balanced Incomplete Block Designs, *J. Am. Statistics Assoc.*, **47:** 151–184 (1952).

8 Bott, R., and J. P. Mayberry: Matrices and Trees, in O. Morgenstern, "Economic Activity Analysis," pp. 391–400, John Wiley & Sons, Inc., New York, 1954.

9 Bryant, P. R.: The Algebra and Topology of Electrical Networks, *Proc. Inst. Elec. Engrs.* (*London*), **C108:** 215–229 (1961). MR 22-1801

10 Carteblanche, F. de: Pile of Cubes, *Eureka*, April, 1947.

11 Cayley, E.: Über die Analytischen Figuren, Welche in der Mathematic Bäume genannt werden, und ihre Anwendung auf die Theorie chemischer Verbindungen, *Ber.*, **8:** 1056 (1875).

12 Chien, R. T.: Synthesis of a Communication Net, *IBM J. Res. Develop.*, **4:** 311–320 (1960) MR 22-2276

13 Cherry, E. Colin: Generalized Concepts of Networks, Proc. Symp. Inform. Networks, Polytechnic Institute of Brooklyn, 1955, pp. 175–184. MR 16-1077

14 Clarke, L. E.: On Cayley's Formula for Counting Trees, *J. London Math. Soc.*, **33:** 471–474 (1958). MR 20-1193

15 Clarke, L. E.: On Otter's Formula for Enumerating Trees, *Quart. J. Math. Oxford, Ser.* 2, **10:** 43–45 (1959). MR 21-623

16 Crowell, R. H.: Graphs of Linear Transformations over Finite Fields, *J. Soc. Ind. Appl. Math.*, **10**(1): 103–112 (1962).

17 Dulmage, A. L., and N. S. Mendelsohn: The Characteristic Equation of an Imprimitive Matrix, *J. Soc. Ind. Appl. Math.*, **11:** 1034 (1963).

18 Edmonds, J.: *J. Res.,Natl. Bur. Std.* B, **69:** (1965).

19 Edmonds, J.: Maximal Matching and a Polyhedron with 0, 1-Vertices, *Natl. Bur. Std. Rept.*, 1963. (Mimeographed.)

20 Edmonds, J.: Paths, Trees, and Flowers, *Natl. Bur. Std. Rept.*, 1964. (Mimeographed.)

21 Feller, W.: "An Introduction to Probability Theory and Its Applications," John Wiley & Sons, Inc., New York, 1957.

22 Ford, L. R. Jr., and D. R. Fulkerson: "Flows in Networks," Princeton University Press, Princeton, N. J., 1962.

23 Ford, L. R. Jr., and S. M. Johnson: A Tournament Problem, *Am. Math. Monthly*, **66:** 387–389 (1959).

24 Fulkerson, D. R.: "Expected Critical Path Lengths in PERT Networks," *Operations Res.*, **10:** 808–818 (1962).

25 Fulkerson, D. R., and O. A. Gross: Incidence Matrices and Interval Graphs, *RAND Project* 1057.

26 Gaffney, M.: *European Sci. Notes, Office Naval Res., London*, 17-4, 22 April 1963.

27 Gardner, M.: Mathematical Games, *Sci. Am.*, pp. 124, 129, October, 1958.

28 Gardner, M.: Mathematical Games, *Sci. Am.*, pp. 150, 152, July, 1961.

29 Gilbert, E. N.: Enumeration of Labelled Graphs, *Can. J. Math.*, **6:** 405–411 (1956).

30 Gill, A.: "Introduction to the Theory of Finite-state Machines," McGraw-Hill Book Company, New York, 1962.

31 Gilmore, P. C., and A. J. Hoffman: A Characterization of Comparability Graphs and of Interval Graphs, *Can. J. Math.*, **16:** 539–548 (1964).

32 Ginsburg, S.: "An Introduction to Mathematical Machine Theory," Addison-Wesley Publishing Company, Inc., Reading, Mass., 1962.

32a Grossman, I., and W. Magnus: "Groups and Their Graphs," Random House, Inc., New York, 1964.

33 Grünbaum, B., and T. S. Motzkin: Longest Simple Paths in Polyhedral Graphs, *J. London Math. Soc.*, 152–160 (1962).

34 Guillemin, E. A.: How to Grow Your Own Trees from Given Cut-set or Tie-set Matrices, *IRE Trans. Circuit Theory*, **CT-6** (spec. suppl.): 110–126 (1959).

35 Guy, R. K.: A Combinatorial Problem, *Bull. Malayan Math. Soc.*, **7:** 68–72 (1960).

36 Harary, F.: "On Local Balance and n-Balance in Signed Graphs," *Mich. Math. J.*, **3:** 37–41 (1955).

37 Harary, F., and Ian C. Ross: A Procedure for Clique Detection Using the Group Matrix, *Sociometry*, **20**(2): 205–215 (1957).

38 Harary, F.: Graph Theoretic Methods in the Management Sciences, *Management Sci.*, **5**(4): 387–403 (1959).

39 Harary, F.: Some Historical and Intuitive Aspects of Graph Theory, *SIAM Rev.*, **2**(2): 123–131 (1960).

40 Harary, F.: Note on the Polya and Otter Formulas for Enumerating Trees, *Mich. Math. J.*, **3**: 109–112 (1956). MR 17-1231

41 Harary, F.: Graph Theory, in "Encyclopedia of Science and Technology," vol. 6, pp. 253–256, McGraw-Hill Book Company, New York, 1960.

42 Harary, F.: A Matrix Criterion for Structural Balance, *Naval Res. Logistics Quart.*, **7**(2): 195–199 (1960).

43 Harary, F.: Unsolved Problems in the Theory of Graphs, *Publ. Math. Inst. Hung. Acad. Sci., Ser. A*, **5**: 63–95 (1960).

44 Harary, F.: The Number of Linear, Directed, Rooted, and Connected Graphs, *Trans. Am. Math. Soc.*, **78**: 445–463 (1955).

45 Harary, F., and A. Hill: On the Number of Crossings in a Complete Graph, *Proc. Edinburgh Math. Soc.*, **13**: 333–338 (1963).

46 Heise, G. H., and G. A. Miller: Problem Solving by Small Groups Using Various Communication Nets., *J. Abnormal Psych.*, **46**: (1951).

47 Hohn, F.: Some Mathematical Aspects of Switching, *Am. Math. Monthly*, **62**: 75–90 (1955).

48 Hoffman, A. J.: On the Polynomial of a Graph, *Am. Math. Monthly*, **70**(1): 30–36 (1963).

49 Hoffman, A. J.: Some Recent Applications of the Theory of Linear Inequalities to Extremal Combinatorial Analysis, *Proc. Symp. Appl. Math.*, **10**: 315–319 (1959).

50 Hoffman, A. J., and R. Gomory: Finding Optimum Combinations, *Intern. Sci. Tech.*, pp. 26–33, July, 1962.

51 Huggins, W. H.: Signal Flow Graphs and Random Signals, *Proc. Inst. Radio Engrs.*, **45**: 74–86 (1957).

52 Ingram, W. H., and C. M. Cramlet: On the Foundations of Electrical Network Theory, *J. Math. Phys.*, **23**: 134–155 (1944). MR 7-403

52a Jullien, P.: Essai sur la Theorie des Puzzles, *Rev. Franc. Recherche Operationelle*, **33**: 375–384 (1964).

53 Kelley, J. E. Jr.: Critical Path Planning and Scheduling: Mathematical Basis, *Operations Res.*, **9**: 296–320 (1961).

54 Kelley, J. E. Jr., and M. R. Walker: Critical Path Planning and Scheduling, *Proc. Eastern Joint Computer Conf.*, Boston, 1959.

55 Kemeny, J., and J. Snell: "Finite Markov Chains," D. Van Nostrand, Company, Inc., Princeton, N.J., 1960.

56 Kruskal, J. B. Jr.: On the Shortest Spanning Subtree of a Graph and the Traveling Salesman Problem, *Proc. Am. Math. Soc.*, **7**: 48–50 (1956).

56a Kwan, Mei-ko: Graphic Programming Using Odd or Even Points, *Chinese Math.*, **1**: 273–277 (1962).

57 Leavitt, H.: Some Effects of Certain Communication Patterns on Group Performance, *J. Abnormal Psych.*, **46**: (1951).

58 Lehman, A.: A Solution of the Shannon Switching Game, *U.S. Army Math. Res. Ctr. Tech. Summary Rept.* 308, July, 1962.

59 Lekkerkerker, C. G., and J. C. Boland: *Fund. Math.*, pp. 45–64, 1962.

60 MacCrimmon, K. R., and C. A. Ryavec: An Analytic Study of the PERT Assumptions, *Operations Res.*, **12**: 16–37 (1964).

61 Malcolm, D. G., J. H. Roseboom, C. E. Clark, and W. Fazar: Application of a Technique for Research and Development Program Evaluation, *Operations Res.*, **7**: 646–670 (1959).

62 Mason, S. J.: Feedback Theory: Some Properties of Linear Flow Graphs, *Proc. IRE*, September, 1953.

63 Mayeda, W.: Terminal and Branch Capacity Matrices of a Communication Net, *IRE Trans. Circuit Theory*, 1961.

64 Mayeda, W.: Synthesis of Switching Functions by Linear Graph Theory, *IBM J. Res. Develop.*, **4:** 320–328 (1960).

65 Montroll, E. W., and R. B. Potts: Correlation and Spontaneous Magnetization of the Two-dimensional Ising Model, *J. Math. Phys.*, **4**(2): (1963).

66 Ore, Oystein: "Graphs and Correspondences," Festschrift Zum 60 Geburtstag von Prof. Dr. Andreas Speiser, Orell Füssli Verlag, Zurich, 1945.

67 Otter, R.: The Number of Trees, *Ann. Math.*, **49:** 583–599 (1948). MR 10-53

68 Otter, R.: The Multiplicative Process, *Ann. Math. Statistics*, **20:** 206–224 (1949). MR 1950 P41

69 Parker, F. D.: Matrices, Relations, and Graphs, *Math. Mag.*, **34**(1): 5–9 (1960).

70 Parker, F. D.: Boolean Matrices and Logic, *Math. Mag.* **37**(11): 33–38 (1964).

71 Polya, G.: Kombinatorische Anzahlbestimmungen für Gruppen, Graphen, und chemische Verbindungen, *Acta Math.*, **68:** 145–254 (1937).

72 Rényi, A.: Some Remarks on the Theory of Trees, *Publ. Math. Inst. Hung. Acad. Sci.*, **4** (1959).

73 Rosenblatt, D.: On the Graphs and Asymptotic Forms of Finite Boolean Relation Matrices and Stochastic Matrices, *Naval Res. Logistics Quart.*, **4**(2): (1957).

74 Rosenblatt, D.: On the Graphs of Finite Idempotent Boolean Relation Matrices, *J. Res. Natl. Bur. Std.*, **67B**(4): (1963).

75 Rosenblatt, D.: On Linear Models and the Graphs of Minkowski-Leontief Matrices, *Econometrica*, **25:** 325–338 (1957).

76 Roth, J. P.: Algebraic Topological Methods for the Synthesis of Switching Systems, I, *Trans. Am. Math. Soc.*, **88:** 301–326 (1958). MR 20-619

77 Roth, J. P.: Algebraic Topological Methods in Synthesis, *Ann. Harvard Computation Lab.*, **29:** 57–73 (1959).

78 Roth, J. P., and E. G. Wagner: Algebraic Topological Methods for the Synthesis of Switching Systems, III, Minimization of Nonsingular Boolean Trees, *IBM J. Res. Develop.*, **3:** 326–344 (1959). MR 22-920

79 Saaty, T. L.: On the Minimum Number of Intersections for Complete Graphs, *Proc. Natl. Acad. Sci. U.S.*, September, 1964.

80 Saaty, T. L.: A Model for the Control of Arms, *Operations Res.*, pp. 586–609, September–October, 1964.

81 Schiff, H.: Zur Statistik Chemischer Verbindungen, *Ber.*, **8:** 1542 (1875).

82 Schwartz, B. L.: An Analytic Method for the "Difficult Crossing" Puzzles, *Math. Mag.*, **34**(4): 187–193 (1961).

83 Sholander, M. C.: The Linear Graph, *Am. Math. Monthly*, **43:** 543–545 (1942).

84 Steinhaus, H.: Mathematical Snapshots, pp. 37–40, New York, 1950.

85 Synge, J. L.: The Fundamental Theorem of Electrical Networks, *Quart. Appl. Math.*, **9:** 113–127 (1951). MR 13-189

86 Trent, H. M.: Isomorphisms between Oriented Linear Graphs and Lumped Physical Systems, *J. Acoust. Soc. Am.*, **27**(3): 500–527 (1955).

87 Trent, H. M.: On the Construction of Schematic Diagrams for Mechanical Systems, *J. Acoust. Soc. Am.*, **30**(8): 795–800 (1958).

88 Trent, H. M.: On the Conceptual Necessity and Use of Perfect Couplers in Schematic Diagrams, *J. Acoust. Soc. Am.*, **31**(3): 326–332 (1959).

89 Trent, H. M.: A Note on the Enumeration and Listings of All Possible Trees in a Connected Linear Graph, *Proc. Natl. Acad. Sci. U.S.*, **40**: 1004–1007 (1954).

90 Tutte, W. T.: The Factorization of Linear Graphs, *J. London Math. Soc.*, **22**: 107–111 (1947).

91 Western, D. E.: Graphs of Composite Relations, *Am. Math. Monthly*, **69**(5): 418–421 (1962).

92 Whitin, T. M.: An Economic Application of "Matrices and Trees," in O. Morgenstern, "Economic Activity Analysis," pp. 401–408, John Wiley & Sons, Inc., New York, 1954.

93 Zarankiewicz, K.: On a Problem of P. Turan Concerning Graphs, *Fund. Math.*, **41**: 137–145 (1954).

SEVEN

NETWORK FLOWS

7-1 INTRODUCTION

If we assign to each arc of a directed graph a rate of flow of a hypothetical substance, then the structure of the graph, together with the values of the arc flows, becomes a useful model for certain problems in transportation, communications, and other areas concerned with the actual or conceptual movement of commodities, information, or people. In this chapter the concept of flow through a network (as a graph is called in this context) is formalized and the structure of flows is studied. The development centers around two general problems. The first is the problem of maximizing the total rate of flow between specified vertices when the flow in each arc is constrained to lie between given bounds, e.g., maximization of traffic flow when each road segment has a limiting "capacity." The second involves finding constrained flows whose costs are minimal when a cost per unit of flow is associated with each arc. A general example is given in Sec. 7-9. Reference 10 of Chap. 1 gives a comprehensive treatment of this subject.

7-2 BASIC TERMINOLOGY

It is sufficient when studying flows to consider directed graphs which are connected and have no loops. These are called *networks* in this chapter. (Flows in disconnected graphs can be analyzed by considering each component separately, and flow in a loop would contribute nothing to the motion of flow between vertices. So the class of graphs which are networks is sufficiently inclusive for our purposes.) To emphasize the fact that a particular directed graph is a network, we shall represent it by N rather than the usual D. Given a network $N = (V,A)$, a *flow* in N is an integer-valued function φ defined on A. The integer $\varphi(a)$ is called the *flow in arc a*. More specifically, if $a \simeq (v,w)$, the flow is said to be *from v to w* if $\varphi(a) \geq 0$ and *from w to v* if $\varphi(a) \leq 0$.

Most of the concepts and results that will be presented here are particularly transparent if one considers a geometric realization of the network and interprets $|\varphi(a)|$ as the constant rate of flow of a homogeneous substance through a. The direction of flow is determined by the sign of $\varphi(a)$, using the above convention.

The vertices of N are customarily classified, relative to φ, on the basis of whether they create, conserve, or absorb flow. To formalize this, let $v \rightarrow V$ denote the set of arcs having v as initial vertex and let $V \rightarrow v$ denote the set of arcs having v as terminal vertex. Then the integer $Q(v,\varphi)$, defined by the relationship

$$Q(v,\varphi) = \sum_{v \rightarrow V} \varphi(a) - \sum_{V \rightarrow v} \varphi(a)$$

is called the *net output at v, relative to* φ. (We shall write $Q(v)$ if φ is clear from the context.) If $\varphi(a) \geq 0$ in every arc, the first summation is simply the total flow directed away from vertex v and the second is the total flow toward v. If some arc flows are negative, the individual summations do not have these interpretations, but their difference still represents the net flow output. For example, in Fig. 7-1, we have $v \rightarrow V = \{a_1,a_2\}$, $V \rightarrow v = \{a_3,a_4\}$, and

$$Q(v) = [(2) + (-1)] - [(4) + (-7)] = 4$$

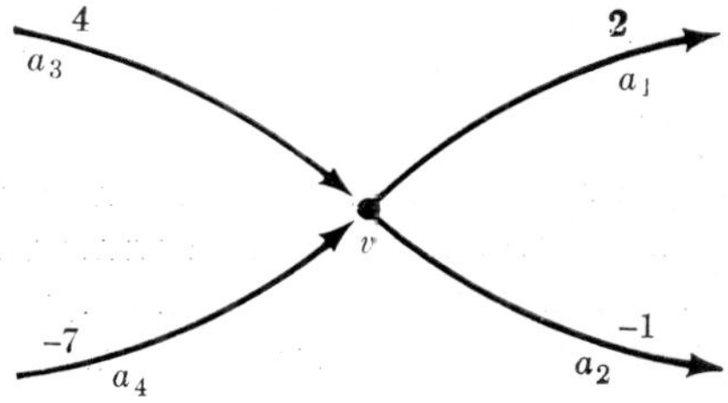

Figure 7-1

Note that the net output at a vertex remains the same if we reverse the direction of an arc and change the sign of its flow. We could, therefore, make all arc flows nonnegative without disturbing the net output at any vertex. However, it is more useful here to let the arc directions remain unchanged, thus providing a fixed frame of reference, and to allow negative flows in certain situations.

We now group the vertices of N into sets V^+, V^-, and V^0 as follows:

$$V^+ = \{v \in V | Q(v) > 0\}$$
$$V^- = \{v \in V | Q(v) < 0\}$$
$$V^0 = \{v \in V | Q(v) = 0\}$$

The elements of V^+, V^-, and V^0 will be called *sources, sinks,* and *intermediate vertices,* respectively. Intuitively, these vertices respectively create, consume, and conserve flow.

The network as a whole conserves flow, relative to any flow φ, in the sense that

$$\sum_{v \in V} Q(v) = 0$$

This can be seen by writing

$$\sum_{v \in V} Q(v) = \sum_{v \in V} \sum_{a \in (v \to V)} \varphi(a) - \sum_{v \in V} \sum_{a \in (V \to v)} \varphi(a)$$

and noting that every arc of the network appears precisely once in each of the double summations. In Fig. 7-2, for example, we have the following classification of vertices given in the accompanying table. If there is a single source v_i and a single sink v_j, the flow is said to be *from* v_i *to* v_j, and $Q(v_i)$, or, equivalently, $-Q(v_j)$, is called the *value* of the flow. [For convenience, this terminology will be extended to include flows of value zero, for which $Q(v) = 0$ for all vertices.]

i	$Q(v_i)$	Type
1	5	source
2	6	source
3	0	intermediate
4	0	intermediate
5	-10	sink
6	-1	sink

Any flow in a network can be transformed into a flow having a single source and sink by augmenting the structure of the network. Specifically, a new vertex w_1 is added, together with an arc $b_i \simeq (w_1,v_i)$ leading from w_1 to each source v_i. Such an arc is assigned the flow

$$\varphi(b_i) = Q(v_i)$$

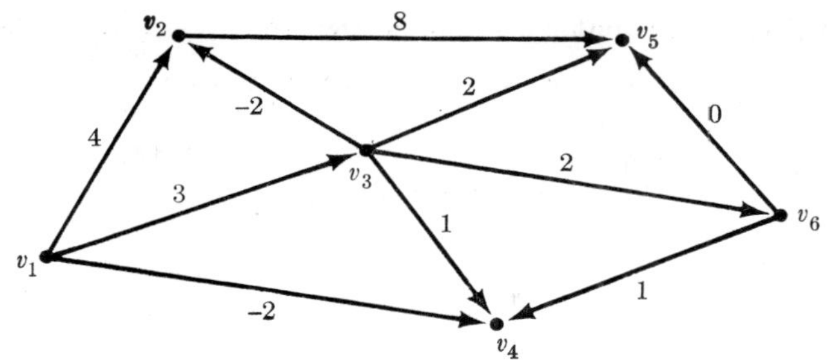

Figure 7-2

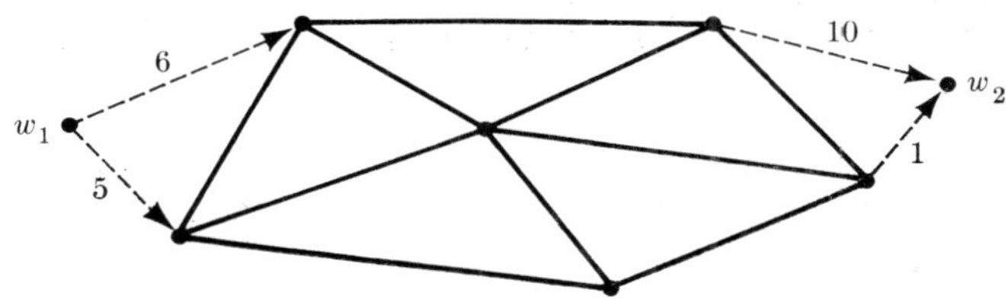

Figure 7-3

Similarly, a second vertex w_2 is added, together with an arc $c_j \simeq (v_j, w_2)$ from each sink v_j to w_2, having flow

$$\varphi(c_j) = -Q(v_j)$$

To illustrate this, Fig. 7-3 shows the augmentation required to put the flow of Fig. 7-2 into this standard form. The resulting flow is a flow from w_1 to w_2 having value 11.

Most of the following material is related specifically to flows having a unique source and sink. However, in view of the above transformation, the results are also applicable to flows having multiple sources and sinks.

7-3 FLOW OPERATIONS AND RELATIONS

Let φ_1 and φ_2 be flows in the same network $N = (V,A)$ and let p be an integer. Then flows $\varphi_1 + \varphi_2$, $\varphi_1 - \varphi_2$, and $p\varphi$ are defined for every $a \in A$ by the relationships

$$\begin{aligned}(\varphi_1 + \varphi_2)(a) &= \varphi_1(a) + \varphi_2(a)\\(\varphi_1 - \varphi_2)(a) &= \varphi_1(a) - \varphi_2(a)\\(p\varphi_1)(a) &= \mathrm{p}\varphi_1(a)\end{aligned}$$

It is readily seen that

$$Q(v, \varphi_1 \pm \varphi_2) = Q(v,\varphi_1) \pm Q(v,\varphi_2)$$
$$Q(v,p\varphi_1) = pQ(v,\varphi_1)$$

It follows that if both φ_1 and φ_2 are flows from v to w, having values k_1 and k_2, respectively, then $\varphi_1 + \varphi_2$ is also a flow from v to w and has value $k_1 + k_2$. Similarly, $\varphi_1 - \varphi_2$ is a flow having value $|k_1 - k_2|$, being a flow from v to w if $k_1 \geq k_2$ and from w to v if $k_1 \leq k_2$. (A flow of value zero is considered as a flow from v to w and also from w to v.) Finally, $p\varphi$ has value $|pk_1|$, being a flow from v to w if $p \geq 0$ and from w to v if $p \leq 0$.

Exercise 7-1 In Fig. 7-2 assign a second flow, consisting of unit flow with each arc, and verify the foregoing relations.

We shall write $\varphi_1 < \varphi_2$ if and only if

$$\varphi_1(a) < \varphi_2(a) \qquad \text{for every } a \in A$$

The flow relations $\varphi_1 \leq \varphi_2$, $\varphi_1 > \varphi_2$, and $\varphi_1 \geq \varphi_2$ are defined analogously.

A flow φ is said to be *bounded* by φ_1 and φ_2 if $\varphi(a)$ lies between $\varphi_1(a)$ and $\varphi_2(a)$ for every arc of the network. It is evident that a sufficient condition for φ to be bounded by φ_1 and φ_2 is that either

$$\varphi_1 \leq \varphi \leq \varphi_2 \qquad \text{or} \qquad \varphi_2 \leq \varphi \leq \varphi_1$$

However, it is also permissible for φ to be bounded above by φ_1 for some arcs and by φ_2 for others.

Flows φ_1 and φ_2 are said to be *conformal* if

$$\varphi_1(a) \cdot \varphi_2(a) \geq 0 \qquad \text{for every } a \in A$$

Expressed differently, φ_1 and φ_2 are conformal if there is no arc a such that $\varphi_1(a)$ and $\varphi_2(a)$ are both nonzero and oppositely oriented. Flows $\{\varphi_1,\varphi_2, \ldots ,\varphi_n\}$ in the same network are said to be *jointly conformal* if they are pairwise conformal.

Exercise 7-2 Verify whether the flow assigned in Exercise 7-1 and that of Fig. 7-2 are jointly conformal. If they are not, assign a flow that is conformal with the flow of Fig. 7-2.

As we shall see shortly, any flow in a network can be conveniently regarded as being composed of a collection of particularly simple flows (defined in the next section) associated with chains and circuits in the network. Moreover, these simple flows are jointly conformal, i.e., they all utilize any given arc in the same direction. This is a simple but important fact in many application contexts where conflicting (nonconformal) sets of flows have no usefulness, and possibly no interpretation.

Theorem 7-1 relates the concepts of conformal flows and bounded flows, and it plays a basic role in later discussions.

Theorem 7-1 If $S = \{\varphi_1, \varphi_2, \ldots, \varphi_n\}$ is a set of jointly conformal flows in a network N, ψ is any flow in N, and $T \subset S$, then $\psi + \sum_T \varphi_i$ is bounded by ψ and $\psi + \sum_S \varphi_i$.

Proof: Let a be any arc. If $\varphi_i(a) \geq 0$ for $i = 1, 2, \ldots, n$, then clearly

$$\psi(a) \leq \psi(a) + \sum_T \varphi_i(a) \leq \psi(a) + \sum_S \varphi_i(a)$$

Similarly, if $\varphi_i(a) \leq 0$ for $i = 1, 2, \ldots, n$, then

$$\psi(a) \geq \psi(a) + \sum_T \varphi_i(a) \geq \psi(a) + \sum_S \varphi_i(a)$$

Since one of these two situations applies for each $a \in A$, the result follows.

Exercise 7-3 Let φ_1 be the flow assigned in Fig. 7-2 and let φ_2 be the flow assigned in Exercise 7-2. Take $S = \{\varphi_1, \varphi_2\}$ and let $T = \{\varphi_1\}$. Assign an arbitrary flow ψ and verify the statement of the theorem.

7-4 SIMPLE FLOWS

In a given network $N = (V,A)$ let C be a set of arcs which form either a simple chain or a simple circuit in the associated undirected graph. Assume, moreover, that we orient the edges by traversing this simple chain or circuit in one of the two possible manners. An arc of C will be called *normal* if its direction agrees with the induced orientation and *inverted* otherwise.

Next, let φ_C be defined on A as follows:

$$\varphi_C(a) = \begin{cases} +1 & \text{for normal arcs of } C \\ -1 & \text{for inverted arcs of } C \\ 0 & \text{for arcs not in } C \end{cases}$$

Figure 7-4 shows a simple chain C oriented from v to w, an oriented simple circuit S, and the associated functions φ_C and φ_S.

One may easily verify that if C is a simple chain joining v to w and oriented from v to w, then φ_C is a flow from v to w in N whose value is 1, called a *simple chain flow from v to w.* Similarly, if S is a circuit oriented in one of the two possible ways, φ_S is a zero-valued flow in N called a *simple circuit flow.* Flows of these two types (and only such) are referred to as *simple flows* in the following pages. Our point of view will be to consider simple flows as the building blocks from which arbitrary flows may be synthesized and into which they may be resolved.

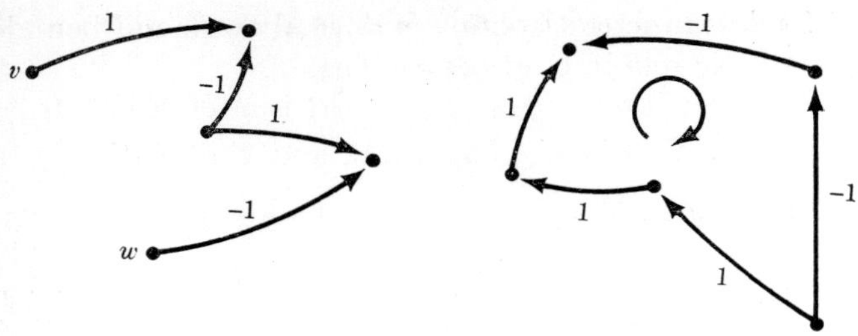

Figure 7-4

7-5 AN ALTERNATIVE FLOW FORMULATION

It is sometimes convenient to consider an alternative representation of a network flow in which both the network configuration and the values of arc flows are reflected in the structure of an associated directed graph. Let φ denote a flow in a network $N = (V,A)$. We construct a directed graph $U(N,\varphi) = (V',A')$ whose vertices are in 1-1 correspondence with those of N. For each arc $a \in A$ such that $\varphi(a) \neq 0$ we include $|\varphi(a)|$ strictly parallel arcs in A' joining the corresponding vertices. If $a \simeq (v,w)$ and $\varphi(a) > 0$, these arcs are taken to be $\simeq(v',w')$ in A'. If $\varphi(a) < 0$, they are taken to be $\simeq(w',v')$. Thus the directions of arcs of $U(N,\varphi)$ indicate the true directions of arc flows in N, while their numbers reflect the magnitudes of arc flows in N. The directed graph $U(N,\varphi)$ is called the *unitary graph* associated with flow φ in network N. Figure 7-5 exhibits the unitary graph associated with a specific N and φ.

Note that $U(N,\varphi)$ may not reflect the complete structure of N, since arcs $a \in A$ for which $\varphi(a) = 0$ are not represented. In particular, $U(N,\varphi)$ may fail to be connected, and hence may not constitute a network in its own right. Except for this technicality, we have essentially eliminated

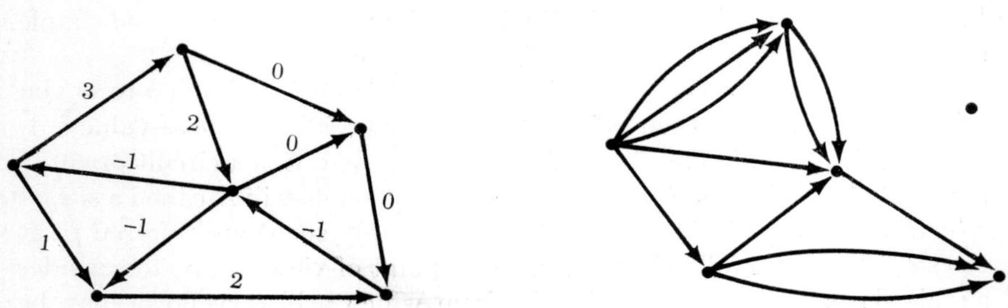

Figure 7-5

the need for the flow function by producing a network with unit flow in every arc.

Exercise 7-4 Give the unitary graph associated with the augmented network of Fig. 7-3 by using the flow given in Fig. 7-2.

If v' denotes the vertex in $U(N,\varphi)$ corresponding to v in N, it is easily seen that

$$\delta^+(v') - \delta^-(v') = Q(v,\varphi)$$

In particular, if φ is a flow of value k from v_i to v_j, then $U(N,\varphi)$ is pseudo-symmetric except possibly at v_i' and v_j', and for these vertices we have

$$\delta^+(v_i') - \delta^-(v_i') = \delta^-(v_j') - \delta^+(v_j') = k$$

Recalling Theorem 3-7, it follows that $U(N,\varphi)$ can be covered by k simple paths from v_i to v_j, together possibly with some simple cycles. This fact plays a central role in the subsequent development, in which more general flows are decomposed into or synthesized from appropriate simple flows. Theorem 7-2 is a restatement of this result in the language of flows.

Theorem 7-2 A flow φ from v to w having value k can be written as a sum

$$\varphi = \sum_{1}^{n} \sigma_i$$

for some appropriate $n \geq k$, where the σ_i's are conformal simple flows, k of which are simple chain flows from v to w, the rest being simple circuit flows.

We shall call such a set of conformal simple flows a *decomposition* of φ. Naturally, a given flow may have more than one decomposition. In Fig. 7-6 two decompositions of the same flow of value 3 from v to w are shown. Note that one of these includes a simple circuit flow and the other does not.

Exercise 7-5 Apply Theorem 7-2 to the flow in Fig. 7-2, augmented as in Fig. 7-3, to obtain a specific set of simple flows σ_i.

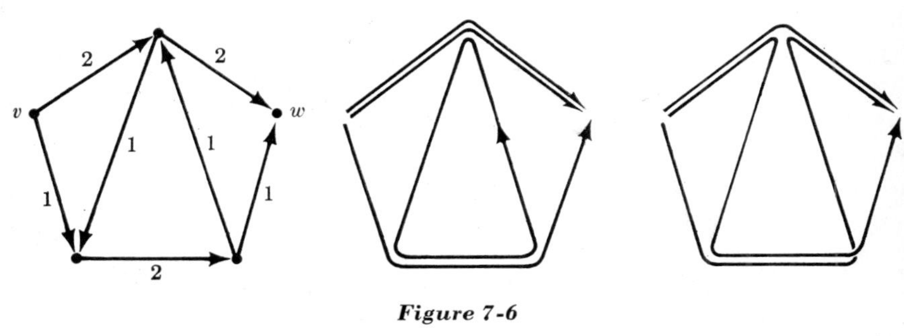

Figure 7-6

7-6 FLOWS CONSTRAINED BY ARC BOUNDS

A network $N = (V,A)$ is said to be bounded if two integer-valued functions α and β are defined on A and satisfy the relationship

$$0 \leq \alpha(a) \leq \beta(a) \qquad \text{for every } a \in A$$

The integers $\alpha(a)$ and $\beta(a)$, called the *lower* and *upper bounds* on arc a, are interpreted as representing the minimum and maximum permissible values of flow in each arc. When $\alpha(a) = 0$ for all $a \in A$, $\beta(a)$ is usually called the *capacity* of arc a, and the network is said to be *capacitated.* A flow φ in N is said to be *feasible* if and only if

$$\alpha(a) \leq \varphi(a) \leq \beta(a) \qquad \text{for every } a \in A$$

If φ_1 and φ_2 are feasible flows, and if φ is any flow bounded by φ_1 and φ_2, then clearly φ is also feasible. The basic questions to be considered here are these:

1. Under what circumstances do feasible flows from v to w exist in a given bounded network?
2. If any feasible flows exist, what is the character of the set of values realized by such flows?
3. If feasible flows from v to w having value k exist, how can we find a specific flow of this type?

In the remainder of this section we assume that the bounded network $N = (V,A)$ under consideration has n vertices indexed as $v_1, v_2, \ldots, v_n$ and that feasible flows from v_1 to v_n are the only flows of interest (if we include zero-valued flows).

As a first step toward answering the above questions, we establish the fact that the arc bounds associated with every cut-set separating v_1 from v_n determine lower and upper bounds on the possible values of feasible flows. If X and Y are (not necessarily disjoint) sets of vertices, $X \to Y$ denotes the set of arcs of the form $a \simeq (x,y)$, where $x \in X$ and $y \in Y$. Also, X' denotes the complement of X relative to V. Recalling the terminology related to cut-sets, if X and X' are both nonempty, then $X \to X'$ and $X' \to X$ are the two directed cut-sets associated with the vertex partition $\{X,X'\}$.

Theorem 7-3 If φ is any feasible flow from v_1 to v_n and W is any set of vertices such that $v_1 \in W$ and $v_n \in W'$, then the value k of φ must satisfy the inequalities

$$\sum_{W \to W'} \alpha(a) - \sum_{W' \to W} \beta(a) \leq k \leq \sum_{W \to W'} \beta(a) - \sum_{W' \to W} \alpha(a)$$

Proof: Since $Q(v_1) = k$ and $Q(v) = 0$ for all other $v \in W$, we have

$$k = \sum_{W} Q(v) = \sum_{W \to V} \varphi(a) - \sum_{V \to W} \varphi(a)$$

This can be simplified to

$$k = \sum_{W \to W'} \varphi(a) - \sum_{W' \to W} \varphi(a)$$

Since φ is assumed to be feasible, the relationship

$$\alpha(a) \leq \varphi(a) \leq \beta(a)$$

holds for every arc. By appropriate substitution of these bounds in the expression for k, the desired inequalities are obtained.

Note that W was arbitrary in Theorem 7-3, except for the requirement that $v_1 \in W$ and $v_n \in W'$. If we define

$$\bar{M} = \min_{W} \left\{ \sum_{W \to W'} \beta(a) - \sum_{W' \to W} \alpha(a) \right\}$$

and

$$\underline{M} = \max_{W} \left\{ \sum_{W \to W'} \alpha(a) - \sum_{W' \to W} \beta(a) \right\}$$

where W ranges over all such sets of vertices, we conclude that a necessary condition for the existence of any feasible flows from v_1 to v_n is that $\underline{M} \leq \bar{M}$.

Consider the network of Fig. 7-7, where the pairs of integers shown with certain arcs a represent the flow bounds $\alpha(a)$ and $\beta(a)$. Taking $W = \{v_1,v_2,v_3,v_4\}$, note that

$$\sum_{W \to W'} \alpha(a) - \sum_{W' \to W} \beta(a) = 10 - 2 = 8 \leq \underline{M}$$

On the other hand, taking $W = \{v_1\}$, we have

$$\sum_{W \to W'} \beta(a) - \sum_{W' \to W} \alpha(a) = 7 - 1 = 6 \geq \bar{M}$$

so the minimal (net) flow which may feasibly cross the first of these cut-sets is not compatible with the maximal flow which may feasibly cross the second. Consequently, no flow in this network can be feasible. (For this reason, the values of α and β on the two vertical arcs are immaterial.)

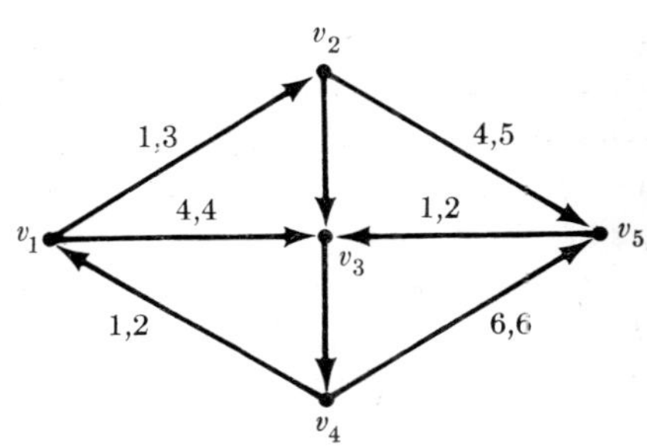

Figure 7-7

Exercise 7-6 Use Fig. 7-7 to take every possible partitioning of the vertices into W and W' when $v_1 \in W$ and $v_3 \in W'$. Evaluate the summations in Theorem 7-3 and find $\bar{M}$ and $\underline{M}$. (Assume that $\alpha = \beta = 3$ on the two vertical arcs.)

Naturally, in a complex network having a large number of vertices it would not be practical to test all partitions $\{W,W'\}$ in order to evaluate $\bar{M}$ and $\underline{M}$. Besides, even if this were practical, what could we conclude if $\underline{M} \leq \bar{M}$? So far, we have only established that this inequality is necessary for the existence of feasible flows. We need a corresponding theoretical result which asserts that this inequality, or something like it, is also sufficient for the existence of feasible flows. The remainder of this section is devoted to developing such a result. In the course of doing so, a practical procedure for realizing flows of all possible values will emerge.

Theorem 7-4 establishes the fact that if an initial feasible flow from v_1 to v_n is known, then feasible flows having all possible values can be obtained by adding or subtracting appropriate simple chain flows.

Theorem 7-4 Let φ be a feasible flow from v_1 to v_n having value k. If there exists a feasible flow φ^* having value $k^* \neq k$, then there exist $m = |k^* - k|$ simple chain flows $\{\sigma_1,\sigma_2, \ldots ,\sigma_m\}$ from v_1 to v_n such that $\varphi \pm \sum_1^m \sigma_i$ is a feasible flow having value k^*. (The plus sign is appropriate if $k^* > k$; the minus sign if $k^* < k$.)

Proof: Suppose that $k^* > k$. Then $\varphi^* - \varphi$ is a flow from v_1 to v_n having value m. By applying Theorem 7-2, we can write

$$\varphi^* - \varphi = \sigma_1 + \sigma_2 + \cdots + \sigma_m + \bar{\sigma}_1 + \bar{\sigma}_2 + \cdots + \bar{\sigma}_p$$

where the σ_i's and $\bar{\sigma}_j$'s are conformal simple chain and circuit flows, respectively. (The circuit flows are indefinite in number and may be absent altogether.) Then, by using Theorem 7-1, we find that $\varphi + \sum_1^m \sigma_i$ is bounded by φ and $\varphi + (\varphi^* - \varphi) = \varphi^*$. Being bounded by two feasible flows, $\varphi + \sum_1^m \sigma_i$ is necessarily feasible, and it clearly has value $k + m = k^*$. This disposes of the case in which $k^* > k$. The case in which $k^* < k$ can be treated in an analogous manner, the details of which are left to the reader.

Exercise 7-7 Use the notation of Theorem 7-4 to draw a network and two flows φ and φ^* (between the same pair of vertices) having values $k = 2$ and $k^* = 5$. Determine the flow $\varphi^* - \varphi$ and write it as the sum of conformal simple flows.

Remark: We can now rigorously verify the not-very-surprising fact that the set of values (if any) realized by feasible flows is a set of consecutive integers. That is, if feasible flows having values k and $k^* > k$ exist and if r is an integer such that $k < r < k^*$, then a feasible flow having

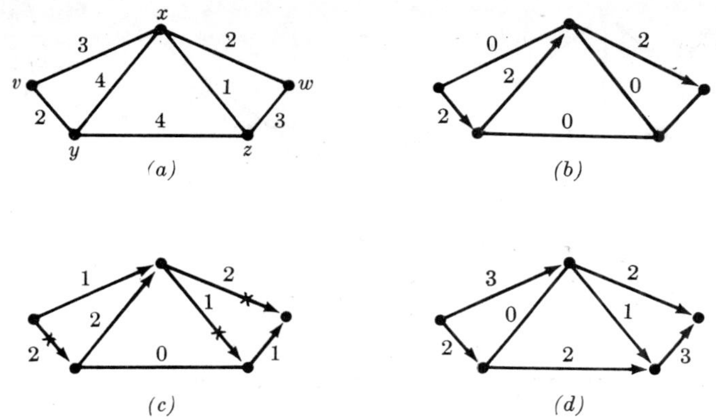

Figure 7-8

value r exists. To see this, consider the flow $\varphi + \Sigma\sigma_i$, where the summation includes any $r - k$ of the simple chain flows of the proof of Theorem 7-4. This flow has value r, and it is feasible for the same reasons as in that proof.

It is not only theoretically possible but also quite practical to synthesize flows having prescribed values by adding (or subtracting) appropriate simple chain flows to some starting feasible flow. Before describing this process of synthesis precisely, we shall illustrate its main features informally and use the capacitated network shown in Fig. 7-8 as an example. The numbers shown in Fig. 7-8*a* represent the arc capacities, and v and w are the intended source and sink, respectively. It is assumed, moreover, that the network is symmetric, so that flow may be in either direction in any arc, provided that it is not greater than the corresponding capacity.

Our object is to maximize the flow from v to w. The maximal value is clearly at most 5, because of the capacities of the arcs incident with v. To begin, we take any path joining v and w such as the one determined by the vertex sequence

$$v,\ y,\ x,\ w$$

and assign to each arc of the path a flow of 2. (Any larger value would violate the capacity of some arcs.) Arcs not in this path are initially assigned a value of zero. The resulting flow is shown in Fig. 7-8*b*.

We now seek any second path which avoids saturated arcs [i.e., arcs like (v,y) and (x,w) for which flow already equals capacity]. Take, for example, the path determined by the vertex sequence

$$v,\ x,\ z,\ w$$

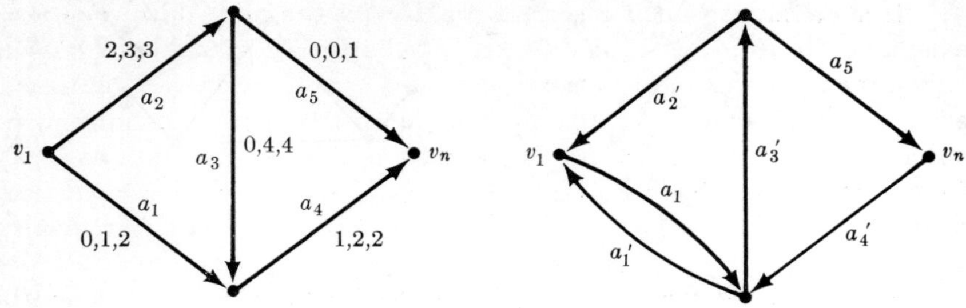

Figure 7-9

and assign one unit of flow to each of its arcs. The resulting flow, when added to that of Fig. 7-8*b*, is shown in (*c*). In this figure, three arcs (those marked with ×'s) are saturated. Note that since we must avoid them, the only way in which we can increase the value of flow from v to w is to negate our earlier decision to send flow in the direction from y to x. To do this, we add a flow of two units in the chain determined by the vertex sequence

$$v,\ x,\ y,\ z,\ w$$

resulting in the flow shown in (*d*). The effect is to retract or cancel the flow in the edge joining x and y and to obtain a total flow whose value is clearly maximal.

We now return to a formal discussion of this approach of synthesizing flows by means of chain flows. The procedure is most easily described in terms of an auxiliary directed graph. If φ is a feasible flow in a network $N = (V,A)$, then the *associated incremental graph* $I(N,\varphi)$ is a directed graph having the same vertices as N and arcs determined as follows: For every arc $a \simeq (v,w)$ in N, $I(N,\varphi)$ includes arc a if $\varphi(a) < \beta(a)$ and a "reverse arc" $a' \simeq (w,v)$ if $\varphi(a) > \alpha(a)$. If a or a' is present in $I(N,\varphi)$, then it is possible to add or subtract, respectively, one unit of flow in a without destroying feasibility in this arc. Figure 7-9 (right) shows the incremental graph associated with the network and feasible flow shown in Fig. 7-9 (left). [Here the triple of numbers x, y, z associated with each arc represent $\alpha(a)$, $\varphi(a)$, and $\beta(a)$, respectively.]

Every simple path from v_1 to v_n in $I(N,\varphi)$ determines a simple chain flow from v_1 to v_n in N which can be added to φ without disturbing feasibility. In Fig. 7-9 (right) the path $\{a_1,a_3',a_5\}$ in $I(N,\varphi)$ determines the simple chain flow σ in N having a_1 and a_5 as normal arcs and a_3 as an inverted arc. Similarly, every simple path from v_n to v_1 in $I(N,\varphi)$ determines a simple chain flow σ^* from v_1 to v_n such that $\varphi - \sigma^*$ is feasible in N. The presence of the path $\{a_4',a_1'\}$ indicates that the chain flow having a_1 and a_4 as normal arcs can be subtracted from φ without destroying feasibility.

It is apparent that if φ is a maximal-valued feasible flow, then there exist no paths from v_1 to v_n in $I(N,\varphi)$. Similarly, if φ has minimal possible value, then there exist no paths from v_n to v_1, except that such paths may exist if φ has value zero. [Recall that, according to our definition, the value of a flow is required to be nonnegative. If $I(N,\varphi)$ contains a path from v_n to v_1 and φ has value zero, then $\varphi - \sigma$ will be feasible for the corresponding chain flow σ, but it will fail to be a legitimate flow from v_1 to v_n.]

We saw earlier that a necessary condition for the existence of feasible flows from v_1 to v_n is that $\underline{M} \leq \bar{M}$. If we let $\underline{M}^*$ denote max $\{\underline{M},0\}$ to avoid flows having negative values, Theorem 7-5 shows that $\underline{M}^*$ and $\bar{M}$ are precise bounds on values realized by feasible flows, if any feasible flows exist.

Theorem 7-5 If there exist any feasible flows from v_1 to v_n, then $\underline{M}^*$ and $\bar{M}$ are respectively the minimal and maximal values realized by feasible flows.

Proof: Let φ be a feasible flow from v_1 to v_n such that $I(N,\varphi)$ contains no paths from v_1 to v_n. (We can always produce such a flow by starting with any feasible flow and adding a succession of simple chain flows until a point when the incremental graph has no paths from v_1 to v_n is reached.) Let W consist of v_1 and all vertices reachable from v_1 by paths. Then $v_n \in W'$ by assumption. Moreover, for every arc in $W \to W'$ we must have $\varphi(a) = \beta(a)$, and for every arc in $W' \to W$ we must have $\varphi(a) = \alpha(a)$. (Why?) But we know that the value of φ, say, k, is given by

$$k = \sum_{W \to W'} \varphi(a) - \sum_{W' \to W} \varphi(a) = \sum_{W \to W'} \beta(a) - \sum_{W' \to W} \alpha(a)$$

Since

$$k \leq \sum_{W \to W'} \beta(a) - \sum_{W' \to W} \alpha(a)$$

for every partition $\{W,W'\}$, it follows that this particular W must realize $\bar{M}$ and that the value of φ is $\bar{M}$.

An analogous line of reasoning establishes the other half of the theorem. In this case, starting with any feasible flow, we can subtract successive chain flows until either a feasible flow of value zero is reached or else no path from v_n to v_1 exists in the current incremental graph $I(N,\varphi)$. In the former case $\underline{M}^* = 0$ and a zero-valued flow exists, so we are finished. In the latter case, let W be the set of vertices not reachable from v_n by paths in $I(N,\varphi)$. Then $\varphi(a) = \alpha(a)$ for arcs in $W \to W'$, $\varphi(a) = \beta(a)$ for arcs in $W' \to W$, and the value k of φ satisfies the relationship

$$k = \sum_{W \to W'} \varphi(a) - \sum_{W' \to W} \varphi(a) = \sum_{W \to W'} \alpha(a) - \sum_{W' \to W} \beta(a)$$

so a feasible flow realizing the value $\underline{M}^*$ exists. This completes the proof.

When $\alpha(a) = 0$ for every arc, feasible flows necessarily exist and Theorem 7-5 is known as the max-flow min-cut theorem.[4]

7-7 MAXIMAL FLOWS IN CAPACITATED NETWORKS

In a capacitated network, a starting feasible flow having minimal value, namely, the flow which is identically zero, is always available. Let φ_0 denote this flow. We can then use the incremental procedure indicated in the proof of Theorem 7-5 to find a succession of flows $\varphi_1, \varphi_2, \varphi_3, \ldots$ having values 1, 2, 3, ... Ultimately, we shall arrive at a flow φ_k having maximal value. According to Theorem 7-5, this will occur when $k = \bar{M}$, at which time one or more cut-sets $\{W,W'\}$ will be saturated, meaning that $\varphi_k(a) = \beta(a)$ for every $a \in W \to W'$ and $\varphi_k(a) = 0$ for every $a \in W' \to W$. To describe an algorithm for finding a succession of feasible flows, let us modify slightly the incremental graph $I(N,\varphi)$ described earlier. For every arc $a \simeq (v,w)$ in N, we now assume that both a and $a' \simeq (w,v)$ are present in $I(N,\varphi)$. However, we associate a length with every arc in $I(N,\varphi)$ as follows:

$$\lambda(a) = \begin{cases} 0 & \text{if } \varphi(a) < \beta(a) \\ \infty & \text{if } \varphi(a) = \beta(a) \end{cases}$$

$$\lambda(a') = \begin{cases} 0 & \text{if } \varphi(a) > \alpha(a) \\ \infty & \text{if } \varphi(a) = \alpha(a) \end{cases}$$

[Note that we could use some positive number rather than ∞ here. However, our convention will allow us to state the algorithm for minimizing cost (discussed later) with a minimum of modification.] Thus arcs which were formerly deleted from $I(N,\varphi)$ are now retained, but have infinite length, and φ is maximal if there are no paths of zero length from v_1 to v_n.

In the following algorithm, I denotes the associated incremental graph (whose structure now does not vary with φ) and λ_k denotes the distance function related to the k-valued flow φ_k.

Maximal-flow Algorithm

1. Initially, set $i = 0$ and take as φ_0 the flow which is identically zero in every arc.

2. Determine the shortest distance from v_1 to v_n in I, relative to the distance function λ_i. (This step can be taken by using the labeling technique described in Chap. 3.)

3. If the distance determined in step 2 is finite, let C denote any simple path from v_1 to v_n with shortest length and let σ_c denote the corresponding simple chain flow in N. Define $\varphi_{i+1} = \varphi_i + \sigma_c$ and repeat step 2, with $i + 1$ in place of i.

4. If the shortest distance from v_1 to v_n is ∞, then φ_i is a flow having maximal value, and the algorithm may be terminated.

There is, of course, one serious drawback to this procedure, namely, that the value of flow is increased by only one unit for each repetition of step 2. In practice, we can expedite matters considerably. Having found a path C in step 3, we can perform a test to see how many units of flow may be routed this way. Expressed differently, we can determine the largest integer t such that $\varphi_i + t\sigma_c$ is a feasible flow. To determine t, note that, for every arc $a \simeq (v,w)$ such that $\lambda_i(a) < \infty$, we can add up to $\beta(a) - \varphi_i(a)$ units of flow to $\varphi_i(a)$ without disturbing feasibility. Similarly, if $\lambda_i(a') < \infty$, a maximum of $\varphi_i(a) - \alpha(a)$ may be subtracted from $\varphi_i(a)$. The integer t is taken to be the minimum of these incremental quantities associated with arcs in C, and step 3 of the algorithm is modified to the extent that we now define not φ_{i+1} but $\varphi_{i+t} = \varphi_i + t\sigma_c$.

Exercise 7-8 Consider Fig. 7-8, for which we have already informally obtained a maximal-valued flow. Solve this problem formally by following the sequence of steps specified by the maximal-flow algorithm.

7-8 MAXIMAL FLOWS IN GENERAL BOUNDED NETWORKS

If $\alpha(a) > 0$ for one or more arcs, but a feasible flow φ_i having value i is known, the procedure described in the preceding section may still be employed to generate feasible flows with values increasing from i to $\bar{M}$ or with values decreasing from i to $\underline{M}^*$. (In this case, step 2 must be modified to find the shortest distance from v_n to v_1 in the current incremental graph.) However, difficulty arises if an initial feasible flow is not known. Note that in this case the flow which is identically zero is specifically nonfeasible.

To find a starting feasible flow, we define an auxiliary problem in a related network N' (Ref. 5 of Chap. 1). The network $N' = (V',A')$ is obtained from $N = (V,A)$ as follows: The vertex set V' consists of the vertices $v_1, \ldots, v_n$ of V, together with two additional vertices v_0 and v_{n+1}. (These will be the source and sink, respectively, of flows in the auxiliary network.)

For every arc $a \simeq (v,w) \in A, A'$ has three corresponding arcs:

$$\begin{aligned} a' &\simeq (v,w) \\ a'' &\simeq (v_0,w) \\ a''' &\simeq (v,v_{n+1}) \end{aligned}$$

These arcs are assigned flow bounds as follows:

$$\begin{aligned} \alpha'(a') &= \alpha'(a'') = \alpha'(a''') = 0 \\ \beta'(a') &= \beta(a) - \alpha(a) \\ \beta'(a'') &= \beta'(a''') = \alpha(a) \end{aligned}$$

Note that for arcs such that $\alpha(a) = 0$, construction of a'' and a''' is superfluous, since $\alpha = \beta = 0$ for such arcs. In practice, these arcs are omitted.

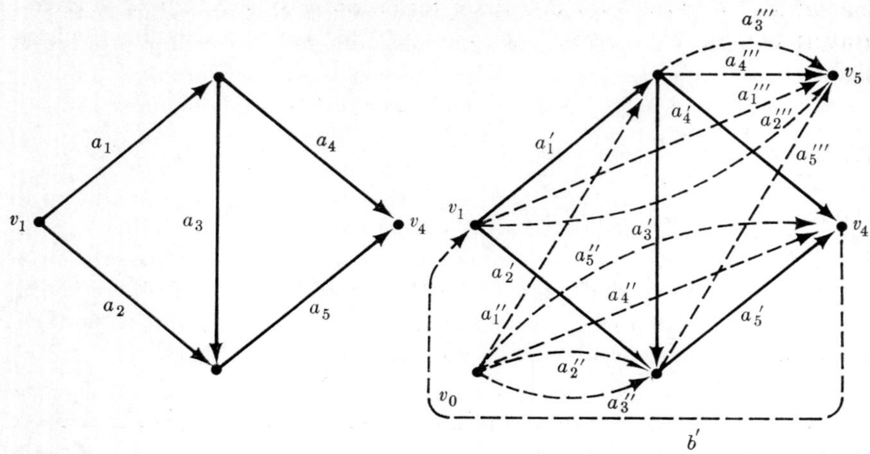

Figure 7-10

In theory, however, it is simplest to suppose that the three arcs a', a'', and a''' are invariably included.

One final arc b' having the following characteristics is added to N':

$$b' \simeq (v_n,v_1) \qquad \alpha'(b') = 0 \qquad \beta'(b') = K$$

where K is an integer larger than the value of any possible feasible flow in the original network N. Figure 7-10 shows the structure of the auxiliary network N' associated with a given network N.

Remark: It should be noted that N' is completely determined by N and its constraining functions α and β. If N is a complex network, it would clearly be very troublesome, at best, to actually construct N'. However, a digital computer can be used to good advantage to construct the auxiliary network and to implement the computational algorithm.

Exercise 7-9 In Fig. 7-7 complete the specification of α and β by setting $\alpha = 1$ and $\beta = 3$ for each of the two vertical arcs. Construct the auxiliary network N' by taking v_1 as source and v_5 as sink. Indicate the value of α' and β' for every arc of N'.

The importance of N' stems from two facts. First, N' is a capacitated network, and the preceding section provides a means for finding a maximal flow in such a network. Second, we can show that a maximal flow φ' in N' can be easily transformed into a feasible flow φ in N, provided that any feasible flows exist. If none exist, this too can be easily deduced from the characteristics of φ'.

A feasible flow φ' from v_0 to v_{n+1} in N' is called a *saturating flow* if $\varphi'(a'') = \beta'(a'')$ for every arc a'' from v_0. Such a flow (if it exists) is necessarily maximal in N'. (Why?) An equivalent characterization is that $\varphi'(a''') = \beta'(a''')$ for every arc terminating at v_{n+1}. If φ' is a saturating flow in N', consider the following function φ defined on the arcs of N:

$$\varphi(a) = \varphi'(a') + \alpha(a)$$

With φ defined in this manner, one can verify the following theorem:

Theorem 7-6 If φ' is a saturating flow in N' and φ is defined as above, then φ is a feasible flow from v_1 to v_n in N. Moreover, the value of φ is $\varphi'(b')$.

Exercise 7-10 Prove Theorem 7-6. Note that three things must be established:

1. $\alpha(a) \le \varphi(a) \le \beta(a)$ for all a
2. $Q(v_i,\varphi) = 0$ for $i = 2, 3, \ldots, n-1$
3. $Q(v_1,\varphi) = \varphi'(b')$

So if, in maximizing flow through N', we obtain a saturating flow φ', we can easily obtain a feasible flow φ in N. On the other hand, if a maximal flow in N' fails to be a saturating flow, Theorem 7-7 asserts that no feasible flows in N exist.

Theorem 7-7 If φ' is a maximal flow from v_0 to v_{n+1} in N' but is not a saturating flow, then no feasible flows from v_1 to v_n exist in N.

Proof: Let W consist of those vertices, other than v_0, which can be reached from v_0 by paths of finite length in the incremental graph $I(N',\varphi')$. Let W' denote the vertices not reachable, other than v_{n+1}. Since φ' is not a saturating flow, both W and W' are nonempty. (Verify this in detail.) The capacity (i.e., sum of upper flow bounds) of the cut-set $\bar{W} \to \bar{W}'$ determined by the partitioning of V' into $\bar{W} = \{v_0\} \cup W$ and $\bar{W}' = \{v_{n+1}\} \cup W'$ is given by

$$k = \sum_{v_0 \to W'} \beta'(a'') + \sum_{W \to W'} \beta'(a') + \sum_{W \to v_{n+1}} \beta'(a''')$$

By using the fact that $\beta' = \beta - \alpha$ for arcs of type a', while $\beta' = \alpha$ for arcs of types a'' and a''', k can be written as

$$k = \sum_{W \to W'} \alpha(a) + \sum_{W' \to W'} \alpha(a) + \sum_{W \to W'} \beta(a) - \sum_{W \to W'} \alpha(a) + \sum_{W \to W} \alpha(a) + \sum_{W \to W'} \alpha(a)$$

Since φ' is not a saturating flow, we also have

$$k > \sum_{W \to W} \alpha(a) + \sum_{W \to W'} \alpha(a) + \sum_{W' \to W} \alpha(a) + \sum_{W' \to W'} \alpha(a)$$

Upon combining the last two relationships, we find that

$$\sum_{W \to W'} \beta(a) - \sum_{W' \to W} \alpha(a) < 0$$

It follows that for any feasible flow in N we must have

$$\sum_{W \to W'} \varphi(a) - \sum_{W' \to W} \varphi(a) < 0$$

(Why?) On the other hand, if v_1 and v_n are both in W or both in W', any flow φ from v_1 to v_n must satisfy the equation

$$\sum_{W \to W'} \varphi(a) - \sum_{W' \to W} \varphi(a) = 0$$

while if $v_1 \in W$ and $v_n \in W'$, φ must satisfy the equation

$$\sum_{W \to W'} \varphi(a) - \sum_{W' \to W} \varphi(a) = k \geq 0$$

where k is the value of φ. We conclude that no feasible flows from v_1 to v_n in N exist, except possibly for the case not considered: when $v_n \in W$ and $v_1 \in W'$. But this case never applies, because if we can reach v_n via an unsaturated path in N', we can also reach v_1, since the arc b' is never saturated. This completes the proof.

Exercise 7-11 Apply the maximal-flow algorithm to the capacitated network N' constructed in Exercise 7-9. Show that the maximal flow is not a saturating flow and hence according to Theorem 7-7 no feasible flows in N exist.

Recapitulation

To summarize the principal findings thus far in the chapter, we have seen that we can determine a maximal flow in a capacitated network by starting with the identically zero flow and adding a succession of simple chain flows from source to sink. The last flow in the sequence has maximal value, and this is detected when, in the associated incremental graph, no further paths having finite length join v_1 to v_n. Every intermediate flow determined during this procedure is also feasible, so we have in fact generated feasible flows realizing all possible values. Moreover, it is possible to accelerate the procedure by taking appropriate multiples of each chain flow. For other networks, for which $\alpha(a) > 0$ for some or all arcs, it is necessary first to construct an associated capacitated network N' and maximize flow from v_0 to v_{n+1} in the above manner. The maximal flow φ' then either leads to a corresponding feasible flow φ in N or else definitely indicates that no such flows exist. In the former case φ can be taken as the starting point for a sequence of increasing (decreasing) flows each obtained from the preceding one by adding (subtracting) an appropriate chain flow. The same device may be employed to accelerate finding the maximal (minimal) flow, once φ has been determined. Thus, for any bounded network, we have a practical means for generating feasible flows between a fixed pair of vertices having all possible flow values.

7-9 MINIMAL–COST FLOWS

If N is a bounded network and $\underline{M}^* \leq k \leq \bar{M}$, there are generally numerous feasible flows from v_1 to v_n having value k. In that case it is frequently desirable to select from among them one which minimizes a quantitative measure which we shall call "cost" but whose true interpretation varies with the application.

We continue to assume that $N = (V,A)$ is a given network with arc bounds $\alpha(a)$ and $\beta(a)$. In addition, we now assume that a real-valued *cost function* γ is associated with N, which assigns to each arc $a \in A$ a *unit cost* $\gamma(a) \geq 0$. If φ is any feasible flow, the *total cost of* φ is denoted by $\gamma(\varphi)$ and defined by the relationship

$$\gamma(\varphi) = \sum_{a \in A} \gamma(a) \cdot \varphi(a)$$

If $a \simeq (v,w)$, $\gamma(a)$ can be interpreted as the cost associated with sending one unit of flow from v to w through a. Then $\gamma(\varphi)$ is merely the total cost of flow in the network. It is important to note that this formulation is an appropriate model for evaluating the cost of flows only when the cost of flow in each arc can be assumed proportional to the magnitude of flow.

For the moment, we shall restrict attention to capacitated networks (for which $\alpha \equiv 0$). The basic problem we shall consider is this: given a capacitated network N and a cost function γ, how can we find a feasible flow φ from v_1 to v_n whose value is a prescribed integer k and whose cost is minimal?

Since $\gamma(\varphi) \geq 0$ for every feasible flow φ, the feasible flow φ_0 which is identically zero is clearly a minimal-cost feasible flow having value zero. Moreover, we shall see in this section that if φ_k is a feasible flow of value k whose cost is minimal, we can find an appropriate chain flow σ such that $\varphi_{k+1} = \varphi_k + \sigma$ is a minimal-cost feasible flow of value $k + 1$ (except, of course, if φ_k is maximal). This being the case, we can start with φ_0 and generate a succession of minimal-cost flows $\varphi_1, \varphi_2, \ldots$ until a flow having the desired value k is reached. For an alternate procedure based on linear programming concepts, see Fulkerson.[7]

If φ is a feasible flow from v_1 to v_n having value k, let $I(N,\varphi)$ denote the associated incremental graph. For each arc a in the original network, define a length $\lambda(a)$ as follows:

$$\lambda(a) = \gamma(a) \qquad \text{if } \varphi(a) < \beta(a)$$
$$\lambda(a) = \infty \qquad \text{if } \varphi(a) = \beta(a)$$

Thus $\lambda(a)$ represents the unit cost of a, with the convention that the unit cost is infinite if a is already saturated.

Similarly, for each arc a' in $I(N,\varphi)$ which corresponds to the "reverse" of an arc a in N, define $\lambda(a')$ by

$$\begin{aligned} \lambda(a') &= -\gamma(a) \qquad && \text{if } \varphi(a) > 0 \\ \lambda(a') &= \infty && \text{if } \varphi(a) = 0 \end{aligned}$$

If P is a simple path from v_1 to v_n or a simple cycle in $I(N,\varphi)$, then $\lambda(P) = \sum_P \lambda(a)$ is readily seen to be the incremental cost associated with adding to φ the simple chain or cycle flow σ_P determined by P. Expressed symbolically,

$$\gamma(\varphi + \sigma_P) = \gamma(\varphi) + \lambda(P)$$

This will be true even for infeasible flows $\varphi + \sigma_P$ if we adopt the convention that all infeasible flows have infinite cost.

Assuming that φ is not maximal, then a shortest path P_0 from v_1 to v_n in $I(N,\varphi)$ clearly determines the best possible chain flow σ with which to augment φ; since if $\lambda(P_0) \leq \lambda(P)$, then

$$\gamma(\varphi + \sigma_{P_0}) = \gamma(\varphi) + \lambda(P_0) \leq \gamma(\varphi) + \lambda(P) = \gamma(\varphi + \sigma_P)$$

There is one difficulty, however: it may happen that $\lambda(Z) < 0$ for some simple cycles Z in $I(N,\varphi)$. We saw in Chap. 3 that our technique for finding shortest simple paths through a directed graph breaks down in this case. Fortunately, if φ is a minimal-cost flow, this cannot happen. In fact, Theorem 7-8 characterizes minimal-cost flows in terms of cycles of negative length.

Theorem 7-8 A feasible flow φ having value k minimizes cost over all feasible flows of value k if and only if $\lambda(Z) \geq 0$ for every simple cycle Z in $I(N,\varphi)$.

Proof: If $\lambda(Z) < 0$ for some simple cycle Z, then the corresponding simple circuit flow σ_z is such that

$$\gamma(\varphi + \sigma_z) = \gamma(\varphi) + \lambda(\sigma_z) < \gamma(\varphi)$$

and hence φ is not a minimal-cost flow. So if φ minimizes cost, then $\lambda(Z) \geq 0$ for every cycle Z.

To prove the converse, assume that $\lambda(Z) \geq 0$ for every cycle Z but that φ is not a minimal-cost flow. (We shall arrive at a contradiction.) Let φ' be a minimal-cost flow having value k. Then we can write

$$\varphi' - \varphi = \sigma_1 + \sigma_2 + \cdots + \sigma_m$$

where the σ_i's are conformal circuit flows associated with cycles Z_1, Z_2, . . . , Z_m in $I(N,\varphi)$. Relative to the distance function λ associated with $I(N,\varphi)$, we have (by assumption)

$$\lambda(Z_1) \geq 0$$

and hence

$$\gamma(\varphi + \sigma_1) = \gamma(\varphi) + \lambda(Z_1) \geq \gamma(\varphi)$$

Now let λ_1 denote the distance function associated with $I(N, \varphi + \sigma_1)$. Clearly, $\lambda(a)$ and $\lambda_1(a)$ are equal on arcs not members of Z_1, since their flow is not altered. For arcs in Z, which are in N, $\lambda_1(a) = \infty$ or else $\lambda_1(a) = \lambda(a)$, depending on whether a is now saturated or not. Similarly, for arcs a' in Z_1 which are the "reverse" of arcs in N, we have $\lambda_1(a') = \infty$ or $\lambda_1(a') = \lambda(a')$, depending on whether arc a is now empty or not. The only types of arcs in the incremental graph for which $\lambda_1 < \lambda$ are those which are the opposites of arcs in Z_1 (that is, arc a if $a' \in Z$ and arc a' if $a \in Z_1$). But since the σ_i's are conformal, such arcs cannot appear in any of the remaining Z_i's, and their new lengths are irrelevant in the remaining argument.

As a result of the above relationship between λ_1 and λ, we have

$$\lambda_1(Z_2) \geq \lambda(Z_2) \geq 0$$

and so

$$\gamma(\varphi + \sigma_1 + \sigma_2) = \gamma(\varphi + \sigma_1) + \lambda_1(Z_2) \geq \gamma(\varphi)$$

By repeating this process m times, we eventually find that

$$\gamma(\varphi + \sigma_1 + \cdots + \sigma_m) \geq \gamma(\varphi)$$

But since

$$\varphi + \sigma_1 + \cdots + \sigma_m = \varphi'$$

this contradicts the assumption that

$$\gamma(\varphi') < \gamma(\varphi)$$

and completes the proof.

Exercise 7-12 Prove that if φ is a minimal-cost feasible flow in N, then there exists at least one other feasible flow φ' in N having the same value and also minimizing cost if and only if $\gamma(C) = 0$ for some cycle in $I(N,\varphi)$.

In a capacitated network, the statement of an algorithm which will find minimal-cost feasible flows of all possible values from 0 to $\bar{M}$ is formally the same as the maximal-flow algorithm of Sec. 7-7, with the understanding that now the appropriate length function λ to use at each stage is the one defined here in terms of incremental arc costs.

Figure 7-11 illustrates the algorithm for finding minimal-cost flows for a very small network. After studying this example, the reader can best gain real familiarity with the procedure by applying the algorithm to various networks of a somewhat more complex variety. For very complex networks, manual implementation becomes impractical. However, networks with several thousand arcs can be readily handled by means of a digital computer.

The top network in Fig. 7-11 shows the structure of the network and gives the unit costs and capacities of all arcs. The left-hand column of graphs is the succession of incremental graphs which must be considered. The numbers associated with the arcs are their lengths, arcs with infinite length being omitted, and the arcs of a shortest path from source to sink are shown as bold arcs. The right-hand column of graphs shows the succession of minimal-cost flows, of increasing value, obtained by successively adding the chain flows associated with the shortest paths in the incremental graphs. The numbers associated with the arcs are the corresponding arc flows. In all, four chain flows, which had associated values of 5,2,3, and 1, respectively, were found. The final flow of value 11 is clearly maximal, since both the arcs leading into the sink are saturated.

Note that the first incremental graph $I(N,\varphi_0)$ is structurally identical with the given network. (Here φ_0 is, of course, the flow which is identically zero in all arcs.) This is the case because no arcs are yet saturated and no positive flows which could potentially be canceled exist. All of the successive incremental graphs have certain forward (solid) arcs missing, corresponding to arcs which are saturated. In addition, they have certain "reverse" (dashed) arcs, corresponding to arcs having positive flow, which could be canceled—by using the arc in the reverse direction—without destroying feasibility.

The simple nature of this example made it possible to find a shortest path through each incremental graph by inspection. In a more complex case, one must use a systematic technique, such as that of Chap. 3, to find each shortest path, since selection of a non-shortest path will result in a flow whose cost is not minimal and which is not a valid starting point for determining the next increment of flow.

Note that the last chain flow does in fact involve a "reverse" arc of the incremental graph $I(N,\varphi_{10})$, so that the optimal flow of value 11 uses this arc to a lesser extent than the optimal flow of value 10.

In this example, there happened to be a unique shortest path in each successive incremental graph. It should be noted that this is not always the case. When there are several shortest paths, any one of these may be used as the basis for the next chain flow.

Exercise 7-13 Referring to the preceding example, suppose that we start with the flow φ_3 of value 3 obtained by assigning three units of flow to the bottom three arcs and 0 to the rest. Construct the corresponding incremental graph $I(N,\varphi_3)$ and observe that it contains at least one cycle whose length is negative. (The theory has shown that this must be the case, since φ_3 is not a minimal-cost flow of value 3.)

It is important to note that, while we have bypassed flows of values 1, 2, 3, 4, 6, 8, and 9 in the above example, we can produce minimal-cost

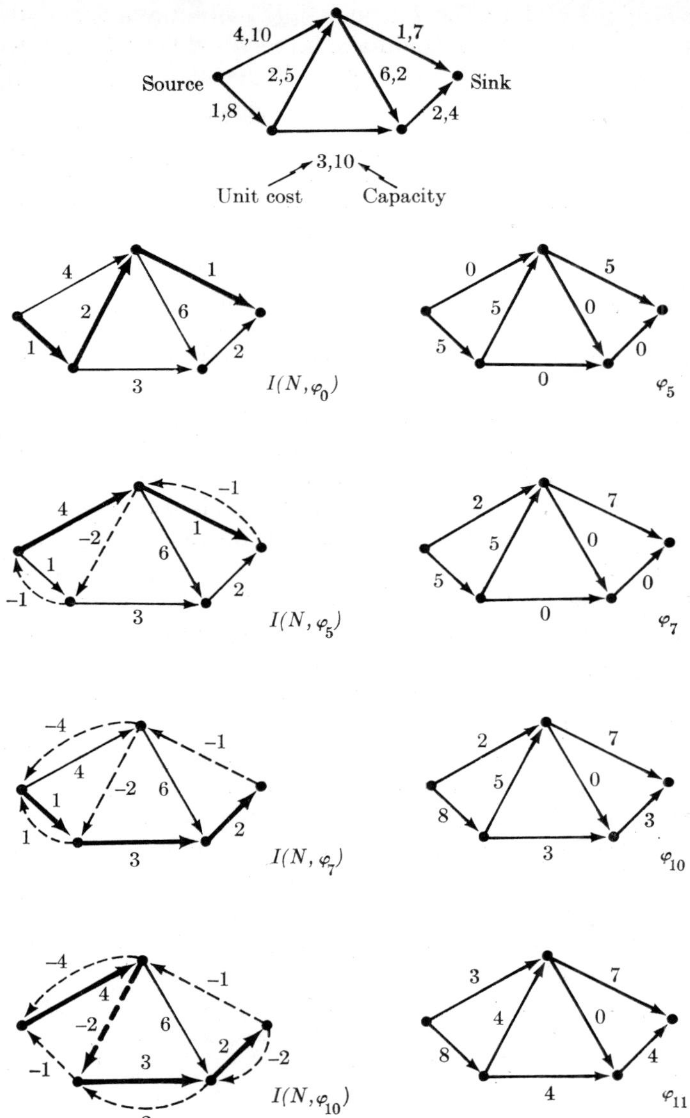
4,10
1,7
Source
2,5
6,2
Sink
1,8
2,4
3,10
Unit cost
Capacity
$I(N, \varphi_0)$
φ_5
$I(N, \varphi_5)$
φ_7
$I(N, \varphi_7)$
φ_{10}
$I(N, \varphi_{10})$
φ_{11}

Figure 7-11

flows having these values by interpolation. For example, to obtain an optimal flow of value 8, we would add to φ_7 one unit of flow in the chain used to produce φ_{10}.

Exercise 7-14 In Fig. 7-8*a*, direct each arc from left to right, assume that the number associated with each arc represents its unit cost $\gamma(a)$ as well as its capacity $\beta(a)$, and assume that $\alpha(a) = 0$, that is, the network is capacitated. Use the method of this section to find a sequence of minimal-cost flows the last of which has maximal value.

7-10 SOME SPECIALLY STRUCTURED FLOW PROBLEMS

As a first example of a class of problems that can be treated as flow problems, we consider the so-called *transportation problem*, which is much studied in the literature of operations research.

Let $S_1, \ldots, S_m$ and $T_1, \ldots, T_n$ be a number of origins and destinations, respectively, between which some homogeneous substance must be distributed. Each origin S_i has a limited supply A_i, and each destination T_j has a specified demand B_j. Moreover, there is a cost of U_{ij} per unit associated with supplying T_j from S_i. In the simplest instance there is no restriction on the amount shipped from S_i to T_j—except, of course, that it must not violate the constraints imposed by A_i and B_j. The problem is to distribute the material from the sources to the destinations at minimal cost, without oversupplying any destination or exceeding the capability of any origin.

This problem can be considered as the problem of finding a minimal-cost flow of maximal value in the network of Fig. 7-12. [The ordered pair (x,y) associated with each arc has the interpretation that x denotes the capacity and y the unit cost.] If movement from S_i to T_j is not permissible for certain pairs i, j, the corresponding arcs are merely omitted. Similarly, if there is an upper limit on the quantity shipped from S_i to T_j a finite capacity is associated with the appropriate arc. In this case, of

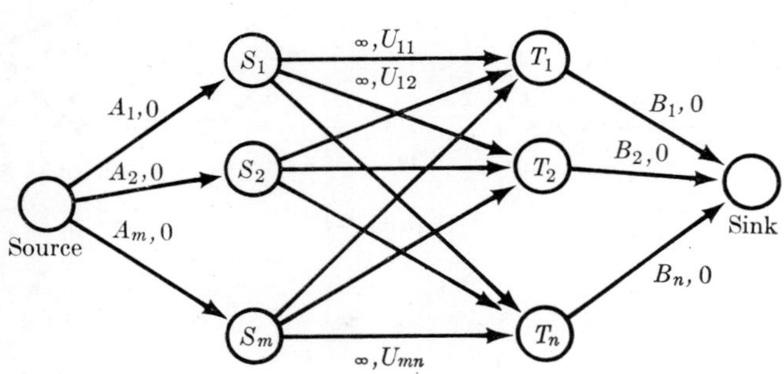

Figure 7-12

course, it may happen that not all destination demands can be met. However, the solution to the flow problem will move the maximum amount possible, and it will do so at minimum cost.

If $m = n$ and if the A's and B's represent individuals and jobs, respectively, then the same model is appropriate for the basic *assignment problem*, in which m men are to be assigned to the m jobs in such a way that the optimal utilization is achieved. (In this case U_{ij} is a measure of the value of assigning the ith man to the jth job.) The capacities A_i and B_j are set to unity, so that each man is assigned to at most one job and each job is performed by one man. If certain men are not qualified for certain jobs, the corresponding arcs are omitted. Here again, this may result in a solution in which not all jobs are performed. (In this problem, one can negate arc values, add a large constant to each, and minimize.)

For bipartite graphs the maximal matching problem discussed in Chap. 6 can also be regarded as a flow problem in a network structured as in Fig. 7-12. In this case, all arcs have unit capacity, costs are not a factor, and the problem is one of maximizing flow from source to sink. Note that the problem of increasing flow by finding a path in the incremental graph is really the same as that of finding an alternating chain joining two exposed vertices, as described in Sec. 6-14.

Exercise 7-15 In Fig. 7-12 assume that $m = n = 8$, that every S_i is joined to every T_j by an arc, that each $A_i = 1$ and each $B_j = 2$, and that $U_{ij} = i + j$. Find a minimal-cost feasible flow from source to sink having maximal value.

7-11 MULTIPLE - COMMODITY FLOW PROBLEMS

Until now, we have assumed that a single homogeneous substance or "commodity" is flowing in the network under consideration. If the network has several sources and/or sinks, we have assumed that the solution may join any source with any sink via a chain flow. Suppose, however, that we now have K commodities and that the vertices of the network include $2K$ distinguished vertices y_i, z_i, for $i = 1, 2, \ldots, k$, such that we seek k simultaneous flows each of which is a flow from y_i to z_i for some i and has a given value Q_i.

To distinguish between commodities, let f_{ij}^k denote the flow of the kth commodity from vertex v_i to vertex v_j, with the added convention that $f_{ji}^k = -f_{ij}^k$. Let C_{ij} denote the capacity (in either direction) of the arc joining vertices v_i and v_j.

The problem, then, is to assign an integer f_{ij}^k to every arc (i,j) for every commodity k in such a way that, for any fixed k, the set of arc flows

$$F^k = \{f_{ij}^k\}$$

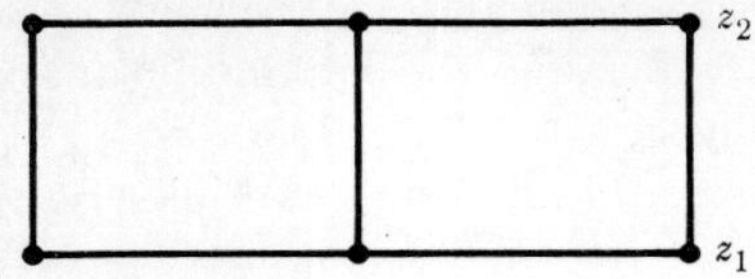

Figure 7-13

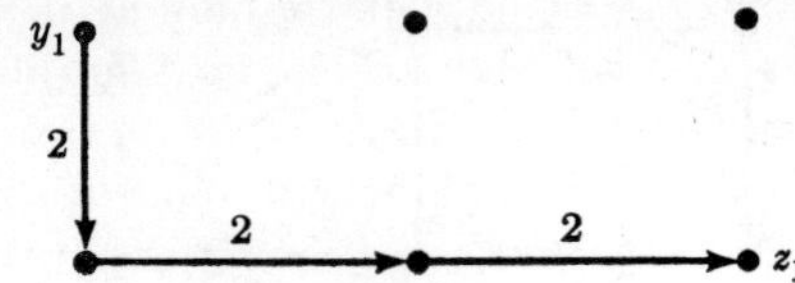

Figure 7-14

is a flow of value Q_k from y_i to z_i and such that, for each arc (i,j), we have

$$\Sigma|f_{ij}^k| \leq C_{ij}$$

(It may happen that different commodities must flow in opposite directions in the same arc in order to achieve a solution.)

For the two-commodity case (i.e., when $K = 2$), Hu[9] has obtained a theoretical result, analogous to the max-flow min-cut theorem, and also a computational procedure for solving the above problem. To illustrate the procedure, consider the network of Fig. 7-13. It is assumed here that each arc has (symmetric) capacity 2, and that

$$Q_1 = Q_2 = 2$$

We first disregard commodity 2 and seek any feasible flow of value $Q_1 = 2$ of commodity 1 from y_1 to z_1, using the single-commodity techniques developed earlier. We might, for example, obtain the flow shown in Fig. 7-14.

If no such flow exists, the problem clearly has no solution. If, as in this case, one does exist, we next reduce all arc capacities by the amount

$$|f_{ij}^1|$$

and repeat the process for commodity 2 alone. In this example, the reduced capacities are shown in Fig. 7-15. Clearly no flow of commodity 2 is possible, since the arcs incident with source y_2 have zero residual capacity. We conclude that we must first reroute the flow of commodity 1 before we can accommodate any flow of commodity 2.

We now give Hu's technique for achieving a suitable rerouting. We first find (if possible) some chain C_1, oriented from y_2 to z_2, which has the

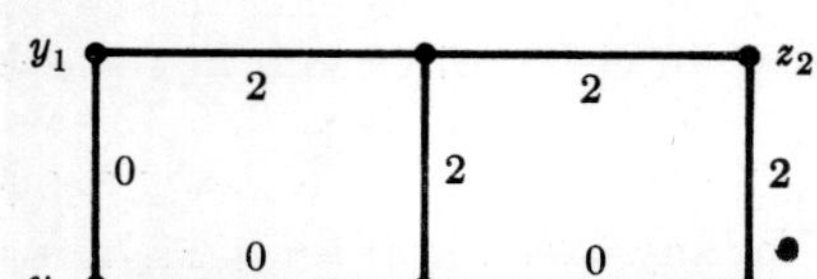

Figure 7-15

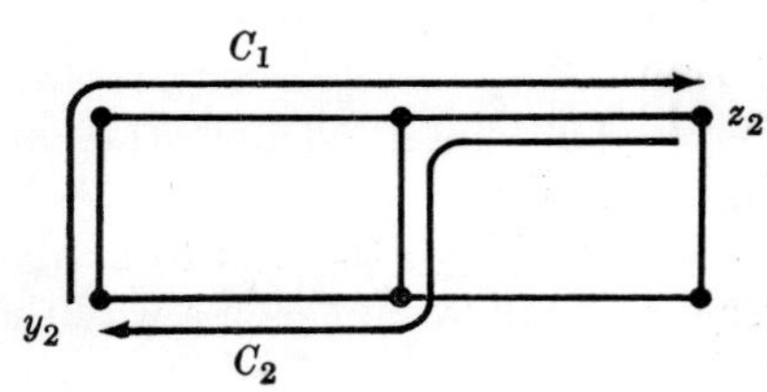

Figure 7-16

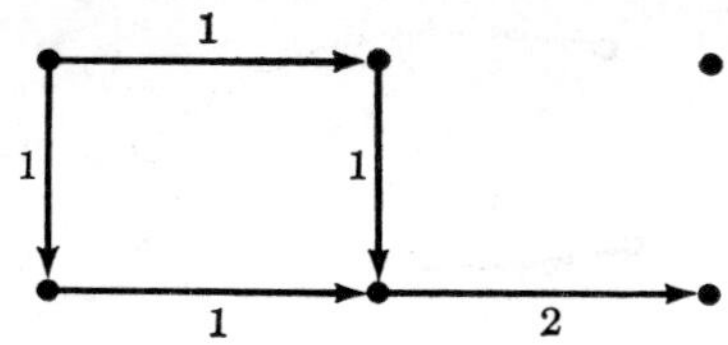

Figure 7-17

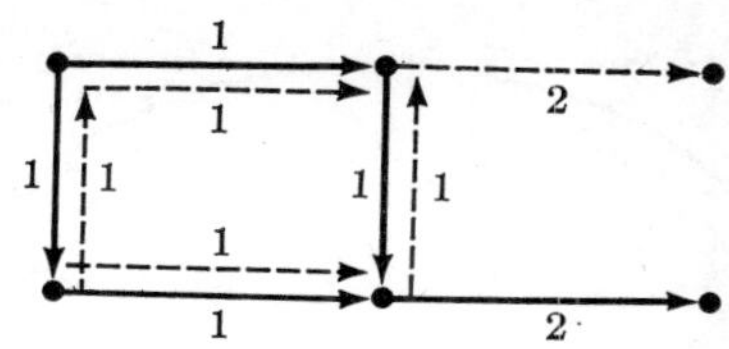

Figure 7-18

property that at least one unit of commodity 1 may be added to each arc in the direction induced by the chain orientation. We next find (if possible) some chain C_2, oriented from z_2 to y_2, having the same formal property. (C_1 and C_2 may involve some of the same arcs.) Appropriate chains for our example are shown in Fig. 7-16. Then we increase the flow of commodity 1, in both chains, by one unit. (In practice, larger flow increments may be used in some cases.) This results in the adjusted flow of commodity 1 as shown in Fig. 7-17. Finally, we increase the flow of commodity 2 along C_1 and along the *reverse* of C_2 by one unit and obtain the (dashed) flows shown in Fig. 7-18.

In the present example we are through, since we have achieved simultaneous flows, of value 2, of both commodities. If the flow of commodity 2 were still less than Q_2, we would proceed to find another pair of chains C_1 and C_2 and a corresponding rerouting of commodity 1 and increase in flow of commodity 2. It can be shown that we can continue to find pairs of chains and ultimately increase the value of flow of commodity 2 to Q_2 (while holding the value of flow of commodity 1 at Q_1) unless the problem has no solution.

Let A_1 and A_2 denote, respectively, the capacities of minimal-capacity cut-sets separating y_1 from z_1 and y_2 from z_2. Also, let A_{12} be the capacity of a minimal-capacity disconnecting set which meets all chains joining y_1 and z_1 and also all chains joining y_2 and z_2. Then the two-commodity analog of the max-flow min-cut theorem is as follows:

Theorem 7-9 There exist jointly feasible flows F_1 and F_2 from y_1 to z_1 and from y_2 to z_2, respectively, having values Q_1 and Q_2, respectively, if and only if the following three conditions are all satisfied:

$$Q_1 \leq A_1 \qquad Q_2 \leq A_2 \qquad Q_1 + Q_2 \leq A_{12}$$

Necessity of the conditions is clear. For sufficiency, see Ref. 9.

7-12 STOCHASTIC FLOW IN NETWORKS

In this section we give a brief summary of ideas which pertain to network flows but have their origin in the theory of queues.

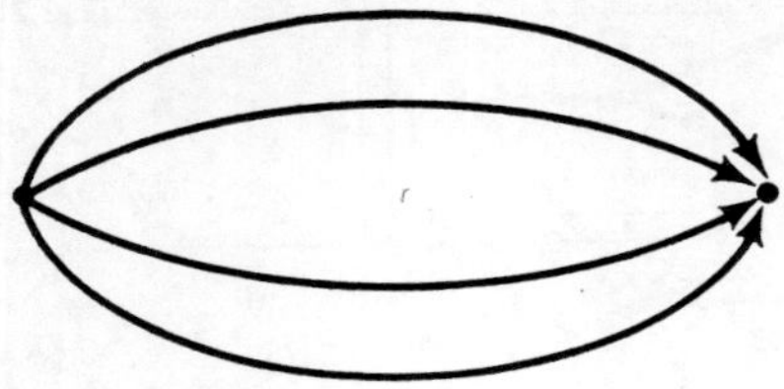

Figure 7-19

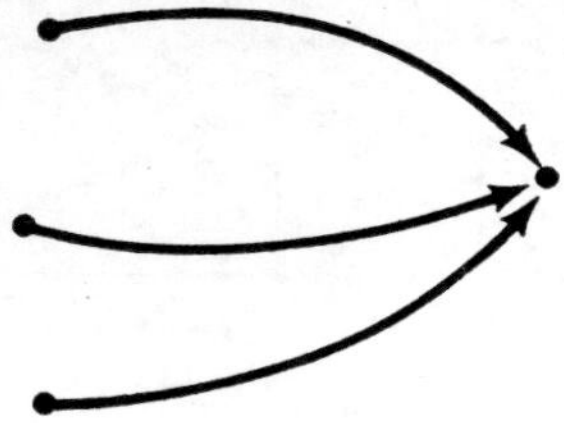

Figure 7-20

Consider a network whose arcs have capacities prescribed according to given probability distributions that are functions of time (i.e., stochastic processes), and let it be desired to pass flow from the source to the sink of such a network. The problem is to determine the total average flow at any time through the network. In general, the amount of flow present at the source is also given according to a stochastic process. Thus at any given moment there may or may not be any flow passing through the source to an initial vertex of an arc. If the amount of flow available exceeds the capacity of an arc through which it must pass, it is delayed at the initial vertex of that arc.

A queueing network consists of sets of interconnected specialized service channels connected in series and in parallel. Before each service channel (or set of service channels if they cooperate in parallel) there is a waiting line (whose length may be zero) of items ready to be served at the service channels. The output from a server may comprise the input of another. If we associate a vertex with the arrivals to a queue and another with its departures and if we join these vertices with an arc corresponding to the service, we obtain a graph. The graph is simple if each queue consists of a single channel and a single line. If there are several channels in parallel, all with the same waiting line and all leading to a single output, we have a figure of the form of Fig. 7-19 to represent that part of the network. If, on the other hand, each of a set of parallel channels has its waiting line and all channels lead to a single output, we have a figure of the form of Fig. 7-20. The situation in which some of the parallel channels may lead to other channels in series or to sinks can be represented by a figure of the form of Fig. 7-21. For the source vertex we take the local population in need of service and for the sink vertex the same population (with satisfied demand).

In the theory of queues the flow consists of discrete items, e.g., customers; whereas in the usual network flow problems flow is considered as continuous, but more frequently one is interested in integer-valued flow which can be interpreted as discrete flow. Note that in the stochastic arc capacity problem there are two alternatives. Either the

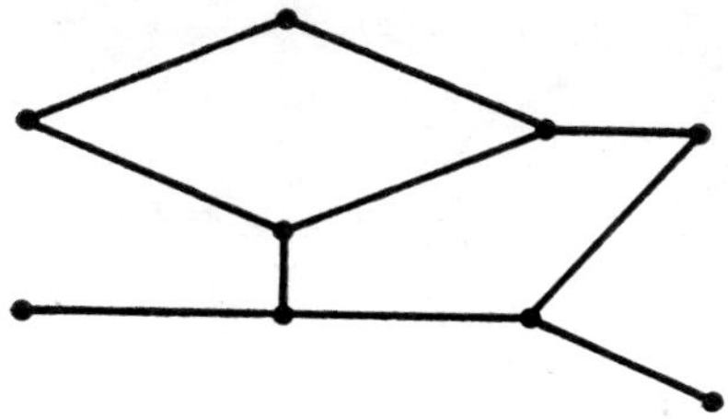

Figure 7-21

flow is lost at the initial vertex when the arc is entirely occupied with flow in process or it is delayed there awaiting passage. An interesting theorem for discrete flow applies to queueing networks in which flow has been in process for a long time and one is interested in the steady state, i.e., in the asymptotic behavior as the time $t \to \infty$.

In a certain queueing network flow which is common to a number of operations, arrivals are observed to occur at random according to the Poisson process

$$\frac{(\lambda t)^n e^{-\lambda t}}{n!} \qquad n = 0, 1, 2, \ldots$$

An important property of this process is that the probability of more than a single arrival in a small period of time is negligible compared with the probability of no arrivals or of one arrival.

The times between arrivals which occur according to a Poisson process have an exponential distribution of the form $\lambda e^{-\lambda t}$. Thus queueing problems in which arrivals occur according to a Poisson process and enter service according to another such process may also be studied by considering the exponentially distributed interarrival intervals and service time intervals.[11]

If the arc has exponential capacity (exponential times of service) and if the arrival intervals are exponential, the flow from the exit has exponential distribution. The requirement here is that the ratio of the input rate to the arc passage rate (service rate) must not exceed unity; otherwise, flows will be delayed for infinitely long time at the initial vertices. With this theorem it is possible to trace flow through a network, because the output distribution from one terminal vertex is automatically a known input distribution with the given terminal vertex as an initial vertex of the next stage.

An alternative representation of a queueing network is to associate a vertex with the waiting line and a second vertex with the service and let the arc incident with the two vertices indicate a transition from waiting into service. One can then associate a third vertex with the departing stream of customers and let the arc incident with this vertex and the service vertex represent the transitions from service to departure or to the next waiting line. Thus we can classify the vertices according to whether they are service vertices, waiting-line vertices, or departure vertices.

Exercise 7-16 Interpret both of these representations for a network of queues in parallel and in series and draw the diagrams.

Exercise 7-17 Mention ten possible conceptual ramifications of network flows such as stochastic flow, flow through porous pipes, and through deformable arcs.

More comprehensive expositions of network flow theory, from various points of view, are to be found in Refs. 5, 10, 12, and 14 of Chap. 1. The reader is also referred to the selected list of papers at the end of this chapter and to the extensive bibliographies contained in the above books.

REFERENCES

1 Berge, C.: Les Problemes de Flot et de Tension, *Cahiers Centre Études Rech. Oper.*, **3:** 69–93 (1961).

2 Dantzig, G. B., and D. R. Fulkerson: "Computation of Maximal Flows in Networks," The RAND Corp., p. 677, 1955.

3 Ford, L. R. Jr.: "Network Flow Theory," The RAND Corp. p. 923, 1956.

4 Ford, L. R. Jr., and D. R. Fulkerson: Maximal Flow through a Network, *Can. J. Math.*, **8:** 399–404 (1956).

5 Ford, L. R. Jr., and D. R. Fulkerson: A Simple Algorithm for Finding Maximal Network Flows and an Application to the Hitchcock Problem, The RAND Corp., RM-1604, 1955.

6 Ford, L. R. Jr., and D. R. Fulkerson: Constructing Maximal Dynamic Flows from Static Flows, *Operations Res.*, **6:** 419–433 (1958).

7 Fulkerson, D. R.: An Out-of-kilter Method for Minimal-cost Flow Problems, *J. Soc. Ind. Appl. Math.*, **9:** 18–27 (1961).

8 Gomery, R. E., and T. C. Hu: Multi-terminal Network Flows, *J. Soc. Ind. Appl. Math.*, **9:** 551–570 (1961).

9 Hu, T. C.: Multi-commodity Network Flows, *Operations Res.*, **11:** 344–360 (1963).

10 Jewell, W. S.: Optimal Flow through Networks, *MIT Interim Tech. Rept.* 8, 1958.

11 Saaty, T. L.: "Elements of Queueing Theory," McGraw-Hill Book Company, New York, 1961.

GLOSSARY

The following glossary presents brief, informal definitions of those terms which are basic and which recur frequently in the text.

Acyclic graph A directed graph having no cycles.

Adjacency matrix Same as **Vertex matrix.**

Adjacent edges or **Arcs** Two edges or arcs having at least one common end point.

Adjacent vertices Two vertices that are joined by an edge or arc.

Arc See **Directed graph.**

Arc progression A finite sequence of (not necessarily distinct) arcs such that the terminal vertex of each arc (except the last) is the same as the initial vertex of the succeeding arc. The progression is said to be *closed* if the initial vertex of the first arc is the same vertex as the terminal vertex of the last arc, and *open* otherwise.

Articulation point A vertex of a connected graph whose removal disconnects the graph.

Asymmetric graph A directed graph that has no arc from w to v if it has an arc from v to w.

Bipartite graph A graph whose vertices can be partitioned into two sets in such a way that every edge has an end point in each set.

Branch An edge of G which is contained in a specified tree T that is a subgraph of G; said to be a branch relative to T.

Capacitated network A network in which flow in each arc is constrained to lie between zero and an upper bound called the capacity of the arc.

Chain A set of edges which, if properly ordered, form a chain progression. In a geometric graph, a set of edges which form an open curve.

Chain progression An open edge progression having no repeated edges.

Chord An edge of G which is not contained in a tree T that is a subgraph of G; said to be a chord relative to T.

Chromatic number of a graph The smallest k such that the vertices of the graph can be partitioned into k sets each of which is an independent set of vertices.

Circuit A set of edges which, if properly ordered, form a circuit progression. In a geometric graph, a set of edges which form a closed curve.

Circuit matrix A matrix whose rows and columns correspond to the simple circuits and edges of a graph, respectively, and each of whose elements is 1 or 0, depending on whether or not the corresponding circuit contains the corresponding edge.

Circuit progression A closed edge progression having no repeated edges.

Complementary graph A simple graph G obtained by deleting the edges of a given simple graph G from a complete graph having the same vertices.

Complete graph One in which every two distinct vertices are adjacent.

Component of a graph A connected subgraph which is not contained in any larger connected subgraph.

Connected graph A graph such that every pair of distinct vertices are joined by at least one chain.

Cut-set A disconnecting set consisting of all the edges that join a specified set of vertices with the complementary set of vertices.

Cut-set matrix A matrix analogous to a circuit matrix, with proper cut-sets replacing simple circuits.

Cycle A set of arcs which, if properly ordered, form a cycle progression. In a geometric graph, a closed curve composed of consistently directed arcs.

Cycle progression A closed arc progression having no repeated vertices.

Cyclic graph A directed graph having at least one cycle.

Degree of a vertex The number of edges (or arcs) with which the vertex is incident, with loops counted twice.

Directed Euler graph A directed graph such that the negative degree and positive degree of each vertex are equal.

Directed graph A mathematical system consisting of two sets V and A, together with a mapping Δ of A into $V \times V$. The elements of V and A are called the *vertices* and *arcs* of the directed graph, respectively, and Δ is called its *directed incidence mapping*. If $\Delta(a) = (v,w)$, v is called the *initial vertex* of arc a and w is called its *terminal vertex*.

Directed incidence mapping See **Directed graph.**

Disconnecting set A set of edges of a connected graph whose removal disconnects the graph.

Dominating set of vertices A set W of vertices such that every vertex not in W is adjacent to a vertex in W.

Edge See **Graph.**

Edge progression A finite sequence of (not necessarily distinct) edges such that one end point of the first edge is also an end point of the second, the remaining end point of the second is also an end point of the third, etc. The edge progression is *closed* if the "free" end point of the first edge is the same vertex as the "free" end point of the last, and *open* otherwise.

End points The vertices with which an edge or arc is incident.

Euler graph A graph having no vertices of odd degree.

Finite graph A graph having a finite number of vertices and edges (or arcs, in the case of a directed graph).

Flow An assignment of integers to the arcs of a network, these integers being interpreted as the rates of flow of a substance through the arcs.

Forest A graph that has no circuits, so that each component is a tree.

Geometric graph A graph whose vertices are selected points in two- or three-dimensional space and whose edges are nonintersecting simple curves each of which joins two vertices (or, in the case of a loop, closes on a single vertex) without containing any other vertices.

Geometric realization A geometric graph that is isomorphic with a given graph.

Graph A mathematical system consisting of two sets V and E, together with a mapping Γ of E into $V \,\&\, V$ (the set of unordered pairs of elements of V). The elements of V and E are called the *vertices* and *edges* of the graph, respectively, and Γ is called its *incidence mapping*. If $\Gamma(e) = (v \,\&\, w)$, vertices v and w are called the *end points* of edge e.

Graph of a binary relation A directed graph whose vertices correspond to the elements of the set in which the relation is defined and whose arcs join precisely those pairs of vertices which are in the relation.

Hamiltonian chain, circuit, cycle, or path A chain, circuit, cycle, or path that is simple and that is incident with every vertex of the graph under consideration.

Inarticulate graph A connected graph having no articulation points.

Incidence The relationship between an edge (or arc) e and the vertices into which it is mapped by the incidence mapping of the graph.

Incidence mapping See **Graph.**

Incidence matrix A matrix whose rows and columns correspond to the vertices and edges of a graph, respectively. Each element is 1 or 0, depending on whether or not the corresponding vertex and edge are incident.

Independent set of vertices A set of vertices no two of which are adjacent.

Initial vertex See **Directed graph.**

Isomorphic graphs A pair of graphs whose vertices and edges (or arcs) can be placed in 1-1 correspondence in such a way that incidences are preserved. In the directed case, the sense of incidence (i.e., positive or negative) must also be preserved.

Loop An edge (or arc) that is incident with only one vertex.

Matching A set of edges of a graph such that no two edges in the set are adjacent.

Negative degree of a vertex The number of arcs with which the vertex is negatively incident.

Negative incidence The relationship between an arc and its terminal vertex.

Network A directed graph which is connected and has no loops. This term is applied in the context of flow theory.

Parallel edges or **Arcs.** Edges or arcs having the same end points.

Path A set of arcs which, if properly ordered, form a path progression. In a geometric graph, an open curve composed of consistently directed arcs.

Path progression An open arc progression having no repeated arcs.

PERT network A directed graph, used to characterize the interactions of individual activities within an overall complex project. The arcs correspond to the activities, and their end points are chosen in a way which reflects constraints on the order in which the activities can be performed.

Planar graph A graph that is isomorphic with a geometric graph in a plane.

Positive degree of a vertex The number of arcs with which the vertex is positively incident.

Positive incidence The relationship between an arc and its initial vertex.

Primitive graph A directed graph such that, for a suitable integer k, every pair of distinct vertices can be joined by an arc progression having precisely k terms.

Proper cut-set A cut-set which does not contain any proper subset which is a cut-set.

Reflexive graph A directed graph having a loop incident with each vertex.

Regular graph A graph all of whose vertices have the same degree.

Rooted directed tree A directed graph which is a tree and which has a vertex v such that, for every other vertex w, the unique tree chain joining v and w is in fact a path directed from v to w; the tree is said to be *rooted* at the vertex v.

Simple chain A chain having at most two edges incident with any given vertex.

Simple circuit A circuit having at most two edges incident with any given vertex.

Simple cycle A cycle having at most two arcs incident with any given vertex.

Simple graph An undirected graph that has no parallel edges and no loops; a directed graph that has no strictly parallel arcs and no loops.

Simple path A path having at most two arcs incident with any given vertex.

Sink A vertex in a network at which flow input exceeds output.

Source A vertex in a network at which flow output exceeds input.

Spanning tree A subgraph of a connected graph G which is a tree and which includes all vertices of G.

Strictly parallel arcs Arcs having the same initial vertex and the same terminal vertex.

Strongly connected graph A directed graph for every ordered pair of distinct vertices (v,w) of which there is at least one path from v to w.

Subgraph A graph $G' = (V',E',\Gamma')$ so related to a graph $G = (V,E,\Gamma)$ that $V' \subset V$, $E' \subset E$, and $\Gamma'(e) = \Gamma(e)$ for every edge $e \in E'$; the first graph being said to be a subgraph of the second. It is necessary that V' include all end points (relative to G) of edges in E'. A directed subgraph of a directed graph is defined in essentially the same way if E and Γ are interpreted as the set of arcs and the directed incidence mapping, respectively.

Symmetric graph A directed graph whose arcs can be grouped into pairs of parallel but oppositely directed arcs.

Terminal vertex See **Directed graph.**

Total graph A directed graph such that, for every two distinct vertices v and w, there is a path from v to w or one from w to v (or both).

Transitive graph A directed graph which has an arc from u to w whenever it has an arc from u to v and one from v to w.

Tree A connected graph which has no circuits.

Undirected graph Same as **Graph.**

Unicursal graph A graph (directed graph) the totality of whose edges (arcs) forms a circuit (cycle) or chain (path).

Vertex See **Graph** and **Directed graph.**

Vertex matrix A matrix whose rows and columns correspond to the vertices of a graph and whose elements indicate the number of edges joining the corresponding pairs of vertices.

ANSWERS TO EXERCISES

CHAPTER 1

Exercise 1-2 Let the corresponding vertex sequence be $v_1, v_2, \ldots, v_n$, where $v_1 = v$ and $v_n = w$. If all the v_i's are distinct, the edge progression is necessarily a simple chain progression. If $v_i = v_j$ for some $i < j$, delete the edges corresponding to the vertex subsequence $v_i, v_{i+1}, \ldots, v_j$. Repeat such deletions until all vertices in the sequence determined by the remaining edge progression are distinct.

Exercise 1-3 Let G consist of three vertices v_1, v_2, and v_3 and two edges e_1 and e_2, where $e_1 \sim (v_1 \& v_3)$ and $e_2 \sim (v_1 \& v_2)$. The edge progression e_1, e_1, e_2, joining v_1 and v_2, is the desired example.

Exercise 1-4 Three or more edges, all of which have one common end point and whose other end points are all distinct, satisfy the stated characterization but fail to have the quality of "traversability," which is vital for the concept of an edge progression.

Exercise 1-5 One can obtain an open edge progression from v_1 to v_3 by simply adjoining the chain progression from v_2 to v_3 to the chain progression from v_1 to v_2. If this progression is not a simple chain progression, use Exercise 1-2.

Exercise 1-6 The example given for Exercise 1-3 will suffice.

Exercise 1-7 It is sufficient to show that the edges of a simple chain C joining v and w can be traversed in just two ways (one from v to w and one from w to v). Note first of all that precisely one edge of C is incident with v or w and precisely two edges of C are incident with any other vertex which meets C. Otherwise, the sequence of vertices induced by traversing C would include a repeated vertex. Consequently, if we start at v and wish to traverse each edge precisely once, we have at each stage a unique choice for the next edge to be traversed.

Exercise 1-8 If a loop is added at one or more of the vertices along a simple chain, the resulting nonsimple chain can still be traversed in just two ways.

Exercise 1-9 Note first that every vertex which is incident with the circuit has even degree relative to the circuit. Remove loops at the outset. Starting at any vertex v, select an edge incident with v and traverse it. From the vertex reached in this way, select and traverse a new edge (which must necessarily exist). Repeat this process until a vertex previously reached is encountered. Remove the simple circuit obtained in this way. Vertices incident with the remaining edges (if any) have even degree relative to these, and the process can be repeated.

Exercise 1-10 Modify the construction indicated for Exercise 1-9, or else join v and w by means of a new edge, apply Exercise 1-9, and then remove this edge.

Exercise 1-11 Consider the closed progression e, e and the open progression e, e, e involving a single repeated edge.

Exercise 1-12 Arrange the vertices of odd degree in pairs. Join each pair by an edge. Then add enough loops at each vertex to attain the desired degree.

Exercise 1-13 If the edge is not contained in any circuit, then its end points are not connected by any chain when the edge is removed. If it is contained in a circuit, then the circuit, less this edge, is a chain C joining the edge's end points. Thus any edge progression that involves this edge can be replaced by one that does not by substituting C; and so any pair of vertices which were connected before remain connected.

Exercise 1-14 A simple chain (and its vertices) or a Y-shaped graph has this property.

Exercise 1-15 Delete all edges; then replace them, one at a time, in such a way that at each stage (after the first) the edge being added is adjacent to at least one previously added edge. After the first edge, which "covers" two vertices (unless it is a loop), each succeeding edge covers at most one new vertex.

Exercise 1-16 If some vertex has degree less than 2, it is not incident with any simple circuit; while if some vertex has degree greater than 2, no simple circuit can account for all edges incident with it. On the other hand, if every vertex is of degree 2, we can start at any vertex, traverse one of the two edges incident with it, then traverse the unique edge adjacent to the first, and continue in this way. Ultimately, all edges will be traversed precisely once, and the starting vertex will be reached once again. The characterization for chains differs in that precisely two vertices have degree 1.

Exercise 1-17 Start with any edge progression. If not all edges are included, use connectivity to show that a new edge can be included by adjoining an appropriate edge progression to the end of the present one. By continuing this process, all edges can be ultimately included.

Exercise 1-18 The two edges leading to the left from W and the three leading to the right are the desired cut-sets.

Exercise 1-19 It is a general property of improper cut-sets. Consider the cut-set determined by some vertex set W and its complement W'. If, after removal of the cut-set, the vertices of W' determine two or more connected components with vertex sets $W'_1, \ldots, W'_n$, then these vertex sets and their respective complements determine disjoint proper cut-sets whose union is the original cut-set.

Exercise 1-20 No.

Exercise 1-21 A single edge in a connected graph is a cut-set if and only if that edge is not included in any circuit.

Exercise 1-22 Base a proof on the answer to Exercise 1-19.

Exercise 1-23 $q - 1$.

Exercise 1-24 The complete graph having two vertices is bipartite. In general, the complete graph having q vertices is q-partite but not p-partite for any smaller p.

Exercise 1-25 Consider the 3-regular graph corresponding to the vertices and edges of a cube. It is possible to change one end point of two appropriate edges, leaving others fixed, in such a way that the resulting graph is still 3-regular but includes triangles (i.e., 3-edged circuits).

Exercise 1-26 Since a bipartite graph need not be connected, whereas the existence of cut-sets requires connectivity by definition, the answer is no. However, it is true that a connected graph is bipartite if and only if the totality of its edges is a cut-set.

CHAPTER 2

Exercise 2-1 If the arcs could be traversed in two different ways, some vertex would necessarily have two outgoing arcs. But then, in order for the traversal to include both arcs, such a vertex would occur twice in the corresponding vertex sequence. Consequently, the path is nonsimple. If loops are adjoined to some vertices along a simple path, the resulting nonsimple path may also be traversed in only one way.

Exercises 2-2 to 2-4 Analogous to the corresponding proofs for the undirected case.

Exercise 2-5 No. (Join two disjoint cycles by means of a path, having at least two arcs, whose interior vertices are not incident with either cycle.)

Exercise 2-6 To show that the converse is not necessarily true, consider a graph having vertices v, w, x, and y and arcs (v,x), (x,y), (y,w), (w,x), and (y,v). To prove the theorem, adjoin one path progression to the other, forming a closed arc progression. Show that some subset of consecutive arcs that is a cycle progression exists.

Exercise 2-7 Start at any vertex v_1. Since v_1 has at least one outgoing arc, traverse one such arc, arriving at vertex v_2. Apply the same reasoning at v_2, traversing some arc to arrive at v_3. Continue until the first time a vertex is repeated. The arcs traversed between the first and second occurrence of this vertex then form a cycle.

Exercise 2-8 Label every vertex with either A or B, so that no two A or B vertices are adjacent. The desired path must alternate between A and B vertices, but these do not exist in equal numbers.

Exercise 2-9 Similar to Exercise 2-8. Since the rook reaches white and black squares alternately, it cannot start on white and also end on white after 63 moves.

Exercise 2-10 The existence of such a progression clearly implies strong connectedness. To prove the converse, consider a strongly connected graph and a closed edge progression P which starts and ends at vertex v. If some arc (x,y) is not included in P, adjoin to P an arc progression from v to x, then arc (x,y), and then an arc progression from y to v. If the resulting closed progression fails to include all arcs, repeat the process.

Exercise 2-11 If, for some W, there is no arc from W to $V - W$, then there can be no path from any vertex in W to any in $V - W$, so D is not strongly connected. If D is not strongly connected, choose v and w in such a way that there exists no path from v to w. Choose as W the set of vertices consisting of v and all vertices reachable from v by paths. Then there exist no arcs from W to $V - W$.

Exercise 2-12 Two vertices in the same cycle are clearly joined by a path. For two vertices in different cycles, construct an arc progression from one to the other by taking appropriate portions of consecutive cycles; and use the fact that an arc progression contains a path progression as a subprogression.

Exercise 2-13 Note that there is at least one cycle of length 3 containing v and that every vertex is joined with v by a cycle of length 2. These two facts in conjunction can be made the basis of a proof.

Exercise 2-14 $(n + 1)/2^n$.

Exercise 2-15 Show that there is at most one vertex which is not the terminal vertex of any arc. Then show that there exists a path from this vertex to each other vertex. Finally, show that a subset of the arcs in these paths (one to each other vertex) which is a rooted tree exists.

Exercise 2-16 Show that it is possible to arrange directions in such a way that there are two vertices neither of which is the terminal vertex of any arc. Then show that there cannot be a rooted tree spanning the graph under these circumstances.

Exercise 2-17 Given any cycle C and any vertex v incident with C, repeated application of transistivity reveals that there is a loop at v.

Exercise 2-18 All such graphs have isolated vertices. Moreover, every component which is not an isolated vertex has a loop at each vertex and at least one arc from each vertex to every other vertex of the component.

Exercise 2-19 In (a), for example, the arcs are (0,0), (1,1), (4,2), and (9,3).

Exercise 2-20 Corresponding to every path of two arcs, add a new arc joining its end points directly. The set of such arcs represents "is a grandparent of." For every path consisting of a "parent" arc followed by a "grandparent" arc, add a new direct arc. These arcs represent the desired relationship.

CHAPTER 3

Exercise 3.1 In Figs. 1-2, 1-3, and 1-6 the entire edge set is a chain or circuit. In Fig. 1-5 any one of the edges e_2, e_3, or e_6, together with the complementary set, yields a minimal covering of "size" 2.

Exercise 3-2 Let v and w be some (but not all) vertices having odd degree and assume that some minimal covering includes a chain joining v and w. Show that connectivity implies the possibility of producing a modified minimal covering which includes a chain joining v with some vertex other than w.

Exercise 3-3 A circuit and chain illustrate the second and third cases, respectively. For the first case consider a graph with vertices v, w, x, and y, edges (v & w) and (x & y), and two parallel edges (w & x). Removal of one of the parallel edges reduces the size of a minimal cover from 2 to 1.

Exercise 3-4 Given a partitioning of the vertices into W and W', in traversing an Euler circuit it is clear that each crossing from W to W' is matched by a crossing from W' to W and that each edge of the cut-set is traversed precisely once.

Exercise 3-5 One possible cyclic sequence is *aaaabbbbaababbab*.

Exercise 3-6 One possible cyclic sequence is *aadabbddbccdcacb*.

Exercise 3-7 It consists of a single vertex corresponding to a $(q-1)$-letter word, all of whose letters are a, and a loop at this vertex.

Exercise 3-8 Given any $(q-1)$-letter word, p different q-letter words are formable by prefixing any of the p letters of the alphabet. Since each of the latter words occurs once, the original word occurs p times. Repetition of this argument establishes the general result.

Exercise 3-9 In Figs. 2-2, 2-4 (left), 2-4 (right), and 2-7 the number of paths and/or cycles in a minimal covering is 1, 2, 2, and 13, respectively. Obtain specific coverings by inspection.

Exercise 3-10 Simply traverse an Euler circuit (or chain) and assign to each edge the direction induced by the traversal.

Exercise 3-11 Yes.

Exercise 3-12 Parallel the proof of the corresponding undirected result (Theorem 3-1).

Exercise 3-14 If there exists a hamiltonian circuit, it necessarily includes the 10 edges incident with the 5 vertices of degree 2. But these edges form a circuit, and they cannot be a subset of a larger simple circuit.

Exercise 3-15 See answer for Exercise 3-16.

Exercise 3-16 Since the edges of a chain or circuit in a bipartite graph alternate between two vertex sets, the existence of a hamiltonian circuit implies that these vertex sets have the same size. Existence of a hamiltonian chain implies that they have the same size or differ in size by 1.

Exercise 3-18 Visit the vertices in the order *AFLKJEDCIHBGA*.

Exercise 3-19 Complete graphs necessarily have this property. So do graphs corresponding to simple circuits and chains having an even number of vertices.

Exercise 3-21 For the graph corresponding to a simple circuit (other than a loop) having an odd number of vertices, every dominating set necessarily contains a pair of adjacent vertices.

Exercise 3-22 The radius is n, the diameter is $2n$, and one vertex is a center.

Exercise 3-23 For a chain having $2n + 1$ edges, the quantities are $n + 1$, $2n + 1$, and 2, respectively. For a circuit having $2n$ edges they are n, n, and $2n$, respectively.

Exercise 3-24 A simple graph is complete if and only if its diameter is 1.

Exercise 3-25 For examples, see answers to Exercises 3-22 and 3-23.

Exercise 3-26 Let the number of edges in a longest chain be n. The end points of such a chain are extreme vertices of the graph (i.e., vertices of degree 1), and $T = n$. Show that the distance from a vertex which is adjacent to one of these extreme points to any vertex of the graph is at most $n - 1$. Hence, $R < T$.

Exercise 3-29 The distances range from 1 to 5, two vertices having distance 4 from v_1. One vertex is inaccessible from v_1.

Exercise 3-30 The distances from v_1 vary from -3 to 4.

Exercise 3-31 Fourth-stage distances (top to bottom) are 9, 8, and 11. Fifth-stage distances are 12 (top) and 10. The shortest distance to v is 10.

CHAPTER 4

Exercise 4-1 Place the south pole of a sphere inside the square and map the graph onto the sphere. Then rotate the sphere so that the image of the square contains the north pole and map back onto the plane.

Exercise 4-2 The existence of such a subgraph means that one can take one of the two Kuratowski subgraphs, subdivide some of its edges by introducing vertices of degree 2, and arrive at a graph isomorphic to this subgraph. Alternatively, we could start with the given subgraph and eliminate its vertices of degree 2 by regarding chains of edges whose end points are of degree greater than 2 and whose interior points are of degree 2 as single edges.

Exercise 4-3 In Fig. 4-7, if the dashed graph is regarded as the original graph, the solid graph may be regarded as its dual, since it can be constructed by means of the construction described in the text.

Exercise 4-4 The dual graph would be the complete simple graph having five vertices, which is nonplanar.

Exercise 4-5 (*a*) Observe that any edge of the complete graph is adjacent to $n - 2$ other edges at each end. (*b*) Let the two nonadjacent vertices correspond to edges $(v \,\&\, w)$ and $(x \,\&\, y)$. Then the four vertices corresponding to $(v \,\&\, x)$, $(v \,\&\, y)$, $(w \,\&\, x)$, $(w \,\&\, y)$, and only these, are adjacent to both vertices. (*c*) Let the two adjacent vertices correspond to edges $(v \,\&\, w)$ and $(w \,\&\, x)$. Then each is adjacent to the $n - 2$ other vertices corresponding to edges of the form $(w \,\&\, y)$, where $y \neq v$ or x.

Exercise 4-6 According to part (*a*) of Exercise 4-5, each of these two vertices is adjacent to $2(n - 2)$ vertices. According to part *b*, four vertices occur in both sets of $2(n - 2)$ vertices. Thus there are $4(n - 2) - 4$ distinct vertices adjacent to one or both of the original two vertices.

Exercise 4-7 The three families are Dick, Jane, and Emily; Tom, Mary, and Michael; and Harry, Susan, and Alan.

Exercise 4-9 Note that every vertex of the resulting graph has even degree, and apply Theorem 4-23.

Exercise 4-10 The result clearly holds for six or fewer regions. Assume it holds for n. Consider a map with $n + 1$ regions and shrink some region which has five or fewer sides to a point. Color the resulting map with six colors. When the shrunken region is restored, there is at least one color not assigned to any of its neighbors.

Exercise 4-11 Show that the number of vertices and edges satisfy $2m \geq 3n$ and substitute in Euler's formula.

Exercise 4-12 Use Euler's formula to show that if every region is bounded by at least 5 sides, then there are at least 30 edges. However, use Exercise 4-11 to show that if there are fewer than 12 regions, then there are fewer than 30 edges.

Exercise 4-13 Use an inductive proof. Assume that the theorem holds for n regions. Every map with $n + 1$ regions has one region having at most four neighbors. If such a region has three or fewer neighbors, shrink and reexpand as in Exercise 4-10. If it has four, there are two neighboring regions which are not neighbors to each other. Form a single region from the original region and these two neighbors. Color the resulting map, and show the possibility of coloring the original map when the boundaries are reinstated.

Exercise 4-14 Each triangular face can be assigned one of two numbers (or orientations) in such a way that for any vertex, the number of faces of one type meeting that vertex is congruent, mod 3, to the number of the other type.

Exercise 4-17 The following figure shows a map of six regions on a Möbius strip (whose left and right sides are so joined that the arrows coincide) such that each region is adjacent to all five others.

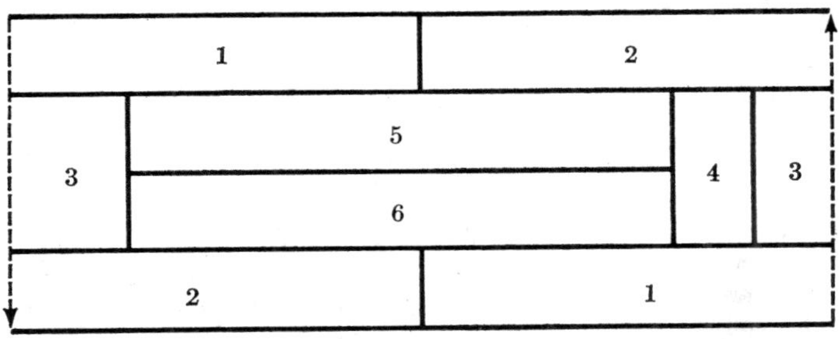

CHAPTER 5

Exercise 5-1 The possibility of permuting rows and/or columns of one incidence matrix in such a way that the matrix is identical with a second incidence matrix is clearly equivalent to establishing 1-1 correspondences between vertices (rows) and also between edges (columns) in such a way that incidences are preserved.

Exercise 5-2 The column corresponding to arc (v_3,v_5), for example, has a 1 in the row corresponding to v_3, a -1 in the row corresponding to v_5, and 0's elsewhere. Note that if the vertices and arcs associated with one component are all listed before (i.e., above and to the left of, respectively) the rest, the matrix is block-diagonal.

Exercise 5-3 Taking the v_3 row of A and the C_2 row of C (i.e., the C_2 column of C'), for example, which are

$$(0 \quad 1 \quad 1 \quad 0 \quad 1 \quad 0 \quad 0 \quad 0 \quad 0 \quad 0)$$

and

$$(0 \quad 1 \quad 0 \quad 0 \quad 1 \quad 1 \quad 0 \quad 0 \quad 0 \quad 0)$$

respectively, the sum of term-by-term products is 2, which is congruent to 0 mod 2. The same holds for every row of A and of C.

Exercise 5-4 If the set of edges incident with some vertex v of a connected graph is not a proper cut-set, then the removal of these edges divides the graph into at least three components. But then v is an articulation point.

Exercise 5-5 Since the circuit matrix and the transposed cut-set matrix are orthogonal, $k_1 + k_2 - q \leq 0$ (where k_1 and k_2 are their respective ranks). Moreover,

$$k_1 = m - n + 1$$

and $q = m$.

Exercise 5-6 Letting the rows of A correspond (top to bottom) to v_1 to v_4 as in Fig. 5-3, we have

$$A_{11} = \begin{pmatrix} 0 & 0 \\ 0 & 1 \\ 1 & 0 \\ 1 & 0 \end{pmatrix} \qquad A_{12} = \begin{pmatrix} 1 & 0 & 0 & 0 \\ 1 & 1 & 0 & 0 \\ 0 & 1 & 0 & 1 \\ 0 & 0 & 1 & 0 \end{pmatrix}$$

hence

$$A_{12}{}^{-1} = \begin{pmatrix} 1 & 0 & 0 & 0 \\ 1 & 1 & 0 & 0 \\ 0 & 0 & 0 & 1 \\ 1 & 1 & 1 & 0 \end{pmatrix} \qquad A_{12}{}^{-1}A_{11} = \begin{pmatrix} 0 & 0 \\ 0 & 1 \\ 1 & 0 \\ 1 & 1 \end{pmatrix}$$

Exercise 5-9 For the complete simple graph having five vertices, numbered 1 to 5, arrange the edges in the order

23 24 25 34 35 45 12 13 14 15

and consider the tree to be that determined by the last four edges. Then

$$A = \begin{pmatrix} 1 & 1 & 1 & 0 & 0 & 0 & 1 & 0 & 0 & 0 \\ 1 & 0 & 0 & 1 & 1 & 0 & 0 & 1 & 0 & 0 \\ 0 & 1 & 0 & 1 & 0 & 1 & 0 & 0 & 1 & 0 \\ 0 & 0 & 1 & 0 & 1 & 1 & 0 & 0 & 0 & 1 \\ 0 & 0 & 0 & 0 & 0 & 0 & 1 & 1 & 1 & 1 \end{pmatrix}$$

where the rows (top to bottom) correspond to 2, 3, 4, 5, and 1. Also,

$$\bar{C}_{12} = \bar{K}'_{11} = \begin{pmatrix} 1 & 1 & 0 & 0 \\ 1 & 0 & 1 & 0 \\ 1 & 0 & 0 & 1 \\ 0 & 1 & 1 & 0 \\ 0 & 1 & 0 & 1 \\ 0 & 0 & 1 & 1 \end{pmatrix}$$

Exercise 5-11 If $\vec{V}^k > 0$ for some k, then $(\vec{V}^t)^n > 0$ whenever $tn \geq k$.

Exercise 5-12 Since the graph is primitive, there is an h such that, for all powers of the incidence matrix $\geq h$, every entry in the row corresponding to v_i is greater than zero.

Exercise 5-13 Since the graph is strongly connected, there is an arc from v_k to v_j for some k. So an arc progression of length $h_i + 1$ from v_i to v_j can be obtained by taking an arc progression of length h_i from v_i to v_k and adjoining the arc from v_k to v_j. Similar reasoning can be applied to $h_i + 2$, $h_i + 3$, etc.

Exercise 5-14 This follows from Exercises 5-12 and 5-13 and the definition of index of primitivity.

Exercise 5-15 Every open arc progression contains a simple path joining the same vertices. Such a path has length at most $n - 1$. If the length less than $n - 1$, adjoin to such a progression a loop taken a sufficient number of times to yield length $n - 1$.

Exercise 5-16 Consider the left component in Fig. 5-3 with arcs indexed as in Fig. 5-5, except that e_7 is regarded as e_1. Then, if the end points of the ith arc are denoted as p_{2i-1}, p_{2i}, we have

$$
\mathfrak{D} = \begin{array}{c} \\ 1 \\ 2 \\ 3 \\ 4 \\ 5 \end{array}
\begin{array}{c}
\begin{array}{cccccccccccc} 1 & 2 & 3 & 4 & 5 & 6 & 7 & 8 & 9 & 10 & 11 & 12 \end{array} \\
\begin{pmatrix}
0 & 0 & 0 & 1 & 0 & 0 & 0 & 0 & 0 & 0 & 0 & 0 \\
0 & 1 & 1 & 0 & 0 & 1 & 0 & 0 & 0 & 0 & 0 & 0 \\
0 & 0 & 0 & 0 & 1 & 0 & 1 & 0 & 0 & 0 & 1 & 0 \\
0 & 0 & 0 & 0 & 0 & 0 & 0 & 1 & 1 & 0 & 0 & 0 \\
1 & 0 & 0 & 0 & 0 & 0 & 0 & 0 & 0 & 1 & 0 & 1
\end{pmatrix}
\end{array}
$$

and

$$
D = \begin{array}{c} \\ 1 \\ 2 \\ 3 \\ 4 \\ 5 \\ 6 \\ 7 \\ 8 \\ 9 \\ 10 \\ 11 \\ 12 \end{array}
\begin{array}{c}
\begin{array}{cccccc} 1 & 2 & 3 & 4 & 5 & 6 \end{array} \\
\begin{pmatrix}
1 & 0 & 0 & 0 & 0 & 0 \\
-1 & 0 & 0 & 0 & 0 & 0 \\
0 & 1 & 0 & 0 & 0 & 0 \\
0 & -1 & 0 & 0 & 0 & 0 \\
0 & 0 & 1 & 0 & 0 & 0 \\
0 & 0 & -1 & 0 & 0 & 0 \\
0 & 0 & 0 & 1 & 0 & 0 \\
0 & 0 & 0 & -1 & 0 & 0 \\
0 & 0 & 0 & 0 & 1 & 0 \\
0 & 0 & 0 & 0 & -1 & 0 \\
0 & 0 & 0 & 0 & 0 & 1 \\
0 & 0 & 0 & 0 & 0 & -1
\end{pmatrix}
\end{array}
$$

CHAPTER 6

Exercise 6-1 If $D(A)$ contains any cycles, shrink some cycle by replacing all its vertices by one pseudovertex, and delete all loops created in this way. Continue to shrink cycles until none exist. (The final graph will be a single vertex if the original graph was strongly connected.) Establish the fact that every end vertex (i.e., a vertex which is not the initial vertex of any arc) of the final graph corresponds to a set of vertices of the original graph which determine a closed strongly connected subgraph. Moreover, since the final graph has no cycles, every vertex of the final graph which is not an end vertex is joined by a path to an end vertex.

Exercise 6-2 Given that $x(I - B) = w$ has an admissible solution, form

$$B^* = \begin{pmatrix} B & p \\ w & 1 \end{pmatrix}$$

where p is a column vector of zeros. Let x' be a variable corresponding to the one new activity. Then

$$(x,x')(I - B^*) = (x(I - B) - x'w,0)$$

If x is an admissible solution of the open model, then $(x,1)$ is a solution of this enlarged closed model.

Exercise 6-3 Let the vertices associated with A be $v_1, \ldots, v_n$ and let w be the vertex associated with the last row and column. There is an arc (v_i,w) and an arc (w,v_i) for every i, since x and y are assumed positive. Hence there exist cycles of length 2. If there is at least one arc of the form (v_i,v_j), then there also exist cycles of length 3. But then the GCD of lengths of cycles is 1, which implies primitivity. To see that the index of primitivity is 4 for some matrices A, suppose that there is only one arc of the form (v_i,v_j). Then there is no arc progression of length 3 from v_j to v_i.

Exercise 6-4 Show that if some activity (v_i,v_j) fails to be completed before some adjacent activity (v_j,v_k) is begun, then these activities could not have been started at $T(v_i)$ and $T(v_j)$, respectively.

Exercise 6-5 Note that $T(v_i) + X(v_i)$ is the length of some path from v_l to v_n. Thus this sum does not exceed $T(v_n)$. If equality holds, this path is necessarily a critical path.

Exercise 6-7 For the odd case, we have $\sum r_i^2 = \dfrac{n(n-1)^2}{2^2}$ and hence the maximum number is $\binom{n}{3} + \frac{1}{2}\binom{n}{2} - \frac{1}{8}n(n-1)^2$.

Exercise 6-8 Use the same technique as in Fig. 6-15.

Exercise 6-9 Use the same technique as in Fig. 6-15.

Exercise 6-11 The graph has the structure shown in the accompanying figure. Note that there are 60 intersections. This agrees with the value of M'_{10}.

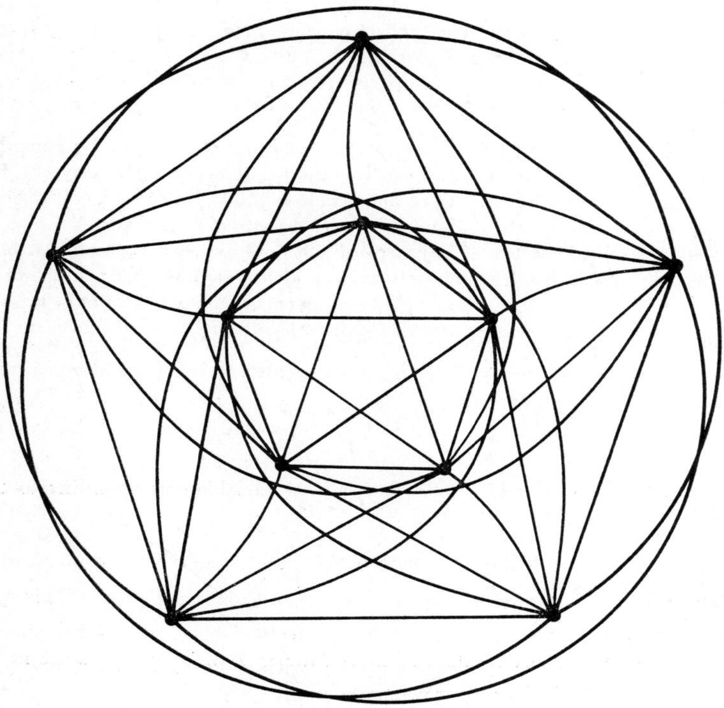

Exercise 6-12 Assume that the cubes are stacked in such a way that, from top to bottom, they are cubes 1, 2, 3, and 4. It is possible to arrange that sides x, y, x', and y' (top to bottom) are colored GYBR, BGRY, BRYG, and YBRG, respectively.

Exercise 6-13 Yes, fewer crossings are required. If the distinction between C and K is removed, by replacing K by C, then in the solution the subsequence MMCK, (MC), MMCC, (CC) can be replaced by MMCC, (CC), eliminating one round trip.

Exercise 6-14 Number the vertices 1 to 8 clockwise, starting at v. The desired path goes through the vertices in the order 143821856. Note that 1 and 8 are repeated.

Exercise 6-15 This is essentially the vertex or adjacency matrix of the graph shown in Fig. 6-20.

Exercise 6-16 This follows from the fact that n elements (the diagonal elements) have fixed values, while the remaining $n^2 - n$ can assume either of two values.

Exercise 6-17 Use as V the directed vertex matrix having 1's in the following positions: (1,2), (2,1), (2,3), (2,4) (3,4), (4,2), (4,5), (5,2), (5,7), (6,5), (7,5). Raise V to appropriate powers to obtain the answers.

Exercise 6-18 If an inner vertex were covered by more than one edge of the selected set, its degree would exceed 2. If an outer vertex were covered by more than one edge, the graph could not be a tree.

Exercise 6-19 Note that if an inner vertex were exposed, then two outer vertices would also be exposed, since every edge joins an inner vertex with an outer vertex.

Exercise 6-20 Consider a chain progression passing through vertices v_0, v_1, . . . , v_k, where v_0 is the exposed outer vertex. The edge joining v_0 with v_1 is not in M, since v_0 is exposed. Since v_1 is of degree 2 and is covered, the edge joining v_1 with v_2 must be in M. Repetition of this reasoning establishes the alternating property.

Exercise 6-21 In the case of Fig. 6-17 (including the dashed part), matchings which leave only one vertex exposed can be found. For Fig. 6-15 perfect matchings exist. In both cases the maximal matching is not unique.

Exercise 6-22 In general, these questions may be answered by summing (*a*) rows or (*b*) columns, computing V^2 (*c*), and computing $V + V^2 + V^3 + V^4$ (*d*). Explain why.

Exercise 6-25 Let t_i denote the ith term. Whenever $t_i - t_{i+1} > 1$, insert $t_i - t_{i+1} - 1$ ones after t_i. Whenever $t_i \leq t_{i+1}$, insert $t_i - 1$ ones after t_i. (Do the latter also when t_i is the last term.)

Exercise 6-26 $t_0 = 4$.

Exercise 6-27 Let v have a loop. If the graph has n vertices, then there exist paths from any v_i to v, and from v to any v_j, having at most $n - 1$ arcs. Thus one can obtain an arc progression of length $2n - 2$ by taking two such paths together with an appropriate number of loops. (Compare with the earlier discussion of primitive matrices.)

Exercise 6-28 There are four automorphisms. The top and bottom vertices may be either permuted or left fixed. Independently of this, the left and right vertices may be either permuted or left fixed.

Exercise 6-29 The graph corresponding to cubing the residues has a loop at each vertex. The graph corresponding to squaring the residues has the following arcs: (0,0), (1,1), (2,4), (3,3), (4,4), (5,1).

Exercise 6-30

x	0	1	2	3	4	5	6
$f(x)$	5	4	0	6	3	*	1

* Undefined.

Exercise 6-31 The arcs of the graph form a simple cycle which passes successively through 0, 5, 4, 3, 2, and 1.

Exercise 6-32 With respect to the first partitioning, rearrange the rows and columns into the order 1, 3, 2, 4, by taking

$$P = P^{-1} = \begin{pmatrix} 1 & 0 & 0 & 0 \\ 0 & 0 & 1 & 0 \\ 0 & 1 & 0 & 0 \\ 0 & 0 & 0 & 1 \end{pmatrix}$$

Compute $B = P^{-1}AP$ and note that its submatrices are as given. Note also that B^2 is block-diagonal, each diagonal block being

$$\begin{pmatrix} 2 & 0 \\ 0 & 12 \end{pmatrix}$$

Find an appropriate P for the other partitioning.

Exercise 6-33 41 comparisons are required. $L(15)$ and $N(15)$ are 41 and 45, respectively.

CHAPTER 7

Exercise 7-1 Let φ_1 be the flow given in Fig. 7-2 and let φ_2 consist of unit flow in each arc of the same network. Then the net flow outputs at v_1 to v_6, relative to $\varphi_1 + \varphi_2$, are 8, 5, 3, -3, -13, and 0, respectively. Relative to $\varphi_1 - \varphi_2$, they are 2, 7, -3, 3, -7, and -2, respectively. Compare with net outputs relative to φ_1 and φ_2 individually.

Exercise 7-2 The flows are not conformal. In two arcs, namely, (v_3,v_2) and (v_1,v_4), the arc flows are both nonzero and oppositely oriented.

Exercise 7-3 The numerical results will depend entirely on the choice of flows in Exercise 7-2 and the choice of ψ. For each arc, verify that $\psi(a) + \varphi_1(a)$ lies between $\psi(a)$ and $\psi(a) + \varphi_1(a) + \varphi_2(a)$.

Exercise 7-4 In the unitary graph, there are, for example, three arcs from v_1 to v_3, ten arcs from v_5 to w_2, two arcs from v_2 to v_3, no arcs between v_5 and v_6, etc.

Exercise 7-5 The following is one decomposition, given in terms of the sequence of vertices through which the oriented chains and circuits pass. (Asterisks indicate arcs traversed oppositely to their inherent directions.)

6 chains $w_1v_2v_5w_2$
2 chains $w_1v_1v_2v_5w_2$
2 chains $w_1v_1\overset{*}{v}_2v_3v_5w_2$
1 chain $w_1v_1v_3v_6w_2$
1 circuit $v_1v_3v_6\overset{*}{v}_4v_1$
1 circuit $v_1v_3\overset{*}{v}_4v_1$

Exercise 7-6 There are eight relevant partitionings. When $W = \{v_1,v_2,v_5\}$, for example, the right-hand expression in Theorem 7-3 has the value $(4 + 3 + 2) - (1 + 6)$.

Exercise 7-9 The arc (v_1,v_2), for example, having lower and upper flow bounds 1 and 3, respectively, is replaced by three arcs: (v_0,v_2), (v_1,v_6), and (v_1,v_2), having capacities 1, 1, and 2, respectively. Their lower flow bounds are all zero (as are those of all arcs in the auxiliary network).

Exercise 7-11 The maximal-flow value through N' is only 7, whereas a saturating flow requires 19 (the sum of lower flow bounds of all arcs in the original network).

Exercise 7-12 Show that if such a cycle exists, then φ plus a unit flow in this cycle is feasible and has the same total cost as φ. On the other hand, show that if a second minimal-cost flow φ' exists, then each unit circuit flow in a decomposition of $\varphi' - \varphi$ determines a cycle of this variety in the incremental graph.

Exercise 7-13 Note, for example, that the outermost clockwise cycle in the incremental graph has total length $4 + 1 - 2 - 3 - 1 = -1$. There is also a second clockwise cycle, including all vertices except the source, with length -2.

Exercise 7-14 Send two units of flow along the chain v, x, w, at unit cost of 5. Then send one unit along the chain v, x, z, w, at unit cost 7. Finally, send two units along the chain v, y, z, w, at unit cost 9. Total cost: 35.

Exercise 7-15 The left-to-right flows in the following figure indicate one solution.

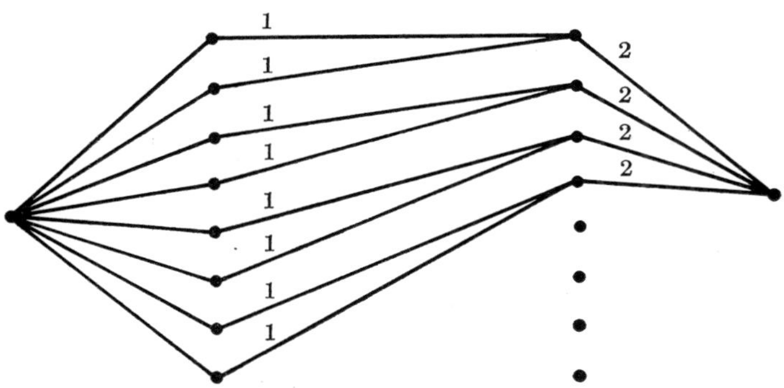

INDEX

Page numbers in *italics* refer to defining pages or pages of special importance.